ROCKY MOUNTAIN CARBONATE RESERVOIRS:

A CORE WORKSHOP

ORGANIZED AND EDITED
BY
MARK W. LONGMAN, KEITH W. SHANLEY,
ROBERT F. LINDSAY, AND DAVID E. EBY

SEPM CORE WORKSHOP NO. 7
GOLDEN, COLORADO
AUGUST 10–11, 1985

© COPYRIGHT 1985 BY SOCIETY OF ECONOMIC PALEONTOLOGISTS AND MINERALOGISTS

PREFACE

The 1985 SEPM Midyear Meeting brings together sedimentary geologists from across the country and around the world. This core workshop was organized to give these geologists the opportunity to see a wide variety of carbonate reservoirs as well as some carbonate source rocks from the Rocky Mountain region. It is hoped that the papers presented here will provide both sedimentologists and explorationists with a useful record of the cores displayed and discussed by the contributors at SEPM Core Workshop #7.

Cores displayed at the workshop range in age from Cambrian to Cretaceous and come from a number of the major oil-producing basins in the Rocky Mountains. Included are cores from the Williston, Paradox, Powder River, Alberta, and Western Interior basins, and the Overthrust Belt in west Wyoming. Depositional facies represented in the cores range from sabkhas and tidal flats through algal and coral buildups to relatively deep water chalks.

Dolomite and evaporite minerals are important in approximately half the cores described; the others are dominantly limestone. Porosity of many different types including intergranular, intercrystalline, framework, moldic, and fracture are displayed and discussed. Diagenesis, or lack of it, has played a major role in forming virtually all the reservoirs. Thus, the workshop offers the chance to observe and study a wide variety of depositional and diagenetic textures in a number of economically important rock units.

This workshop and these notes were made possible through the help of many people and companies. Tenneco, Gulf, and Champlin provided time and support for three of the organizers (K. S., R. L., and D. E., respectively). All manuscripts were retyped by the Information Processing Center of Gulf Oil Exploration Technology Center in Houston and the support of this group in helping turn out a high-quality publication is much appreciated. Use of the U. S. Geological Survey Core Library was willingly offered and critical to the success of the workshop.

Finally, the unsung heroes who indirectly made this workshop possible are the people and companies who paid to cut the cores in the first place and who carefully labelled, slabbed, stored, and made available the cores displayed here. The fact that almost all major oil companies and many smaller ones operated the wells from which these cores come attests to the widespread recognition of cores as a major tool in interpreting carbonate reservoirs. Many thanks to all the people who helped take and make these core available.

TABLE OF CONTENTS

Page

CAMBRIAN - POWDER RIVER BASIN

Origin of Upper Cambrian Flat Pebble Conglomerates in the Northern Powder River Basin, Wyoming
by: Michael D. Wilson. 1

PALEOZOIC EXAMPLES - WILLISTON BASIN

Depositional and Diagenetic Alteration of Yeoman (Lower Red River) Carbonates from Harding County, South Dakota
by: Alan C. Kendall 51

Depositional Environment and Diagenesis of the Red River Formation "C" Interval, Divide County, North Dakota and Sheridan County, Montana
by: Douglas G. Neese 95

Depositional Environments, Paleoecology and Diagenesis of Selected Winnipegosis Formation (Middle Devonian) Reef Cores, Williston Basin, North Dakota
by: Nancy A. Perrin and William F. Precht 125

Deposition, Diagenesis and Paleostructural Control of Duperow and Birdbear (Nisku) Reservoirs, Williston Basin
by: James R. Ehrets and Don L. Kissling 183

Rival, North and South Black Slough, Foothills and Lignite Oil Fields: Their Depositional Facies, Diagenesis and Reservoir Character, Burke County, North Dakota
by: Robert F. Lindsay 217

Deposition and Diagenesis of the Mississippian Charles (Ratcliffe) Reservoir in Lustre Field, Valley County, Montana
by: Mark W. Longman and Kenneth H. Schmidtman 265

DEVONIAN RESERVOIRS - CANADA

Devonian Hare Indian - Ramparts (Kee Scarp) Evolution, MacKenzie Mountains and Subsurface Norman Wells, N.W.T.: Basin-Fill and Platform Reef Development
by: Iain Muir, Pak Wong, and Jack Wendte 311

Sedimentology of a Carbonate Source Rock: The Duvernay Formation of Alberta, Canada
by: Frank A. Stoakes and Stephen Creaney 343

MISSISSIPPIAN - OVERTHRUST BELT

Evidence of Rapid Fluid Migration during Deformation, Madison
Group, Wyoming and Utah Overthrust Belt
by: Joyce M. Budai 377

PENNSYLVANIAN - PARADOX BASIN

Pennsylvanian Phylloid Algal Mound Production at Tin Cup
Mesa Field, Paradox Basin, Utah
by: Wilson H. Herrod, Mike H. Roylance and
    Elizabeth C. Strathouse 409

CRETACEOUS - WESTERN INTERIOR BASIN

Sedimentology and Reservoir Characteristics of the Niobrara
Formation (Upper Cretaceous), Kansas and Colorado
by: Peter A. Scholle and Richard M. Pollastro 447

ORIGIN OF UPPER CAMBRIAN FLAT PEBBLE CONGLOMERATES

IN THE NORTHERN POWDER RIVER BASIN, WYOMING

Michael D. Wilson
Geological Consultant
11440 West 39th Place
Wheat Ridge, Colorado 80033

ABSTRACT

Flat pebble conglomerates consisting of rounded clasts of bioclastic siltstone, sandstone and grainstone are abundant in a 233 foot (71 m) core from an Upper Cambrian sequence in the northern Powder River Basin. Flat pebble conglomerates form 20 percent of a sequence dominated by ripple-laminated and horizontal parallel-laminated lenticular beds of bioclastic siltstone, sandstone and grainstone.

Most flat pebble conglomerate beds are 1 to 7 cm thick, but some multistory units are up to 60 cm thick. Most conglomerates display weak to strong clast imbrication, and several beds exhibit bimodal imbrications. The flat pebbles are derived by segmentation of interbedded thin shale and bioclastic siltstone, sandstone and grainstone beds. The segmentation is produced by subvertical dewatering channels along which relatively fine sediment is injected. In addition to dewatering structures, other features in the finer grained deposits which indicate rapid sedimentation are load casts, piled ripples, ball and pillow structures, microfaults, and climbing ripples.

The character of the clastic component in the conglomerate matrix changes abruptly midway through the core, suggesting a major change in the nature of clastic input. Conglomerate beds in the lower part of the core contain subangular silt to very coarse quartz brought to the coastline by fluvial processes. Younger conglomerates contain smaller amounts of silt- to very fine sand-sized quartz, which most likely was introduced by aeolian processes.

The conglomerates are interpreted to have been deposited below wave base by tidal currents in a shallow marine basin. Work by Palmer (1971), Aitken (1978) and others has shown this basin was separated from the open ocean to the west by a peritidal bank. Glauconite is a minor to abundant component throughout the Upper Cambrian sequence. Its relative abundance in a sequence of rapidly deposited sediments behind a shelf-edge bank may reflect increased rates of glauconite development in the Cambrian. The increase in rate may been favored by lower temperatures and widespread anoxic conditions in the oceans during the Cambrian.

The much greater abundance of flat pebble conglomerates during the Cambrian and Lower Ordovician may be attributable to changes in biologic evolution, as proposed by Sepkoski (1982), or to an increased tidal range during this time period, or to both of these. Available evidence bearing on the tidal range in the Cambrian is sparse and inconclusive.

INTRODUCTION

A well drilled by Shell Oil in the northern Powder River Basin in 1948 was cored through most of the middle and lower Paleozoic sequence. Of the almost 1,600 feet (488 m) of core recovered in this well, 233 feet (71 m) were taken in a Cambrian sequence dominated by thin interbedded bioclastic siltstones and sandstones, silty bioclastic grainstones, and shales. Within this siltstone/grainstone/shale package are numerous beds of flat pebble conglomerate.

Sepkoski (1982) and many other geologists have observed that these conglomerates are unusually abundant in Cambrian and Lower Ordovician deposits. The origin of these flat pebble conglomerates remains in dispute. It is the purpose of this paper to describe the conglomerates encountered in the Cambrian sequence from the Shell Oil #1 Clear Creek and to review possible mechanisms by which they might have formed. To the author's knowledge, the cores from this well constitute the only cores taken in the flat pebble conglomerate-bearing portion of the Cambrian in the Rocky Mountain area; and, as such, they provide a unique opportunity the investigate the origin of these conglomerates.

A concise and insightful paper by Sepkoski (1982) reviews the characteristics and origins of Upper Cambrian flat pebble conglomerates based on outcrop studies in southern Montana and northern Wyoming. He interprets these conglomerates to be the products of storm activity in a subtidal environment, as does Demicco (1983) in his study of Upper Cambrian subtidal grainstone-thrombolite-flat pebble conglomerate deposits in the central Appalachians. Demicco (1983) also encountered flat pebble conglomerates in intertidal-subtidal "ribbon rock" sequences; and, though he does not make any statements regarding their origin, he presumably also considers them to have been generated by storms. Lochman-Balk (1970) describes both angular and rounded flat pebble conglomerates from the Upper Cambrian of the western United States. She suggests they are formed by storm or tidal reworking of desiccated grainstones and mudstones in tidal flat and subtidal lagoon settings. Flat pebble conglomerate grain flow deposits are reported from Cambrian outer shelf-slope environments in Newfoundland (Hubert and others, 1977) and central and eastern Nevada (Cook and Taylor, 1977).

The discussion that follows will first focus on the siltstone-shale sequence within which the conglomerates occur and then on the characteristics of the conglomerates themselves. A synthesis of the data from both types of beds is then utilized to generate a comprehensive analysis of the origins of the conglomerates.

LOCATION AND STRATIGRAPHIC FRAMEWORK

The well of interest, Shell Oil #1 Clear Creek, is located at the north end of the Powder River Basin (Sec. 11, T57N, R78W) in Sheridan County, Wyoming (Fig. 1). This figure also shows the location of this well in relation to the regional lithofacies patterns established by Palmer (1971) for the Upper Cambrian of the western Cordillera. Three lithofacies belts have been documented:

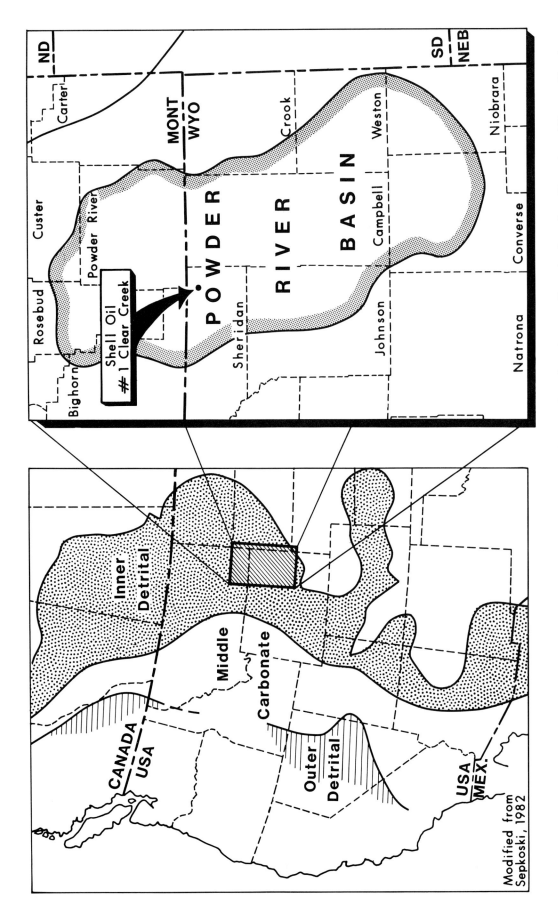

Figure 1. Map (right) showing the location of the Shell Oil #1 Clear Creek in the Powder River Basin, and (left) the position of this well relative to the regional lithofacies belts of Palmer (1971).

(1) Inner Detrial belt - siltstones, sandstones, silty grainstones and shales with subordinate flat pebble conglomerates. Sandstones become the dominant lithology as the shoreline is approached. Glauconite in trace to moderate amounts occurs in most beds.

(2) Middle Carbonate belt - diverse carbonates ranging from shallow subtidal to supratidal. Deposition occurred in a band separating a stable platform on the landward side (east) from a more rapidly subsiding basin to the west.

(3) Outer Detrital belt - outer shelf and slope dark colored siliceous and calcareous mudstones and debris flow deposits. Deep-water siliceous deposits also are present. Aitken (1978) notes that in Canada some deposits of the outer belt cannot be distinguished from rocks of the Inner Detrital belt.

The Cambrian rocks in the Shell Oil #1 Clear Creek are tentatively assigned to the Upper Cambrian based on the presence of the trilobite Aphelaspis sp. (?) at 12,200 to 12,201 feet (3719 m) by Michael Taylor of the U.S. Geological Survey (pers. commun.,). Stratigraphic terminology for the Upper Cambrian in the northern portion of the Powder River Basin is not well established. Names given to the Cambrian sequence of thin siltstones, sandstones, limestones and pebble conglomerates which outcrop in the northern Big Horn Mountains include the Depass Formation (Miller, 1936) and Gros

Ventre Formation (Sepkoski, 1982). Rocks on the east side of the Big Horn Mountains are simply referred to as Gallatin and Gros Ventre formations and equivalent rocks by Love and Christiansen (1980). The Upper Cambrian of the Shell Oil #1 Clear Creek is probably equivalent to the Upper Cambrian shale and intraclast limestones and laminated limestone lithofacies described by Kurtz (1976) in the northern Big Horn Mountains. A formation name will not be used in the paper; instead, the term "Cambrian flat pebble conglomerate sequence" will be used. The sequence of units present in the Shell Oil #1 Clear Creek and the location of cored intervals are shown on the SP and resistivity log profiles (Fig. 2).

CORE DESCRIPTION

General

The sequence of lithologies and structures encountered in the interval 11,960 to 12,216 feet (3645 to 3723 m) in the Shell Oil #1 Clear Creek is shown in Figure 3. The top of the Cambrian flat pebble conglomerate sequence is placed at 11,975 feet (3650 m). In the 241 foot (74 m) interval between the top of the Cambrian flat pebble conglomerate sequence and the base of the last core at 12,216 feet (3734 m), only 8 feet of core missing. Generally, 50 to 80 percent of each foot of core is present. However, in the last 20 feet (6 m) of core, many intervals, particularly shaly zones, are represented by only a few inches of core. Lack of adequate labelling in this relatively old core has caused many core segments to be inverted in the core

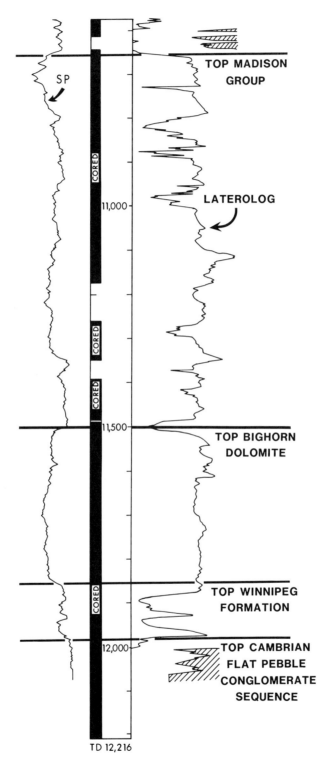

Figure 2. SP and laterolog profiles are shown for the middle and lower Paleozoic section in the Shell Oil #1 Clear Creek. Cored intervals are indicated by black bars. The interval of interest, the Cambrian flat pebble conglomerate sequence, occurs in the lower 240 ft of the well.

boxes. Only where ripple structures, load casts and geopetal structures are present can the correct orientation of many core segments be determined with assurance.

Thin Bedded Siltstone/Grainstone/Shale Lithologies

Twenty percent of the Cambrian flat pebble conglomerate sequence in the Shell Oil #1 Clear Creek consists of flat pebble conglomerates. Of the remaining core, approximately 90 to 95 percent consists of thin interbedded bioclastic siltstone and very fine-grained sandstone, silty or sandy bioclastic grainstone, and dark greenish-gray shales. For brevity, these finer grained lithologies will subsequently be referred to as the siltstone beds. Excepting the upper 11 feet (3.4 m) of the Cambrian flat pebble conglomerate sequence, which is stained pale red by hematite, the siltstone beds are typically colored light olive gray, pale brown, or medium light gray to medium dark gray.

Siltstone units tend to form lenticular beds ranging in thickness from a fraction of a millimeter to 6 cm, though most have thicknesses in the range of 1 to 25 mm (Fig. 4). Detailed analysis of bed lithology and geometry was conducted on a 7 foot (2.1 m) interval from 11,982.5 to 11,989 feet (3652 to 3654 m), of which 4 feet (1.2 m) consist of coarser sandstone and fine pebble conglomerate. This analysis indicates that 87 percent of the 133 siltstone beds in the interval are distinctly lenticular and that 95 percent have maximum thicknesses of 2 cm or less. Although not typical of all siltstone units in the cored interval, the zone analyzed can be considered

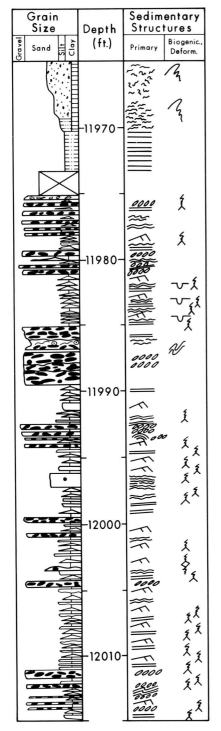
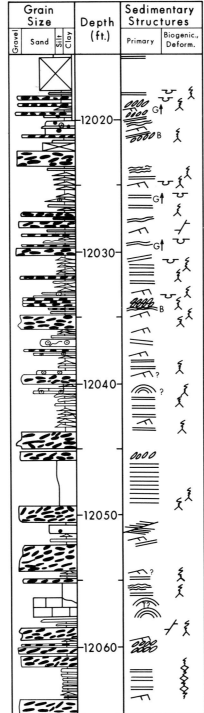
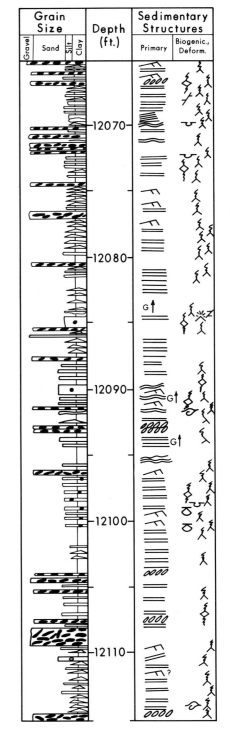

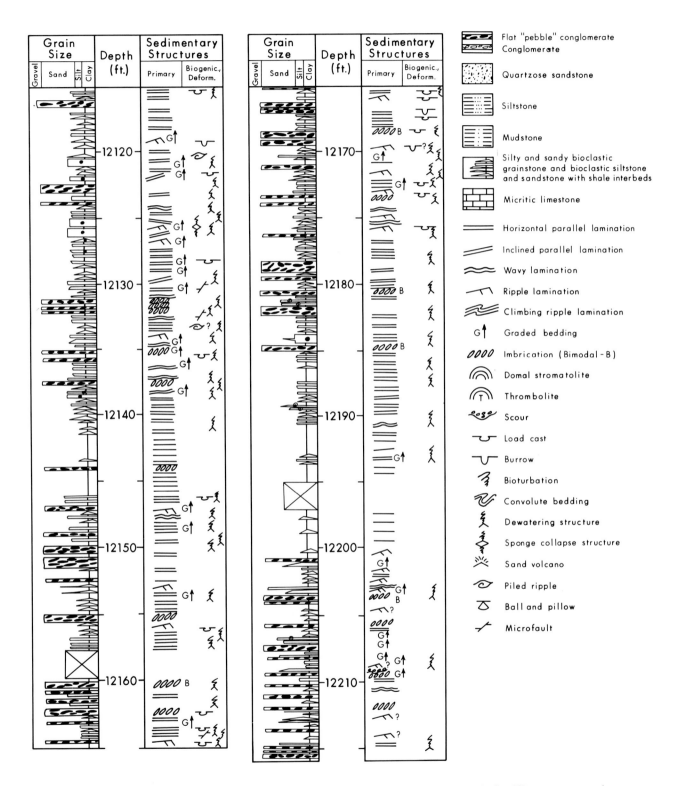

Figure 3. The coregraph shown above and to the left illustrates the lithologies, grain sizes, and sedimentary structures which are present in the Cambrian flat pebble conglomerate sequence in the Shell Oil #1 Clear Creek.

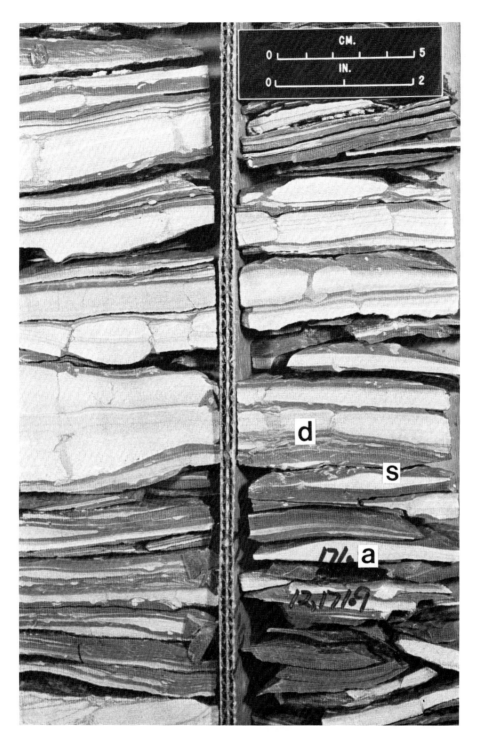

Figure 4. Core photo displaying the thin lenticular beds typical of the siltstone/sandstone/grainstone/shale lithologies which form the bulk of the Cambrian flat pebble conglomerate sequence. At (s) a symmetrical ripple occurs, and at (a) is an asymmetrical ripple. Subvertical penetrations of many of the beds are filled by dikes of shale or siltstone and represent dewatering structures. The dikes are commonly compacted into intestine-like contortions (d).

representative of most such beds in the Cambrian flat pebble conglomerate sequence. Zones containing thicker siltstone and very fine-grained sandstone beds, mostly in the range 1 to 10 cm, occur at 12,093 to 12,100 feet (3686 to 3688 m) and 12,124 to 12,127.5 feet (3695 to 3696 m).

Disregarding distortions due to compaction and load casting or to deposition within troughs of underlying lenticular beds, a very high percentage of the siltstone beds has sharp horizontal bases. The upper contacts are sinuous, primarily due to the development of current or wave ripples on the upper surfaces. Both symmetric and asymmetric ripple forms are present (Fig. 4). None of the ripples have truncated crests. Ripple lamination is recognizable within only a small proportion of the thin beds. In contrast, horizontal parallel lamination is commonly descernible (Fig. 4).

Sedimentary structures indicative of relatively rapid deposition occur in numerous beds. These include load casts (Fig. 5A), climbing ripples (Fig. 5B), ball and pillow structures (Fig. 5C), convolute bedding (Fig. 5D), microfaulting (Fig. 6A), dewatering structures (Figs. 4, 6A, 6B, 6C) and collapse structures produced by the death of sponges buried by rapid sedimentation.

Many of the thin finer grained beds are segmented into slabs or wafers between which straight-sided to jagged streaks or dikes of siltstone, very fine sandstone, bioclastic grainstone, or shale are encountered. These clastic dikes are extensively deformed into intestine-like swirls in the overlying and underlying shale beds (Figs. 4, 6C). These structures are interpreted to be dewatering escape paths into which clay, silt, sand or bioclastic material have been

Figure 5. A. A foot-shaped load cast occurs at the base of the upper bed with horizontal parallel laminations. Fine-grained glauconitic grainstone in the load cast grades upward into very fine-grained sandstone and siltstone. The horizontal laminations have been distorted by compaction over the load cast. Core depth - 12,128.5 ft.

B. Well-developed climbing ripple laminations occur in a very fine-grained sandstone to siltstone bed. Core depth - 12,013 ft.

C. Lenses of light colored coarse siltstone float in a bed of darker finer grained siltstone. This feature, which forms as a result of rapid loading over an unconsolidated bed, is referred to as ball and pillow structure. Core depth - 12,091.2 ft.

D. Convolute bedding has developed in a series of thin interbeds of fine to coarse siltstone and shale. Disturbed bedding of this type is common in sequences where rapid deposition occurs. Core depth - 12,113.1 ft.

14

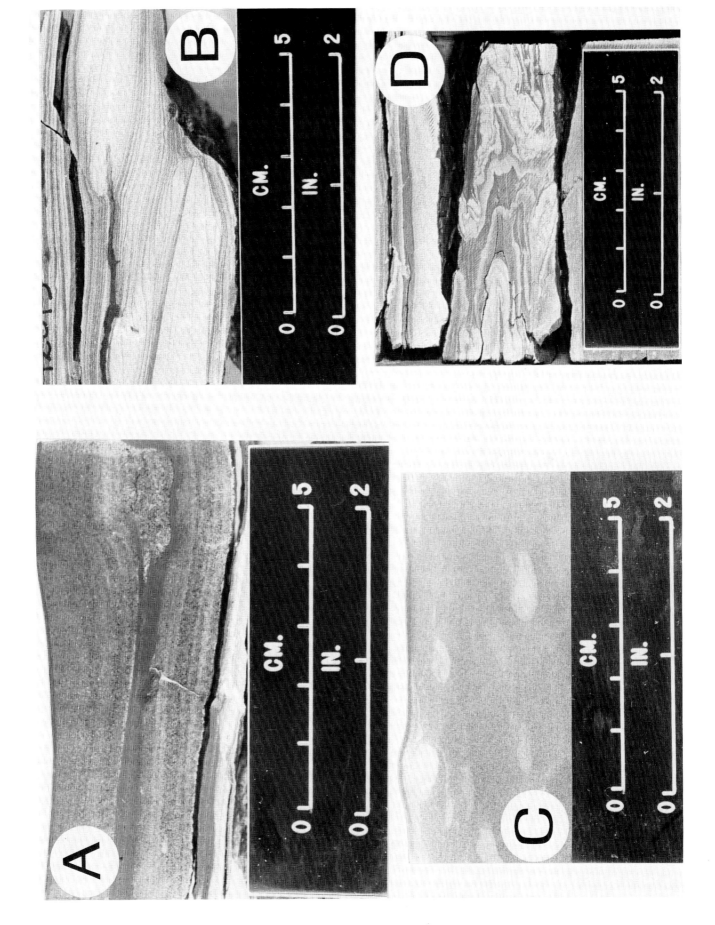

injected. Compaction subsequent to deposition in the shaly interbeds produces the convoluted dikes. The dikes have a subvertical planar geometry. In some beds, the dikes develop parallel to one another, while in others they intersect to form polygonal fences (Fig. 6D). The planar geometry of these dikes precludes their origin by biogenic burrowing. In a few instances, compressed sand volcanos occur at the outlet of the dewatering feature, and overlying beds are fractured due to compaction over these mounds (Fig. 6A). The fact that the edges of these slabs do not curl upwards indicates they are not mudcracks produced by subaerial exposure.

Sponge collapse structures occur as jagged subvertical penetrations of one or more beds filled by thin argillaceous seams and local orange-brown streaks and lobate masses of chert (Fig. 7A). In thin section, the masses of chert contain abundant rod-shaped sponge spicules (Fig. 7B).

In a modest proportion of the finer grained beds, grading is present. In some cases, it takes the form of a thin (a few millimeters) basal zone of fine- to medium-grained sandstone or an upper zone of darker more argillaceous, micaceous medium to fine siltstone (Fig. 5A). The basal graded zone may be entirely contained within load casts. Most graded beds display a relatively continuous fining upwards from fine to medium sand at the base of the bed to fine silt or clay at the top (Fig. 7C). In beds containing abundant skeletal debris, coarser echinoid, brachiopod and trilobite fragments are concentrated in the basal portions of graded beds.

Shale interbeds are a relatively minor component in the upper 100 feet (30 m) of the Cambrian flat pebble conglomerate sequence and are relatively thin, generally being 1 to 5 mm thick. In the lower 140 feet (43 m), shale beds 2 to 6 cm are common, and a few beds up to 20 cm are occasionally encountered.

Burrows present in the siltstone beds tend to be 1 to 3 mm in diameter, have circular to ovoid outlines, and are filled by silt or very fine sand. Burrows are relatively sparse considering the abundance of skeletal material in both the clastic rocks and lime grainstones. Whole and fragmented skeletal material, particularly phosphatic brachiopods and trilobites, is only rarely observed in shale interbeds but is relatively common in the lower 15 feet (4.6 m) of the core.

A domal stromatolite with a maximum relief of 6 cm occurs at 12,057 feet (3675 m) (Fig. 7D). Several thin beds (3 to 15 cm) of mottled, highly stylolitized micritic limestone, which may represent thrombolites (cryptalgal structures lacking lamination but having a clotted fabric; Aitken, 1967), are present at 11,993.2 (3656 m), 12,040.5 (3670) and 12,056.5 (3675 m) feet.

Only a small percentage of the cored sequence contains beds coarser than fine-grained clastic sandstone or bioclastic grainstone. The coarser beds are predominantly medium- to coarse-grained crinoidal grainstone and flat laminated, and possible hummocky bedded, fine- to medium-grained glauconitic bioclastic sandstone and grainstone. These coarser beds typically are 1 to 10 cm thick, and those at the low end of this range display highly lenticular geometries.

Figure 6.   A.   Microfaults occur in a thin bioclastic siltstone bed compacted over a sand volcano (arrow) at the top of the underlying bed. The sand volcano formed at the outlet of a dewatering vent (dark streak) penetrating the bed and two thinner beds below. The vent is better displayed on the opposite side of the core segment. Core depth - 12,084.2 ft.

B.   Oblique dewatering vents in a medium siltstone are defined by faint jagged clay streaks (center). These vents continue into the overlying lighter colored siltstone beds where they become vertical. The uppermost bed appears to have been affected by boudinage as a result of the dewatering penetrations. Core depth - 12,069 ft.

C.   Siltstone and very fine-grained bioclastic grainstone beds have been separated into clasts with rounded borders by injection of clay and silt along dewatering structures. The clay and silt dikes filling the dewatering vents have been severely contorted by compaction in the interbedded shales. Core depth - 12,164 ft.

D.   Polygonal intersections of clastic dikes at the base of a siltstone bed. In some beds, such dikes are oriented parallel to one another. These dikes are produced by dewatering of an underlying bed. Core depth - 12,164 ft.

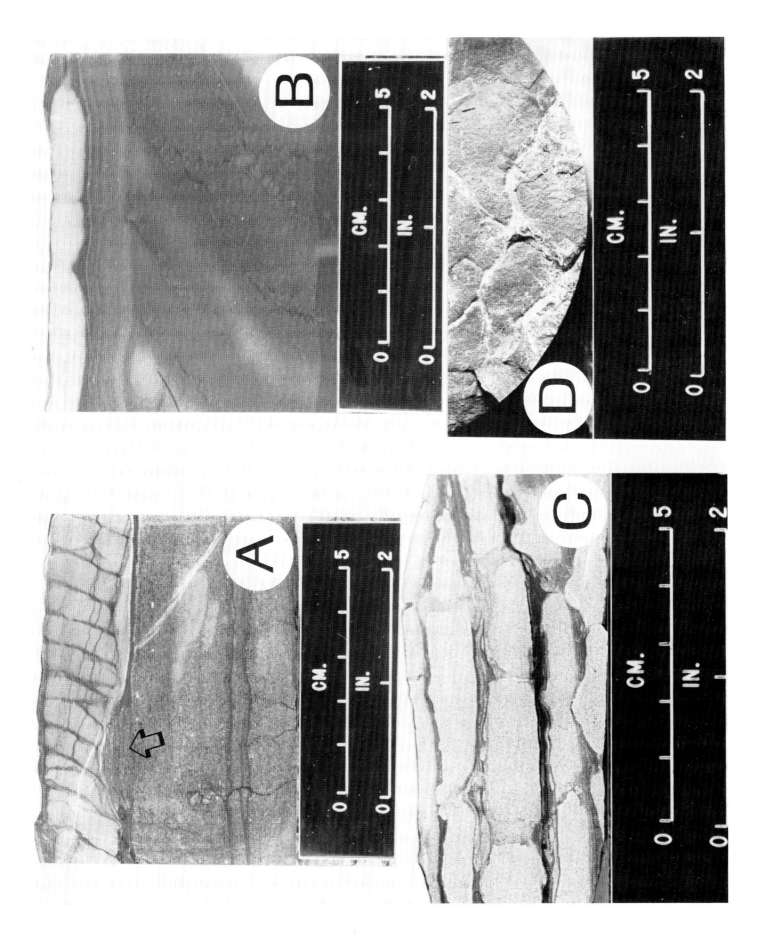

Figure 7. A. Collapse structure interpreted to have formed as a result of the death of a sponge. Swirled material filling the subvertical irregularity is highly siliceous and contains abundant sponge spicules (see photo 7B). The fracture crossing the core is artificial. Core depth - 12,062.5 ft.

B. Swirled mass of rod-shaped sponge spicules (circular in cross section) and glauconitic (black) quartzose silt which fills the sponge collapse structure shown in photo 7A. Scale bar is 0.5 mm in length. Core depth - 12,062.5 ft.

C. Photomicrograph of a thin bed which grades from very fine glauconitic sand at the base to medium silt at the top. Scattered coarser grains of very fine to fine quartz (white) occur at the base of this bed and the overlying graded bed. Scale bar is 4.0 mm in length. Core depth - 12,126.5 ft.

D. Overlapping domal stromatolites (arrows) which occur interbedded with flat pebble conglomerates. The maximum relief recognizable in the core is 6 cm and occurs on the back side of the core segment. Core depth - 12,055.9 ft.

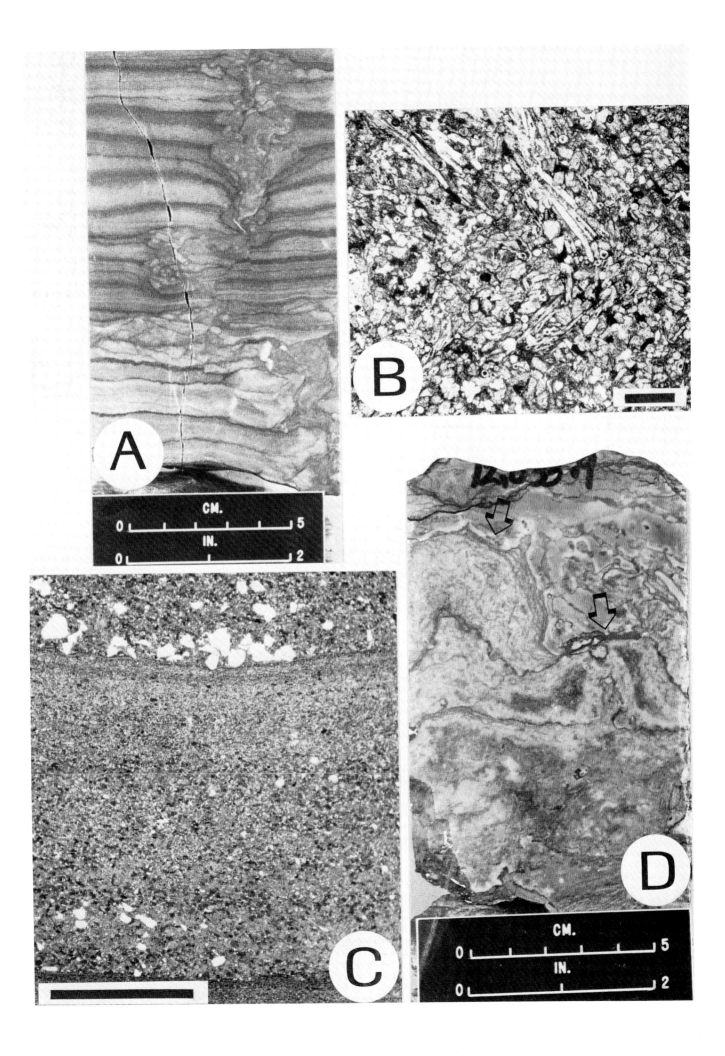

Flat Pebble Conglomerates

In the 233 feet (71 m) of Cambrian flat pebble conglomerate core, 130 flat pebble conglomerate beds were observed. Most beds have thicknesses in the range of 1 to 7 cm (Fig. 8A), but beds up to 60 cm are present. Most beds thicker than 15 cm consist of multiple conglomerate units 3 to 7 cm thick separated by stylolitic clay seams a fraction of a millimeter to a few millimeters thick. The thickest contiguous bed (bed which cannot be segmented into thinner units) is 28.4 cm thick. Excluding beds greater than 16 cm in thickness, the average bed thickness is 5.75 cm.

Most clasts have a disk-type geometry, being 3 to 10 mm thick and 5 to 50 mm in length. The maximum clast length in most beds is 50 to 80 mm, though in many units the largest clast extends beyond the edge of the core in both directions and is greater than 82 mm in length. The maximum clast thickness is generally between 5 and 20 mm (Fig. 8B). A 45 mm clast is the thickest encountered. There is a slight tendency for maximum clast length to increase with increasing bed thickness. Clasts in beds between 1 and 3 cm thickness have maximum lengths ranging from 10 to 40 mm, while those 10 cm or more thick have maximum lengths from 50 to 82+ mm.

Most pebbles are oriented within 20° of the horizontal (Fig. 9A). Clast imbrication (Fig. 9A) occurs in at least parts of about two-thirds of the conglomerate beds. Bimodal imbrication is present in about 10 percent of the beds and produces a weak herringbone pattern (Fig. 9B). In a few beds, the direction of imbrication reverses at least three times. In a small percentage of the beds, clasts assume

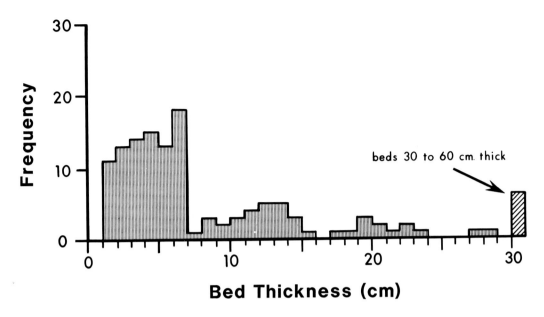

Figure 8. A. Histogram of bed thickness for flat pebble conglomerates. Most beds have thicknesses between 1 and 7 cm. Many of the thicker beds may represent stacked sequences of thinner units.

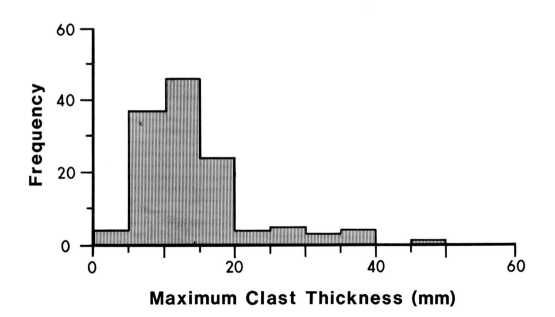

B. Histogram of maximum clast thickness for flat pebble conglomerate beds. In almost 90% of the beds, maximum clast thickness is less than 20 mm.

Figure 9. A. Well-developed imbrication of parallel — laminated glauconitic siltstone, sandstone and grainstone clasts in a flat pebble conglomerate bed 9.5 cm thick. The clasts are oriented at 10 to 25° to the horizontal. Core depth - 12,019.7 ft.

B. Bimodal clast imbrication in a flat pebble conglomerate bed 13.1 cm thick. Imbrication in the lower two-thirds of the bed (clasts dip to right) indicates flow from right to left, while imbrication in the upper one-third (clasts dip to left) indicates flow in the opposite direction. Clasts are rimmed by a hematitic stain. Core depth - 12,184.4 ft.

C. Sigmoidal orientation of flat pebbles in a bed 28.4 cm thick. Many of the clasts are broken by fractures filled by coarse calcite (white). The reorientation of the clasts may have been produced by load casting, wave or current disturbance, or dewatering disruption. Core depth - 12,032.7 ft.

D. Fan-shaped array of flat pebbles in a flat pebble conglomerate bed 5.5 cm thick. Pressure solution suturing produces the interpenetration of pebbles in the center of the bed. Core depth - 12,053.3 ft.

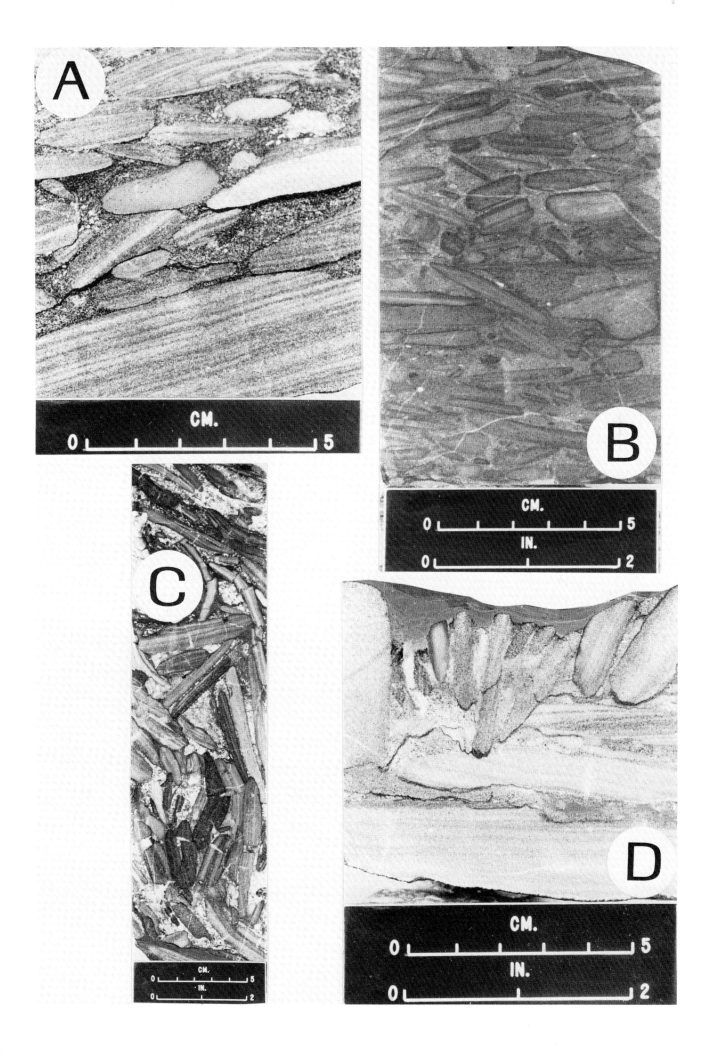

subvertical orientations; and, in a few, a sigmoidal arrangement of clasts is present (Fig. 9C). Some disoriented pebbles occur in fan-like subvertical arrays (Fig. 9D). In some instances, these vertically oriented clasts are associated with irregular streaks of clay which penetrate through a large portion, or all, of a bed.

The tops and bases of most beds are flat or irregular with an overall horizontal orientation. Conglomerate units do not significantly truncate underlying units. Large-scale load casts may be present at the bases of a few beds.

The lateral extent of the flat pebble conglomerates cannot be determined from a 9 cm diameter core. Outcrop studies by Sepkoski (1982) in the Wyoming-Montana area indicate that most beds of flat pebble conglomerate can be traced for tens to hundreds of meters.

Laminations in most clasts parallel the outer planar boundaries of the clasts (Figs. 9A, 9B). Clasts are dominantly subrounded to rounded, but subangular clasts also occur with some frequency. Clasts in the flat pebble conglomerate display the same lithologies as the surrounding thin bedded siltstones, sandstones and grainstones. The most common clast types are bioclastic or peloid siltstone and silty or sandy bioclastic grainstone (Table 1).

Light gray to white lime mudstone, which probably represents fragments of algal stromatolites or thrombolites, forms a modest percentage of the clasts in the upper two pebble counts (12,023 and 12,053.4 feet). Algal stromatolites and thrombolites also are restricted to this same interval. Lime mudstone constitutes only a minor component in lower zones.

	Depth (ft)				
Lithology	12023	12053.4	12104.2	12151	12207.6
Very glauconitic bioclastic siltstone, very fine-grained sandstone, and grainstone					
Light olive gray	55.1	76.0	---	---	---
Light brown	14.6	20.0	---	---	---
Slightly glauconitic bioclastic siltstone, very fine-grained sandstone, and grainstone					
Light olive gray	---	---	---	5.7	30.1
Light brown	---	---	25.0	8.6	27.4
Light brown slightly glauconitic silty bioclastic grainstone	3.4	1.3	68.8	20.0	21.9
Medium gray sandy trilobite grainstone	---	---	---	2.9	1.4
Light brown sandy crinoidal-trilobite grainstone	---	---	---	1.4	8.2
White trilobite wackestone	---	---	---	---	6.8
Light brown crinoidal grainstone	5.6	---	1.6	57.1	---
Light brown peloid bioclastic grainstone	2.2	---	---	---	---
Light gray to white lime mudstone	18.0	16.0	4.7	2.8	2.7
Flat-pebble conglomerate	1.1?	---	---	---	---
Light brown dolomite	---	---	---	---	1.4
TOTAL COUNTS	89	75	64	70	73

Table 1. Lithologic pebble counts for Upper Cambrian flat pebble conglomerates. Values listed are percentages of total clasts.

A small number of the conglomerates contains clasts having rinds of hematite and/or limonite (a general term which encompasses a series of ferric oxides and hydroxides) a fraction to 2 mm thick. The rinds border entire clasts and, thus, were formed after the clasts had developed. Pocket-shaped embayments indicative of borings are rare in most clasts but are relatively common in those clasts which are stained by iron oxides or hydroxides (Fig. 10A). This relationship suggests that the iron-stained clasts were derived from areas, probably tidal flats, which had experienced subaerial exposure.

Thin glauconitic rinds occur along the entire circumference of clasts in several beds. These rinds also occur above iron oxide-hydroxide rims and are concentrated along the borders of borings (Fig. 10A). The distribution of these glauconitic rims indicates they developed subsequent to both deposition of the clasts and formation of the borings. In contrast to the observations of Sepkoski (1982), the glauconitic rinds do not appear to be better developed in the upper portions of conglomerate beds.

PETROGRAPHY

Siltstone/Sandstone/Grainstone/Shale

Detrital Components

Point count analysis (300 points per slide) of ten samples documents a general increase in skeletal grain content as grain size increases (Table 2). Medium siltstones typically contain a high proportion of quartz and micas, and glauconite is a minor component (less than 10 percent).

Grain Type	Grain Size (mm)				
	0.03	0.06	0.07-0.10	0.35	0.50-0.75
Clastic					
Quartz + Feldspar	42.4	58.6	10.4-45.2	2.2	0.6
Glauconite	7.6	4.4	0.0-17.4	0.7	0.0
Micas + Chlorite	47.1	16.6	0.0-0.6	0.0	0.0
Other Clastic	0.9	0.5	0.0-0.6	0.0	0.0
Carbonate					
Echinoid	0.0	9.4	13.1-46.2	64.0	50.0
Trilobite	0.0	0.5	0.0-3.2	5.1	0.0
Brachiopod	0.5	0.5	0.0-3.2	4.4	0.0
Uncertain Fossil (probably trilobite)	1.4	8.3	0.0-14.4	11.0	0.0
Peloids	0.0	0.5	0.0-32.1	0.0	0.0
Micritic Clasts	0.0	0.5	0.0-9.4	12.5	0.0
Glauconitic Dolomite Clasts	0.0	0.0	0.0	0.0	49.4

Table 2. Framework grain content variation with grain size. Grain sizes were estimated visually. Values are in percent of the total framework fraction.

Figure 10. A. Glauconite is concentrated along embayments (arrows) interpreted to have been produced by boring organisms. Borings occur much more commonly on the surfaces of clasts stained by limonite or hematite. Such clasts are presumed to have experienced subaerial exposure. Core depth - 12,203.7 ft.

B. Poorly sorted matrix consisting primarily of echinoid (dark gray) and trilobite (arcuate- and rod-shaped grains) fragments. The pebbles in this bed (large dark gray to black clasts) consist primarily of peloid grainstone. Matrix quartz in this bed is entirely coarse silt to very fine sand in size. Scale bar is 0.5 mm in length. Core depth - 12,058 ft.

C. Poorly sorted matrix containing large amounts of subangular quartz up to very coarse sand size. Such quartz appears abruptly at 12,108 ft and is common in flat pebble conglomerate matrices below this depth. Compare with photo 10B. Scale bar is 4.0 mm in length. Core depth - 12,108.5 ft.

D. Total loss of all visible intergranular porosity in a flat pebble conglomerate bed containing no matrix material. Porosity has been destroyed by a combina- tion of clast fracturing (fractures are filled by white calcite) and pressure solution suturing (dark digitate contacts between clasts). Core depth 10,050 ft.

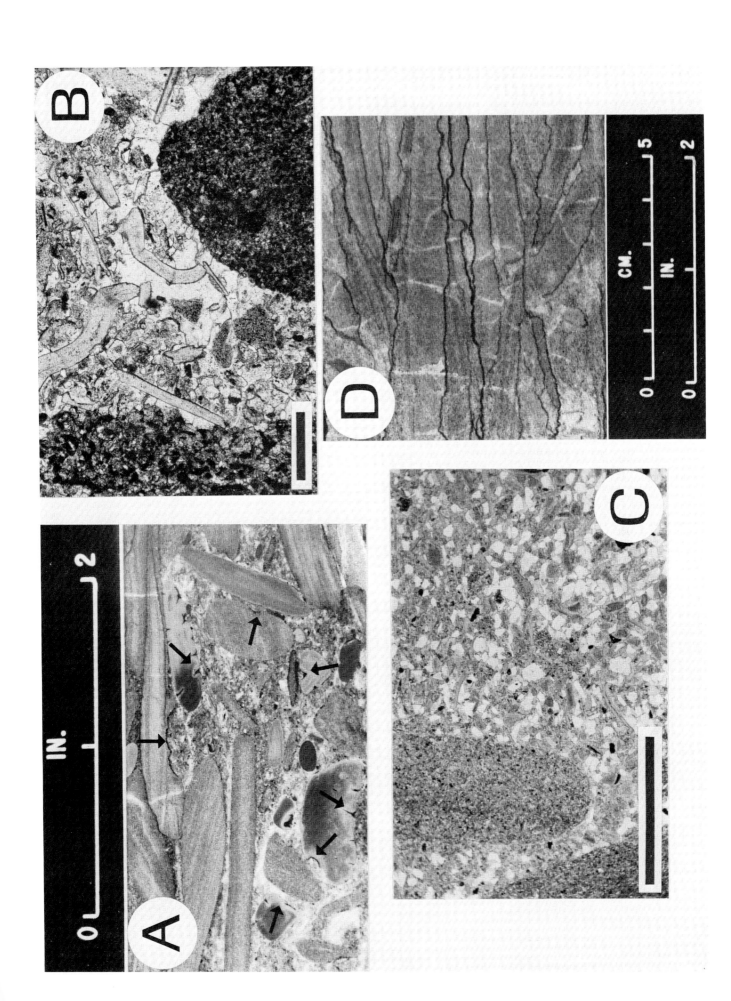

Coarse siltstones to very fine-grained sandstones (Fig. 10B) have subequal amounts of silicic grains (quartz, glauconite) and carbonate grains (echinoderms, trilobites, peloids, micritic fragments). Quartz, echinoid fragments, trilobite fragments, and peloids are the only components whose abundances generally exceed 10 percent, and they may vary significantly between adjacent beds and even between laminae within a bed. Where the content of carbonate framework grains exceeds 50 percent, the term grainstone is applied. Present in very minor amounts in the very fine-grained sandstones and grainstones are micas, microcline, phosphatic and non-phosphatic brachiopod fragments, and phosphate ovules (phosphatic grains displaying no internal structure). Platy grains, such as micas and brachiopod or trilobite fragments, display a strong horizontal alignment in most beds.

Fine- to coarse-grained beds are grainstones dominated by echinoid and trilobite fragments. Rare brachiopod coquinas occur sporadically in the sequence. Large amounts of glauconitic, silty, highly dolomitic clasts occur in some grainstones.

Diagenetic Components

Although all samples were injected with blue-dye epoxy under high vacuum, no visible porosity was observed in any of the siltstone, sandstone or grainstone samples. The dominant cementing agent in all samples examined is fine to coarse spar calcite, most of which formed as syntaxial overgrowths on echinoid fragments (Table 3). These overgrowths extend outwards from the echinoid nuclei, in some cases for distances of up to 0.5 mm, and about or surround adjacent skeletal fragments of other types.

	Siltstones, Sandstones and Grainstones	Flat Pebble Conglomerate Matrix
Syntaxial Calcite and Blocky Calcite	28.7 - 48.7%	22.8 - 43.3%
Bladed Isopachous Calcite	0.0 - 5.7%	0.5 - 5.3%
Micrite (neomorphosed)	grouped with clay matrix	0.0 - 1.0%
Ferroan Dolomite	0.0 - 4.0%	0.3 - 10.0%
Anhydrite	0.0 - 15.3%	0.0 - 1.0%
Iron Sulfide and Oxide	0.0 - 2.7%	0.0 - 0.3%
Clay Matrix	0.0 - 16.0%	0.0 - 4.0%
Quartz Overgrowths	0.0 - 1.7%	0.0 - 0.0%

Table 3. Ranges of abundances of pore-filling agents in siltstones, sandstones, grainstones and conglomerates of the Cambrian flat pebble conglomerate sequence, Shell Oil #1 Clear Creek. Values are percentages of the total rock and are based on point counts of 300 points.

Where, due to the local paucity of echinoderm fragments, syntaxial cements have not severely invaded portions of a sample, bladed isopachous and blocky spar cements fill the bulk of the pores. In most samples, such cements constitute no more than 2 percent of the total rock. The bladed isopachous cements are best developed on brachiopod fragments and may extend radially up to 0.15 mm outwards from the surface of these fragments. Isopachous rims are occasionally encountered on trilobite fragments and, though they may extend up to 0.15 mm away from the grain surface, usually they are much thinner (up to 0.03 mm). The blocky spar cement occurs primarily where pore space has not been destroyed by the previously noted syntaxial and isopachous cements. It is most common in peloidal grainstones.

Cements of secondary importance include ferroan dolomite, pyrite, anhydrite and quartz overgrowths. Ferroan dolomite occurs in may samples and ranges in abundance from a trace to 4 percent. It forms rhombic crystals 0.02 mm in length which replace framework grains and earlier-formed cements. It displays curved crystal faces and wavy extinction typical of saddle dolomite (Radke and Mathis, 1980). A ferroan composition is confirmed by staining (bluish-green color when Dickson's dual carbonate stain is applied) and by X-ray diffraction analysis.

Pyrite forms silt-sized pore-filling ovoid grains and irregular patches. In the upper 11 feet (3.4 m) of the Cambrian flat pebble conglomerate sequence, it has totally altered to hematitle and limonite. In most samples, anhydrite is absent or is present in only trace amounts. In a few samples, it occurs in amounts up to 15 percent. It forms irregular patches or blocky masses up to 2 cm in diameter which indiscriminately replace all other framework and pore-filling components, or it locally may preferentially replace ferroan dolomite. It is most abundant in the upper 15 feet (4.6 m) of core.

Quartz overgrowths occur only where carbonate cements are relatively limited and never comprise more than 2 percent of the total rock. Well developed overgrowths also have been observed on microcline grains. Illitic clay matrix also occurs as a minor pore filler. In the sandstones and grainstones, it occurs in very minor amounts (0 to 2 percent). In most siltstones, it is considerably more abundant (3 to 16 percent).

Pressure solution suturing is prominent where clay streaks and laminae are present. It is probably the major source of carbonate cement in the siltstone, sandstone and grainstone beds. Ductile deformation of glauconite and micas in those sandstones and grainstones probably contributes slightly (2 to 8 percent?) to porosity reduction. Micas are abundant in fine and medium siltstones, and ductile deformation of these causes a porosity loss of approximately 15 to 25 percent.

The very rapid early stage cementation by syntaxial and, to a lesser extent, by bladed isopachous cements very quickly eliminated any significant intergranular porosity in the siltstones, sandstones and grainstones of the Cambrian flat pebble conglomerate sequence. Excluding samples with moderate to abundant ductile grain content, minus cement porosities (visible porosity plus all pore-filling components) indicate initial porosities were probably in the range of 45 to 55 percent. The Cambrian sequence in the Clear Creek area has experienced relatively deep burial (greater than 10,000 to 12,000 feet) since the Upper Mississippian. Without the opportunity for extensive exposure and dissolution since this time, the prolonged burial for the Cambrian flat pebble conglomerate sequence in the central Powder River Basin precludes preservation of more than trace amounts of porosity. A possible exception to this statement might be the generation of porosity through the solution enlargement of fractures associated with major folds or faults.

Flat Pebble Conglomerates

Detrital Components

Most flat pebble conglomerate beds contain a range of clast types. Generally, three or four types dominate an assemblage. The lithologies present are the same lithologies encountered in the surrounding siltstones, sandstones and grainstones. Light olive gray and light brown bioclastic siltstones, very fine-grained sandstones and grainstones, along with light gray to white lime mudstone, are the most common clast types. Glauconite content of these clasts ranges from a trace to 20 percent. Glauconite is most abundant in conglomerates interbedded with glauconitic sandstones that occur at 12,019 to 12,024 feet (3663 to 3665 m) and 12,049 to 12,055 feet (3672.5 to 3674 m). Clasts of coarser (fine- to medium-grained) crinoidal and trilobite grainstones and wackestones seldom form more than 10 percent of a particular assemblage. Within individual clasts, lithology may vary widely from one lamina to another. As noted by Sepkoski (1982), no shale or mudstone clasts are present in the flat pebble conglomerates.

The matrix in the conglomerates consists of primarily fine to very coarse sand-sized echinoderm, trilobite and brachiopod fragements (Fig. 10B). The former two are by far the most abundant. Quartz, in the form of coarse silt-sized grains, forms a decreasing (26 percent falling to 4 percent) proportion of the matrix above 12,108 feet (3691 m). Below this depth, very fine to very coarse quartz (Fig. 10C) forms about 25 to 30 percent of the matrix (Fig. 11). The matrix in most samples is poorly to very poorly sorted. Silt, clay and micrite have infiltrated the matrix of some beds.

The coarser quartz is almost entirely monocrystalline and displays consistently moderate to strong wavy extinction. Grain outlines are commonly very irregular (embayed). Most grains are subangular and have relatively low sphericity. A small percentage of polycrystalline grains occurs. Some are granulated, while a few others are highly quartzose quartz-muscovite schist. Rare grains of quartz with swirls of pyrophyllite were observed in samples at 12,108.5 and 12,215.6 feet (3691 and 3723 m). At 12,215.6 feet (3723 m), a grain of quartz-kyanite or quartz-sillimanite in which the aluminous mineral has been altered to pyrophyllite or muscovite is present. The grain shapes and compositions suggest that this coarser quartz has experienced only a single cycle of weathering and transport and was derived primarily from a granitic source with a minor contribution from low to high grade metasedimentary rocks.

Diagenetic Components

As shown in Table 3, the abundances of the pore-filling components present in the matrix of the flat pebble conglomerates are

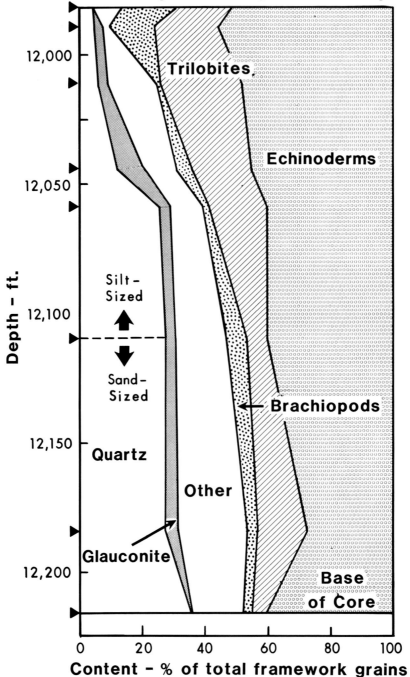

Figure 11. This plot shows a depth profile of matrix composition in flat pebble conglomerates as measured by point count analysis (300 points per sample). There is an abrupt change in the content and type of quartz at 12,108 ft. Above this depth, quartz in the conglomerates is silt or very fine sand in size and forms a relatively low proportion of the matrix. Below this depth, quartz up to very coarse sand size is common, and quartz constitutes a significantly greater part of the matrix.

very similar to those observed in the associated siltstones, sandstones and grainstones. The dominant pore filler is syntaxial calcite (along with much lesser amounts of blocky calcite). Bladed isopachous calcite and ferroan dolomite are the only other cements which form a significant portion of most samples. Ductile deformation of glauconite is a major mechanism of porosity loss only in those zones where glauconite is abundant in the matrix. Bladed isopachous cements are quite prominent in some conglomerates relative to their abundance in most siltstone beds. The more extensive development of this cement is related to the increased grain size of the framework grains in the matrix. The greater distance between the larger grains prevents syntaxial cements from enveloping as many trilobite and brachiopod fragments in the conglomerates. In conglomerates where matrix material was not deposited between the flat pebbles, large voids were present at the time of deposition. As a result of overburden stress, these voids have been totally destroyed by extensive fracturing of clasts and pressure solution suturing along clast boundaries (Fig. 10D).

ENVIRONMENT OF DEPOSITION

The environment of deposition can be reconstructed from a number of lines of evidence. Many features in the Cambrian flat pebble conglomerate sequence indicate rates of deposition were relatively rapid. Sedimentary structures which lead to this conclusion include load casts, graded bedding, climbing ripples, piled ripples, ball and pillow structures, and dewatering structures. Negative evidence for a rapid rate of sedimentation includes the lack of hardgrounds and bored surfaces. The occurrence of embayments, interpreted to be borings, in limonite- and hematite-stained clasts in a few isolated intervals, does suggest that some clasts were derived from subaerially-exposed intertidal environments.

The prevalence of glauconite in both the finer grained deposits as well as the flat pebble conglomerates tends to contradict the previous interpretation. Most specialists regard glauconite as forming most readily in areas where rates of deposition were quite low (see summary in McRae, 1972). In this case, the author feels it is more appropriate to place greater emphasis on the sedimentary structures and to assume that rates of glauconite formation were adequate to overcome the relatively rapid rates of deposition.

As indicated by the predominance of relatively fine grained deposits, the average energy level at the site of deposition is consistently low. These finer grained beds were deposited by relatively modest current and wave activity. Both upper flow regime (horizontal parallel lamination) and lower flow regime (ripple lamination) structures are common.

Water depths are below normal wave base; although, since the site of deposition is protected on the oceanward side by a carbonate bank, the depth to normal wave base may be relatively low. Had deposition occurred above normal wave base, constant wave and current action would have prevented the deposition of the numerous siltstone and shale beds, and the dominant lithologies present would have been ripple-laminated and cross-bedded relatively coarse grainstone or quartzose sandstone. The single occurrence of a domal stromatolite and presence of several possible thrombolites suggest subtidal to, at the shallowest limit, lower intertidal environments (Aitken, 1967).

The large number of flat pebble conglomerate beds in the sequence records frequent pulses of relatively high energy. Sepkoski (1982) and Demicco and Hardi (1981) favor a storm origin for the flat pebble conglomerates. Such an origin is incompatible with the characteristics of the flat pebble conglomerates in cores from the Shell Oil #1 Clear Creek. With one exception, the conglomerates do not exhibit an upward gradation to finer deposits of sandstone, siltstone or grainstone; nor is there any size grading of the flat pebbles within any of the conglomerate beds. Instead, conglomerates are abruptly overlain by shale or mudstone, a relationship also noted by Sepkoski (1982). Grading is common, however, in many of the finer grained deposits of sandstone, siltstone and grainstone.

The development of bimodal imbrication in some of the flat pebble conglomerate beds argues strongly for a tidal current origin for these deposits. In contrast, imbrication was observed only rarely in the flat pebble conglomerates studied by Sepkoski (1982). Grading to finer deposits is not observed at the tops of conglomerate beds. Current and/or wave energies subsequent to deposition of the conglomerates were adequate to maintain traction or suspension transport of the finer fraction. The thin graded siltstone, sandstone and grainstone beds which are interbedded with the conglomerates may represent storm deposits.

The jumbled and swirled clast orientations encountered by Sepkoski also are present in some beds in the Shell Oil #1 Clear

Creek. The disruption of clasts in the conglomerates may have been produced by one or more of the following mechanisms:

1) Water and sediment is injected from below as a result of dewatering initiated by rapid deposition of a conglomerate bed.

2) Reversal of tidal currents may have partially reoriented previously deposited imbricated clasts, or local scour may have removed matrix from around groups of clasts, causing them to "slump" into a subvertical position.

3) Oscillatory currents produced by wave activity, possibly related to storms, may have reoriented clasts. Sepkoski (1982) proposed this mechanism as an explanation for the development of edgewise conglomerates.

4) Development of large scale load casts at the base of rapidly deposited conglomerates might promote reorientation of clasts within the conglomerate.

The rounded clasts in the flat pebble conglomerates are not created by storm disruption of thin siltstone beds and transport of the broken fragments. Rounding occurs syndepositionally when dewatering produced by rapid loading disrupts beds. Water and sediment injected upwards through a thin bed of siltstone rounded the edges of clasts bordering the escape vent. Beds of rounded clasts separated by dewatering vents locally have the appearance of boudinage features.

As indicated by the lack of broken clasts and the low degree of rounding and sphericity of the sand-sized quartz in conglomerate matrices, transport of the clasts and matrix was not extensive. Sepkoski (1982) has hypothesized that the flat pebble clasts were rigid at the time of deposition. As evidence, he cites the lack of plastic deformation of these clasts. Observations made in this study confirm that plastic deformation has not occurred. Close examination of clasts which appear to be plastically deformed reveals that they have been deformed by fracturing.

However, little evidence could be found to substantiate Sepkoski's claim of early cementation. No radial fibrous marine cements or meniscus or pendant vadose cements are present in the flat pebble clasts or in the finer grained interbeds. Very localized and generally very thin rims of bladed isopachous cement do occur but do not appear adequate to lend rigidity to the flat pebble clasts. No hardgrounds or bored surfaces were recognized anywhere in the core. It is expected that such surfaces would be common in the flat pebble sequence if rates of sedimentation were low and significant early cementation were prevalent.

Related to this problem is the relative lack of burrowing in the siltstone, sandstone and grainstone interbeds. The faunal components (echinoids, trilobites, brachiopods) suggest normal marine salinities. Yet burrowing is generally absent, and fossils are seldom encountered in the interbedded shales. An absence of fossils also has been reported by Aitken (1978) for similar deposits in the Cambrian of western Alberta. As suggested by Sepkoski (1982), the lack of burrowing may be simply explained by the lack of an extensive infauna during the Cambrian. However, other authors report extensive burrowing in marine Cambrian deposits (Demicco, 1983; Cook and Taylor, 1977; Kurtz, 1976).

It is possible that hypersaline conditions may have existed within the broad shallow embayment in which the flat pebble conglomerates formed. The low latitude coastal position of western North America during the late Cambrian would favor such a possibility (Ziegler and others, 1981). The abrupt shift from quartz sand-bearing matrices to quartz silt-bearing matrices in the flat pebble conglomerates may reflect a change in climatic conditions. The sand-sized material was probably introduced by fluvial processes to the basin of deposition. The abrupt decrease in size of the clastic grains in the matrix may reflect a shift to an aeolian mechanism of clastic input.

An explanation for the abnormally high incidence of flat pebble conglomerates in Cambrian and Lower Ordovician deposits remains to be resolved. Sepkoski (1982) relates flat pebble conglomerate abundance to evolutionary trends in marine fauna during the early Paleozoic. Expansion of marine infaunal organisms in the Ordovician produced more extensive burrowing of marine carbonates. This burrowing prevented the preservation of thin marine storm deposits which could be disrupted and transported as flat pebbles. This hypotheses, with the exception of the suggestion that storms are the major mechanism of transport and deposition of flat pebble conglomerates, appears to be quite viable.

However, as was discussed previously, tidal currents are interpreted to be the mechanism of transport and deposition, and it is reasonable to inquire whether variations in tidal range may play a role in the irregular distribution of flat pebble conglomerates with geologic age. This possibility was first suggested to the author by G. R. Davies. The occurrence of abnormally high tides during the Cambrian and Lower Ordovician would not only explain the greater abundance of flat pebble conglomerates during this period, but it would also provide a possible rationale for the great thicknesses and abnormally wide distribution of tidal flat and subtidal lithofacies.

Klein (1977) evaluated paleotidal ranges using both the study of thicknesses of fining-upward tidal sequences and the frequency distribution of tidalites with geologic age. Using both types of data, Klein concluded that paleotidal range had not varied significantly since the Precambrian. This author is of the opinion that both the accuracy of the measurement technique and number of measurements are probably inadequate to properly evaluate paleotidal range in the Cambrian.

Severe doubt has been cast on attempts to estimate tidal range by means of growth relief of intertidal stromatolites (Piper, 1978). However, both stromatolite data (Cloud, 1968) and application of Klein's tidal sequence thickness by von Brunn and Hobday (1976) indicate tidal range in at least portions of the Precambrian was significantly higher than it is at present.

As reviewed by Scrutton (1978), stromatolites from Middle Cambrian rocks in Siberia up to 15 m high have been reported (Walter, 1970). These data would suggest, if such growth relief data were meaningful, that the moon's closest approach to the earth would have been in the Middle Cambrian. At present there do not appear to be any reliable techniques for estimating tidal range in the Cambrian.

Resolution of this issue must await the development of new techniques or improvement and wider application of established techniques.

Upper Cambrian paleogeography may have favored oceanic circulation patterns which resulted in a significant reduction of absorbed solar radiation (Ziegler and others, 1981). If the Cambrian was significantly cooler than succeeding periods, the loci of glauconite formation might have been shifted to shallower depths than those in which it is found today. McRae (1972) notes that in the present-day tropics warm surface waters restrict glauconite formation to depths in excess of 250 m, while in non-tropical areas it may occur at depths as shallow as 30 m.

Since glaucontie generation appears to be favored by slightly reducing conditions (McRae, 1972), an increased abundance of this mineral may be associated with widespread anoxic events during the geologic past. Wright and others (1984) have documented cerium anomalies in apatite of Cambrian conodonts which suggest anoxic conditions were prevalent in the lowermost Paleozoic oceans. The combination of both cooler and more anoxic oceanic waters during the Cambrian may have been the key which allowed glauconite to form rapidly in relatively shallow marine waters and in locations where rates of deposition would normally have suppressed glauconite development.

CONCLUSIONS

Flat pebble conglomerate beds of the Upper Cambrian sequence in the Shell Oil #1 Clear Creek were deposited as thin sheets by tidal currents in a relatively shallow low energy marine basin. The basin was protected on the seaward side by a peritidal bank. Evidence supporting this interpretation includes:

(1) Congolmerate beds are generally thin (mostly 1 to 7 cm), do not grade upwards into finer deposits, and do not significantly truncate underlying beds.

(2) Many conglomerate beds display bimodal imbrication.

(3) Interbedded deposits are bioclastic (crinoid-trilobite) siltstones, very fine-grained sandstones and grainstones which exhibit structures indicating rapid sedimentation. These structures include abundant dewatering channels, along with load casts, convolute bedding, ball and pillow structures, climbing ripples, microfaulting, and sponge collapse structures.

(4) Rare domal stromatolites and possible thrombolites occur in the sequence.

The quartz content in the matrix of the flat pebble conglomerates changes abruptly from silt and very fine sand to much coarser sizes below 12,108 feet (3691 m). This change most likely reflects a change in the mechanism of clastic influx. The silt and very fine sand were probably introduced by aeolian action, while the coarser quartz could only have been carried in by fluvial processes. This shift also may be recognizable in equivalent Upper Cambrian deposits of western North America.

The relative abundance of flat pebble conglomerates in the Cambrian and Lower Ordovician may be controlled by one or both of two mechanisms. As proposed by Sepkoski (1982), lack of an extensive infaunal burrowing community may allow thin bioclastic siltstone and grainstone beds to experience sufficient syndepositional cementation to permit transport and deposition as cohesive flat pebble clasts. Alternatively, tidal amplitudes in the Cambrian may have been significantly greater than those in subsequent periods. Evidence for or against higher tides in the Cambrian is both slight and inconclusive, and resolution of this problem must await additional investigations.

ACKNOWLEDGMENTS

The author would like to thank Sabine Corporation and Houston Natural Gas Oil Company for release of the core description used in this paper. Special thanks also are due the staff of the United States Geological Survey Core Library in Denver (Tom Michalski, Director) for providing the needed core layout facilities. Discussions with Michael Taylor of the United States Geological Survey and his dating of trilobite specimens from the core have added significantly to the content of this paper. My wife, Sandra Wilson, typed and helped edit and proof the paper.

REFERENCES

AITKEN, J. D., 1967, Classification and environmental significance of cryptalgal limestones and dolomites with illustrations from the Cambrian and Ordovician of southwestern Alberta: Jour. Sed. Petrol., v. 37, p. 1163-1178.

AITKEN, J. D., 1978, Revised models for depositional grand cycles, Cambrian of the southern Rocky Mountains, Canada: Bull. Canadian Petrol. Geol., v. 26, p. 515-542.

von BRUNN, V., and HOBDAY, D. K., 1976, Early Precambrian tidal sedimentation in the Pongola Supergroup of South Africa: Jour. Sed. Petrol., v. 46, p. 670-679.

CLOUD, P. E., 1968, Atmospheric and hydrospheric evolution of the primitive earth: Science, v. 160, p. 729-736.

COOK, H. E., and TAYLOR, M. E., 1977, Comparison of continental slope and shelf environments in the Upper Cambrian and lowest Ordovician of Nevada: in Cook, H. E., and Enos, P. (eds.), Deep Water Carbonate Environments, Soc. Econ. Paleont. Mineral. Spec. Pub. 25, p. 51-81.

DEMICCO, R. V., and HARDI, L. A., 1981, Patterns of platform and off-platform carbonate sedimentation in the Upper Cambrian of the central Appalachians and their implications for sea level history: in Taylor, M. E. (ed.), Short Papers for the Second Intl. Symp. on the Cambrian System, U. S. Geol. Surv. Open-File Rept. 81-743, p. 67-70.

DEMICCO, R. V., 1983, Wavy and lenticular-bedded carbonate ribbon rocks of the Upper Cambrian Concocheague Limestone, central Appalachians: Jour. Sed. Petrol., v. 53, p. 1121-1132.

HUBERT, J. F., SUCHECKI, R. K., and CALLAHAN, R. K. M., 1977, The Cow Head Breccia: Sedimentology of the Cambro-Ordovician continental margin, Newfoundland: in Cook, H. E., and Enos, P. (eds.), Deep Water Carbonate Environments, Soc. Econ. Paleont. Mineral. Spec. Pub. 25, p. 125-154.

KLEIN, G. deV., 1977, Clastic Tidal Facies: Continuing Education Publication Company, Champaign, Ill., 149 p.

KURTZ, V. E., 1976, Biostratigraphy of the Cambrian and lowest Ordovician, Bighorn Mountains and associated uplifts in Wyoming and Montana: Brigham Young Univ. Geol. Stud., v. 23, pt. 2, p. 215-227.

LOCHMAN-BALK, C., 1970, Upper Cambrian faunal patterns on the craton: Geol. Soc. Am. Bull., v. 81, P. 3197-3224.

LOVE, J. D., and CHRISTIANSEN, A. C., 1983, Preliminary Geologic Map of Wyoming: U. S. Geol. Surv. Open File Rept. 83-802.

MAZULLO, S. J., 1978, Early Ordovician tidal flat sedimentation, western margin of Proto-Atlantic Ocean: Jour. Sed. Petrol., v. 48, p. 49-62.

MAZULLO, S. J., AGOSTINO, P., SEITZ, J. N., and FISHER, D. W., 1978, Stratigraphy and depositional environments of the Upper Cambrian-Lower Ordovician sequence, Saratoga Springs, New York: Jour. Sed. Petrol., v. 48, p. 99-116

McRAE, S. G., 1972, Glauconite: Earth-Sci. Rev., v.8, p. 397-440.

MILLER, B. M., 1936, Cambrian stratigraphy of northwestern Wyoming: Jour. Geol., v. 44, p. 113.

PALMER, A. R., 1971, The Cambrian of the Great Basin and adjacent areas, western United States: in Holland, C. H. (ed.), Cambrian of the New World, John Wiley and Sons, p. 1-78.

PIPER, J. D. A., 1978, Geological and geophysical evidence relating to continental growth and dynamics and the hydrosphere in Precambrian times: A review and analysis: in Brosche, P., and Sundermann, J. (eds.), Tidal Friction and the Earth's Rotation, Springer-Verlag, New York, p. 197-241.

RADKE, B. M., and MATHIS, R. L., 1980, On the formation and occurrence of saddle dolomite: Jour. Sed. Petrol., v. 50, p. 1149-1168.

SEPKOSKI, J. J., Jr., 1982, Flat pebble conglomerates, storm deposits, and the Cambrian bottom fauna: in Einsele, G., and Seilacher, H. (eds.), Cyclic and Event Stratification, Springer-Verlag, Berlin, p. 371-385.

SCRUTTON, C. T., 1978, Periodic growth features in fossil organisms and the length of the day and month: in Brosche, P., and Sundermann, Jr. (eds.), Tidal Friction and the Earth's Rotation, Springer-Verlag, New York, p. 154-196.

WALTER, M. D., 1970, Stromotolite used to determine the time at nearest approach of Earth and Moon: Science, v. 170, p. 1331-1332.

WRIGHT, J., SEYMOUR, R. S., and SHAW, H. F., 1984, REE and Nd isotopes in conodont apatite: Variation with geologic age and depositional environment: in Clark, D. L. (ed.), Conodont Biofacies and Provincialism, Geol. Soc. Amer. Spec. Paper 196, p. 325-337.

ZIEGLER, A. M., PARRISH, J. T., and SCOTESE, C. R., 1981, Cambrian world paleogeography, biogeography, and climatology: in Taylor, M. E. (ed.) Short Papers for the Second Intl. Symp. on the Cambrian System, U.S. Geol. Surv. Open-File Rept. 81-743, p. 252.

DEPOSITIONAL AND DIAGENETIC ALTERATION OF YEOMAN (LOWER RED RIVER) CARBONATES FROM HARDING CO., SOUTH DAKOTA

A. C. KENDALL
Sohio Petroleum Company
50 Fremont St.
San Francisco, California 94105

ABSTRACT

Cores have been examined in detail from two wells located six miles apart in northwest South Dakota. They exhibit different degrees of dolomitization in the Upper Ordovician Yeoman (Upper Red River) Formation.

Depositional characteristics and early diagenetic alteration of these carbonates is similar to that of the Yeoman of the northern Williston Basin. Early diagenetic cementation as previously proposed (Kendall, 1977) is now questioned.

In addition to dolomite-mottled carbonates that arose by preferential dolomitization of burrow networks, the Yeoman sequence contains carbonates that exhibit nodular structures (produced by preferential lithification of burrows) and those that lack conspicuous burrow structures. These different sediment types have markedly affected the pathways taken by later diagenesis. Variations in skeletal carbonate abundance are primary and have not been significantly affected by later dolomitization. Stratigraphic units composed of skeletal-poor wackestones and mudstones parallel thin euxinic organic rich beds and are thus considered isochronous. One unit (the "D" porosity zone) has a basinwide distribution and has been particularly prone to being dolomitized. It may even correlate with the Cat Head Member of the Manitoba outcrop. Variations in Yeoman lithology are ascribed to variations in water oxygenation during deposition and many Yeoman carbonates were deposited under dysaerobic conditions. Yeoman deposition is believed to have occurred in deep shelf environments.

Chemical compaction has had a profound effect and is responsible for the removal of matrix from many dolomites and the creation of diagenetic grainstone textures by concentrating the more resistant components in the matrix. Breccias composed of broken and reoriented dolomite mottles closely resemble depositional features but also result from matrix elimination by pressure solution. Fractures associated with pressure solution seams and stylolites are identified: previously they had been regarded as being early diagenetic. Relations between dolomitization and chemical compaction are ambiguous and require further study.

INTRODUCTION

The initial purpose in studying two Red River cores to be used in this core conference was to seek petrographic evidence for the number of dolomitization episodes represented. This has considerable importance in deciding between the merits of the various dolomitization mechanisms that have been proposed for the Formation (see in particular the discussions of Kendall, 1984 and Longman and others, 1984). During core examination various features of the Red River depositional environment and diagenetic alteration were noted. These had hitherto gone unnoticed or have received only scant attention. Some, in my opinion, have also been misidentified or misinterpreted. They are believed to be important to a correct understanding of the Red River.

Cores come from two wells located in northwest South Dakota and were in the upper part of the burrow-mottled Lower Red River (equivalent to the "C"-burrowed Member of Kohm and Louden, 1978, and the Yeoman Formation of Kendall, 1976). This interval is normally described as being lithologically rather uniform (if not boring!). Closer examination reveals a significant variation in both depositional and diagenetic features (and an inter-relationship between the two), some of which have been noted elsewhere by Carroll (1978) and Derby and Kilpatrick (1985).

Three aspects of the burrow-mottled carbonates will be emphasized:

(1) The similarity of early diagenetic products in the Lower Red River across the entire Williston Basin. The early diagenesis of Red River carbonates has previously been reported (Kendall, 1976, 1977) and here new evidence and a minor change in interpretation are presented.

(2) The lithologic variability of the Lower Red River. This, with the exception of Carroll's (1978) work has largely gone unnoted. In particular, a restriction of the indigenous fauna within a widespread porous dolomite interval is recognized and is probably equivalent to that reported by Derby and Kilpatrick (1985) from Killdeer Field in North Dakota. This interval may have basin-wide distribution and is equivalent to the Cat Head Member in Manitoba outcrops. In addition, differing early-diagenetic overprints, that presumably reflect slight changes in the depositional environment, are recognized. These have provided the templates for the development of rock units that behaved very differently during later diagenesis.

(3) Later diagenetic changes that have affected the burrow-mottled carbonates. These, particularly stylolitization and compactive fracturing, have created rock textures and structures that have been misinterpreted as being of depositional or early-diagenetic origin. Thin intervals of breccia in one of the cores used in this presentation closely resemble depositional features. They

are, however, interpreted to have formed as a result of chemical compaction during late diagenesis.

GEOLOGIC SETTING

The two wells studied are located six miles apart in Harding County, South Dakata. This places them at the southern edge of the lowest Red River anhydrite (the "C" or Lake Alma anhydrite) (Fig. 1) so that each well contains a complete lowermost Red River depositional cycle (Fig. 2). The cores from these wells are of interest because both extend over almost the exact same stratigraphic interval, yet in one well, the Webb Cenex #7-9 Turbiville (NE SE Sec. 7, T20N, R6E), the carbonates are largely dolomitic limestones with only thin fully dolomitized units, whereas in the other well, the Sohio #11-8 Brown (SE NE Sec. 11, T20N, R5E), the greater part of the cored interval is completely dolomitized.

Porosity logs from the two wells (Fig. 2) clearly reveal a greater development of porosity in the more dolomitized Sohio #11-8 Brown well. Nevertheless, almost identical porosity and tight peaks can be recognized in the two wells suggesting that they possess identical microstratigraphies and that differential complete dolomitization has not obscured this. It must be pointed out, however, that detailed core correlation suggests that the microstratigraphy in the two wells is not exactly identical and that intervals in the Turbiville well are thicker than those in the Brown well (Fig. 3). Most of the discrepancy, however, appears to reside in intervals in the Turbiville well that are identified as having no recovery. It is believed that these are absent

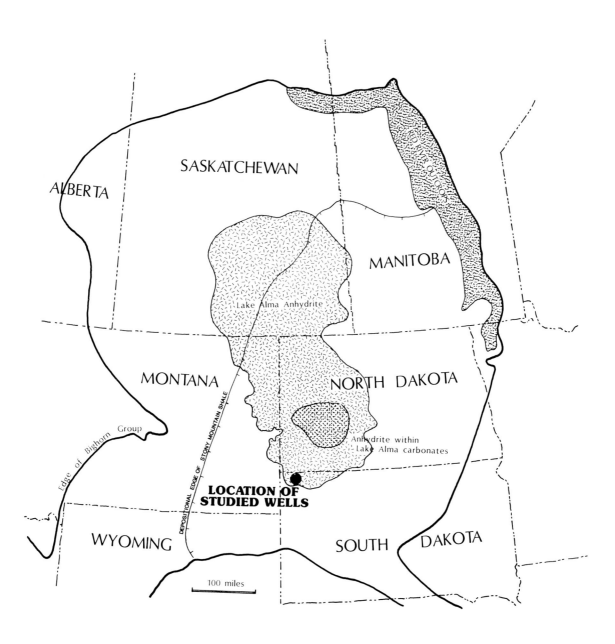

Figure 1. Location map of the Williston Basin. The wells studied lie near the depositional edge of the Lake Alma Anhydrite (Herald or Upper Red River Formation).

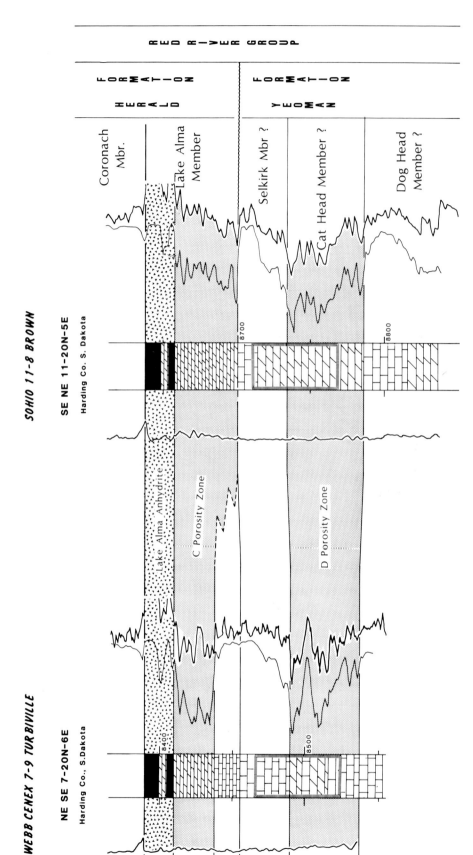

Figure 2. Correlations through the wells studied of note are: (1) the undolomitized nature of the Lower Lake Alma Member in the Turbiville well (with a corresponding thickness reduction of the "C" porosity zone, (2) the greater extent of dolomitization in the Yeoman Formation in the Brown well, (3) the greater development of porosity in the "D" zone of the Brown well but nevertheless, and (4) a similarity in the thicknesses of porosity-defined units in the Yeoman Formation of both wells.

or represent considerably lesser thicknesses than reported, but this cannot be proven. Unfortunately, core-gamma logging is unable to locate portions of the core relative to the gamma ray log of the well. The inability to match core with logs is most unfortunate because, as will be seen, the ability to recognize whether or not thin stratigraphic intervals change thickness with dolomitization is important to timing the dolomitization episode.

The Red River is commonly subdivided into two units. In the center of the basin the Upper Red River is composed of three cyclic units, each of which terminates in a bed of anhydrite. Laminated dolomites beneath the anhydrites are porous and, in the United States, are commonly labelled porosity zones "A" through "C" in descending order. A fourth porosity zone (the "D" zone) is present within the Lower Red River but differs considerably from the higher zones in that it is not composed of laminated sediments. I have suggested elsewhere (Kendall, 1976) that the lithological differences between the Upper and Lower Red River units (together with the history of miscorrelation between surface and subsurface that has affected the Red River) are sufficient to recommend that they be recognized as different formations - the Herald and Yeoman Formations, respectively. Kendall (1976) further subdivided the Yeoman Formation into informally recognized units, and the Herald Formation into three divisions, the lower two of which correspond with the lower two depositional cycles and were identified as members.

The appearance of an interval lying 50 or more feet beneath the top of the Yeoman Formation (= Lower Red River) which is prone to be preferentially dolomitized and develop porosity is apparently

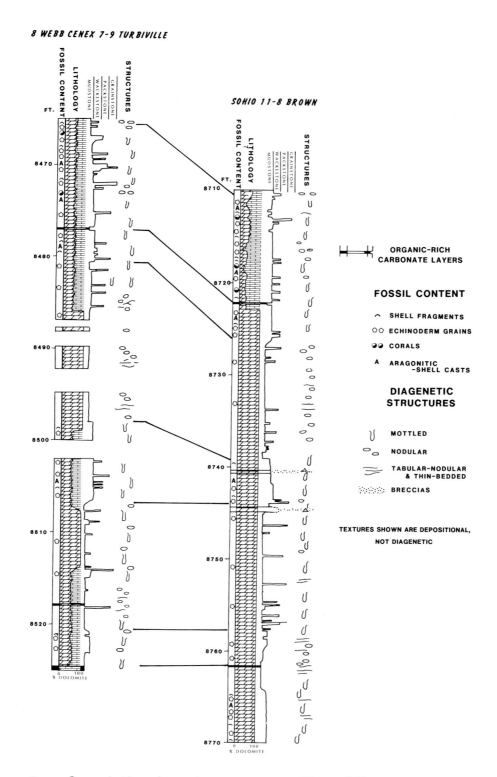

Figure 3. Correlation based on cores. The differences in thickness of the units in the two wells are largely caused by intervals of "no recovery" in the Turbiville well. These variations are not reflected in log correlations (Fig. 2) suggesting the "no recovery" intervals may not exist or are of lesser extent than shown. Core depths based upon those marked onto the cores.

basin-wide, occurring wherever the Yeoman has not been completely dolomitized. It can be correlated with a similar porous dolomite unit that occurs in the shallow subsurface and outcrop belt of Manitoba and which McCae (pers. comm., 1976) suggests may equate with the Cat Head Member (see also Cowan, 1970). The Dog Head, Cat Head and Selkirk Members of the Red River (= Yeoman) outcrop area of Manitoba (Dowling, 1900; Baillie, 1952) have never been convincingly correlated with any subsurface recognized subdivisions of the Formation. If McCabe's correlation is correct then the "D" zone porosity of the subsurface corresponds to the Cat Head Member and the overlying less dolomitic unit corresponds to the Selkirk Member. These correlations are provisionally shown in Figure 4. The uppermost tight dolomitic limestones of the Selkirk Member (?) closely resemble those of the outcrop belt (where they are extensively quarried for building stone) which adds support to the proposed correlation.

It would appear that almost all descriptions of the Yeoman Formation or Lower Red River are those based upon the uppermost (Selkirk?) unit and the underlying more dolomitized beds are commonly considered little more than more dolomitized equivalents. Closer inspection has revealed that these more dolomitized units in northwest South Dakota are significantly less fossiliferous than over- or underlying carbonates.

DEPOSITIONAL AND EARLY DIAGENETIC CHARACTERS OF MOTTLED CARBONATES

Mottled dolomitic limestones from the Turbiville well and lateral equivalents (sometimes fully dolomitized) of the Brown well are lithologically identical with those of the northern Williston Basin

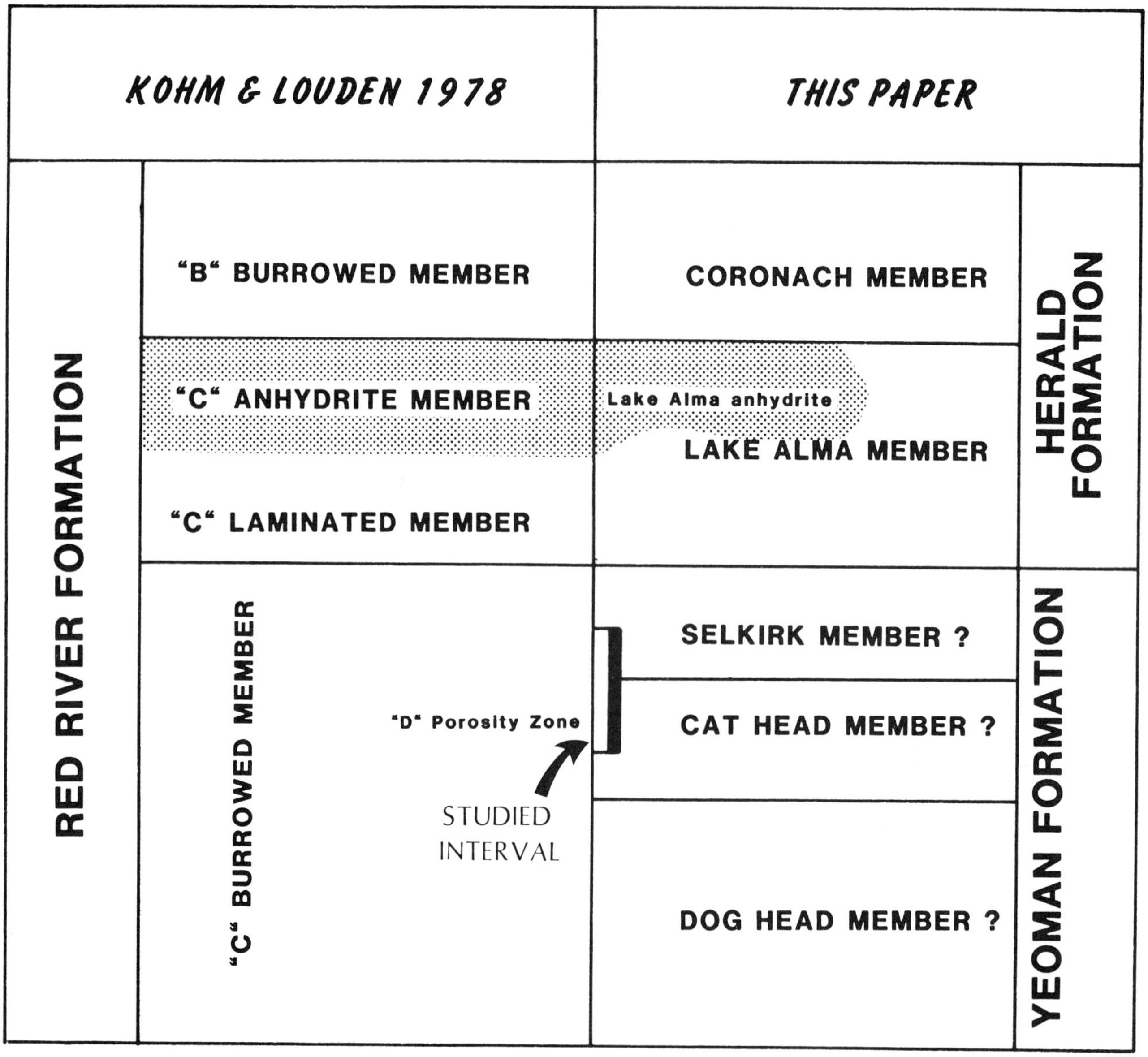

KOHM & LOUDEN 1978		THIS PAPER	
RED RIVER FORMATION	"B" BURROWED MEMBER	CORONACH MEMBER	HERALD FORMATION
	"C" ANHYDRITE MEMBER	Lake Alma anhydrite	
		LAKE ALMA MEMBER	
	"C" LAMINATED MEMBER		
	"C" BURROWED MEMBER ("D" Porosity Zone — STUDIED INTERVAL)	SELKIRK MEMBER ?	YEOMAN FORMATION
		CAT HEAD MEMBER ?	
		DOG HEAD MEMBER ?	

Figure 4. Stratigraphic nomenclature employed in the Williston Basin for the Red River or the combined Yeoman-Herald Formation interval.

(Saskatchewan and Manitoba) where their depositional characteristics and particularly their early diagenetic alteration has been most extensively studied (Kendall, 1976, 1977). Variations from this norm are discussed in the next section.

The matrices of dolomite mottles range from skeletal-rich wackestones (locally packstones) to skeletal-poor wackestones or mudstones. The initial impression is given that the fauna lacks components of fossil groups that would be expected from the Upper Ordovician - particularly brachiopods and trilobites. The fauna appears dominated by relatively large elements, in particular solitary rugose (Fig. 5C) and tabulate (Figs. 5C, 6A) corals, stromatoporoids, Receptaculites (not seen in the South Dakota cores) and thick-walled Macluritid gastropod and nautiloid shells. Even low-power examination of the matrix, however, reveals that the fauna is much more diverse with brachiopods, bryozoans, trilobites, ostracods, trilobites and many unidentified shell fragments. All, however, appear to have been highly comminuted. Given the evidence for significant bioturbation and storm activity (see below), this comminution probably results from a combination of both physical and biological breakage.

All parts of the cored section in both wells contain frequent interruptions by skeletal grainstone or packstone layers (Figs. 5A,5B,5D; 6A,6E; 10C). Commonly only a centimeter thick, these layers have sharp, probably erosive bases and abrupt, sometimes loaded (?) upper contacts. They sometimes exhibit internal grading (Fig. 10C) but most commonly have a uniform clast-size distribution and may possess a mud matrix. Such beds are identified as the products of storms and record episodes when the normal wackestone

Figure 5. Depositional and early-diagenetic features.

    A. Polished core slab of dolomite-mottled limestone. The limestone matrix (lighter gray areas) contains abundant skeletal debris and has stylolitic contacts with dolomitized burrow fills (darker gray areas) that contain lighter colored secondary burrows. Stylolitization has also affected a storm-sheet skeletal packstone layer (st. s). The distinctive shape of one dolomitized burrow suggests that it is a cast of a now-dissolved fossil (F.C.). Turbiville well, 8470 ft.

    B. Storm-sheet skeletal grainstone that cross-cutts the underlying burrowed wackestone and is affected by downward loading(?) of the overlying burrowed sediment. Brown well, 8717 ft.

    C. Polished core slab above fracture, acid-etched below: reburrowed burrow-fills (now dolomitized) avoid solitary coral (s.c.) (note that dolomitization does not enter between the coral septae). In lower part of sample the dolomitized burrows have amalgamated to isolate "islands" of limestone matrix. Also present are (at f.c.) a smoothly curved boundary to a dolomite mottle that suggests the former presence of an aragonitic and concave-upward shell that confined the burrower, and (at c.c.) a fragment of a compound coral (Catenipora) that, like the solitary coral has resisted dolomitization. Turbiville well, 8485 ft.

    D. Mottled dolomite with darker mottles controlled in part by thin storm sheet deposits. Brown well, 8756 ft.

    E. Mottled dolomitic limestone with dolomitization confined to a nautiloid shell (n) and the shell-wall of a thick-shelled gastropod Maclurites (Mac). Note the presence of lighter-colored dolomite filling secondary burrows within the Maclurites shell-filling. Turbiville well, 8506 ft.

Figure 6. Depositional and early diagenetic features.

A. Mottled dolomitic limestone (acid etched) with dolomitized burrow fills that contain darker meniscus-filled secondary burrows. Host sediment is a skeletal wackestone, locally a packstone. Primary burrows were filled by mudstone, not all of which has been dolomitized. Small arrows delineate the boundaries of two such partially dolomitized burrows. In these dolomitization appears to have only affected the secondary burrows and even this is incomplete. Larger arrows point to formerly aragonitic shells that are only partially dolomitized. Remainder of these shells is filled by darker internal sediment that has resisted dolomitization. At d. sh is a dissolved shell, encrusted by a silicified compound coral. Turbiville well, 8478½ ft.

B. Contact between dolomitic limestone above and underlying bituminous limestone. Underlying sediment is a skeletal wackestone and is extensively bioturbated, although unmottled. Overlying sediment is a "normal" dolomite-mottled limestone. The boundary is a flat erosion surface (arrowed) but mostly this is obscured by extensive stylolitization that has concentrated the organic material within the underlying carbonate into a solution seam (S.S.). Brown well, 8722 ft.

C. Dark bituminous dolomite sandwiched between two storm-sheet layers. Lamination in bituminous dolomite results from much-flattened burrow systems. Sediment package occurs within poorly fossiliferous and only slightly bioturbated thin-bedded dolomite mudstones. Turbiville well, 8518 ft.

D. Amalgamated dolomitized burrow fills containing secondary burrows consisting of memiscus-fills of bituminous dolomite similar to beds shown in Fig. 6B and C. Matrix between mottles has been reduced to solution films or to isolated patches of skeletal-rich wackestone or packstone. Oil-stained tension fractures originate from the solution-films. Turbiville well, 8485½ ft.

E. Color-mottled dolomite. Mottles are darker, denser, contain occasional indications of secondary burrows and are locally cut by open, wedge-like tension fractures. Matrix is lighter colored and micro-sucrosic but only locally contains skeletal debris. A dolomitized skeletal-rich storm-sheet deposit (upper right) demonstrates that the absence of skeletal debris from the host sediment is not a product of dolomitization. Brown well, 8748 ft.

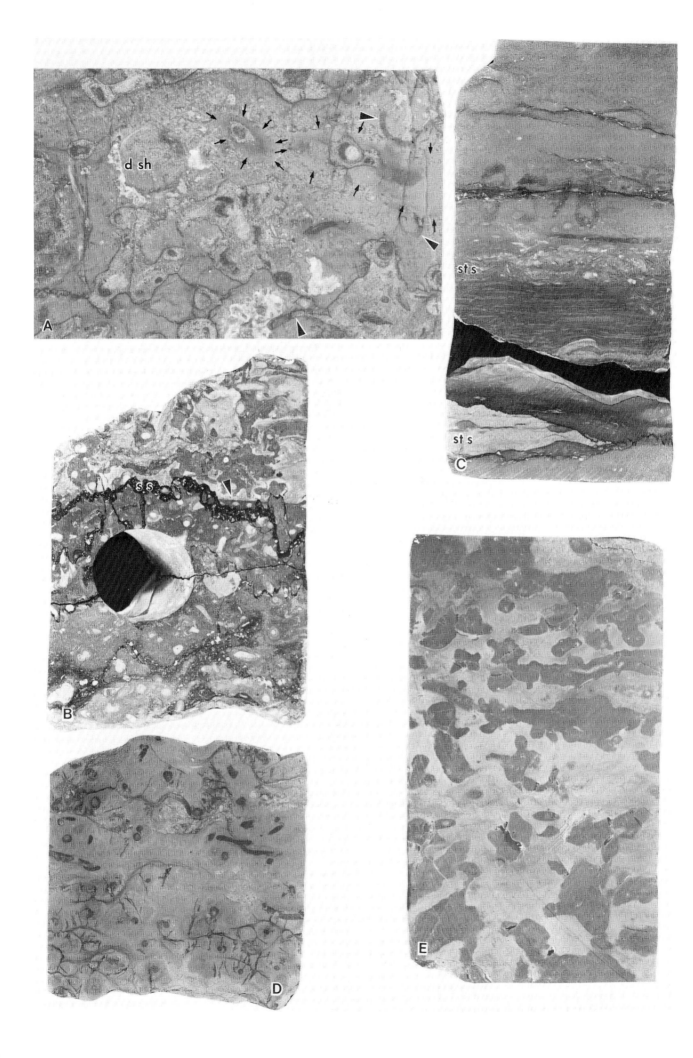

Figure 7. Non-mottled lithologies.

A. Nodular dolomitic limestone. In the upper part of the polished core slab individual small burrows form limestone nodules within a dolomitized matrix that has suffered pressure solution. In lower part, larger and more diffuse nodules of skeletal wackestone are separated by slightly argillaceous and bituminous diagenetic flasers. Microstylolites in the flasers enter and subdivide the margins of the nodules. Flasers and nodule margins have been dolomitized. Chert (ch) has preferentially formed around a nautiloid. Turbiville well, 8465 ft.

B. Thin-bedded to nodular dolomite mudstone with tabular sheets and smaller nodules (after flattened burrows) separated by pressure solution seams and microstylolites. Tension fractures cut the larger dolomite idens. Turbiville well, 8498 ft.

C. Dolomite mudstone exhibiting features transitional between thin-bedded and mottled structures. Mottles differ only slightly from "matrix" and are widely scattered. Matrix contains undulose pressure solution seams that deviate around mottles but otherwise are essentially tabular. Some mottles are cut by open, wedge-shaped tension fractures that have been partially filled with matrix (arrowed). Deflection of neighboring solution seams in the vicinity of the filled fractures indicates matrix was plastic when it was intruded into fractures. Brown well, 8731 ft.

D. Thin-bedded to nodular dolomite mudstone containing well developed pressure solution seams (s.s.). Dolomite adjacent to solution seams is microsucrosic and was oil-stained (darker areas). Local presence of burrow fills may indicate that this is a much diagenetically-modified mottled carbonate that passed through a stage similar to that seen in Fig. 2d. Turbiville well, 8499½ ft.

E. Stylolitic boundary between bituminous, thin-bedded to laminated skeletal wackestone (dolomitized) and overlying, essentially homogeneous bioturbated sparce wackestone to mudstone containing slightly darker and denser areas that represent poorly developed mottles. Brown well, 8762 ft.

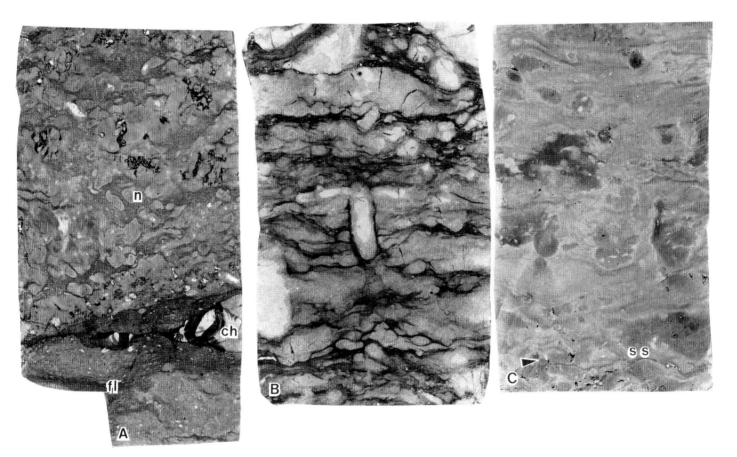
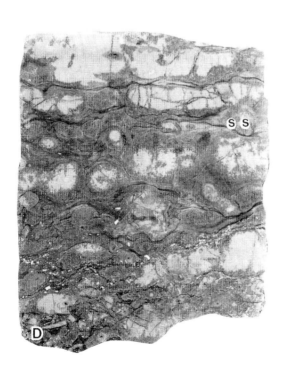

sediment had the greater part of its mud matrix washed away to concentrate the coarser skeletal particles into a tabular sheet deposit. The frequency of these deposits, their thinness and presence of mud matrices suggests that they show greater resemblance to the "distal" tempestites of Aigner (1982) than those identified as being "proximal". The frequency of storm sheets appears to be greater in these South Dakota cores than I have seen elsewhere but this requires confirmation by making counts on cores from different areas.

Thin beds of bituminous mudstone or skeletal wackestone (Figs. 6B,6C; 7E) are also present and compare with the beds of oil-shale or kerogenite recorded from northern and central Williston basin by Kendall (1976) and Kohm and Louden (1978). These appear less organic rich and some are bioturbated. This suggests conditions during their formation were less euxinic than elsewhere but nevertheless they probably record the most restricted conditions during deposition of the lower Red River (Yeoman) Formation. Paradoxically, brachiopod and trilobite remains are best preserved in, or near these beds. This differential preservation is attributed to the absence or scarcity of scavaging organisms that under normal conditions reduced skeletons of these forms to fragments.

Little can be added to the early-diagenetic history of these rocks that was not presented earlier for equivalent rocks from Saskatchewan and Manitoba (Kendall, 1977). The interested reader should refer to that study for details and discussion of the evidence only mentioned here.

The dolomite mottles are burrow systems which have been preferentially reburrowed by a second generation of smaller organisms.

Fluids that caused dolomitization preferentially moved through these burrow systems and this implies that burrow fills were more permeable than sediment matrices during the dolomitizing event. Differential permeability (and sediment plasticity) between burrows and matrix is also shown by the preferential occurrence of displacive halite (now dissolved) within burrow fills. In the South Dakota cores halite casts are uncommon, having only been found at one location (Fig. 9A). This sparcity reinforces a conclusion already made (Kendall, 1977, p. 500) that halite in the Yeoman is more common in, or is confined to, the more central portions of the Williston Basin.

Dolomite commonly preferentially replaces skeletal aragonite (Figs. 5E; 6A; 8A), with calcitic shells being unaffected Figs. 5C; 6A). The presence of burrows, identical with those in the burrow mottles, within the dolomite shell walls (Fig. 5E) indicates that dolomitization has not replaced the shell carbonate directly. Instead, skeletal aragonite dissolved within the sediment and the shell molds were filled by sediment (introduced concurrently with the filling of the primary burrow networks and by the same secondary burrowers). It is this internal sediment that has been preferentially dolomitized. At one horizon in the Webb Cenex #7-9 Turbiville well the later dolomitization event is incomplete and burrows and neighboring shell casts partially retain their original, undolomitized internal sediment fills (Fig. 6A).

In my original interpretation of the early diagenesis of the mottled dolomitic limestones (Kendall, 1977) I argued that the absence of any compaction or distortion of the shell molds or their later sediment casts implied that the host sediment must have been lithified at the time of shell dissolution. I no longer wish to maintain this argument for I

Figure 8. Later diagenetic features.

A. Dolomite-mottled limestone. Matrix, where unaffected by pressure solution is a skeletal wackestone with echinoderm grains preferentially replaced by dolomite (locally by silica). Along solution seams the skeletal debris is concentrated to form diagentic grainstones (i.e., below dol s). Stylolites (with associated tension fractures) cross-cut all other rock components, including solution seams. A dolomite-filled shell cast (dol s) occurs and dolomite burrow mottles contain bituminous dolomite filled secondary burrows. Turbiville well, 8502 ft.

B. Mottled dolomite composed of overly-compacted mottles set within sucrosic matrix that is confined to solution seams. Matrix appears to consist largely of skeletal debris but this grainstone texture is of diagenetic origin. Mottles have stylolitic contacts with the matrix and their shapes have been modified by pressure-solution to allow the close packing. Mottles are also markedly fractured: some are confined to single mottles and represent tension fractures created as mottles accommodated their shapes to each other; other fractures cross-cut several neighboring mottles and represent a fracturing stage after mottles had come into vertical contact so that the rock acted as a whole. Brown well, 8743 ft.

C. Nodular structure originating from mottled dolomite. Boundaries between mottles and matrix are diffuse but rock structure is now dominated by pressure solution seams (and accompanying tensional fractures: f) that preferentially pass through the matrix areas. This is believed to be an arrested stage in the development of nodular dolomites like those of Fig. 2D. Stylolites post-date all other diagenetic events and locally have reactivated some earlier solution stringers. Turbiville well, 8499½ ft.

D. Rapid transition between rock types that appear to differ only in their later diagenesis. At the top is a color-mottled dolomite in which mottles (lighter colored) are separated by a microsucrosic matrix (oil stained at burrow-matrix boundaries) that has been only slightly affected by pressure solution. The matrix here is a skeletal wackestone. In the middle part of the slab, mottles are more closely spaced and the matrix has a grainstone texture. In the lower part, mottles have amalgamated and matrix is confined to pressure solution seams. The grainstone texture in mid-core may represent a bioturbated storm sheet or be the product of alter diagenetic matrix dissolution (or both). Turbiville well, 8508 ft.

Figure 9. Diagenetic breccias

    A. Dolomite breccia composed of irregular to tabular large clasts of bioturbated dolomite (burrows with slightly bituminous, meniscus fills) and smaller, rounded clasts of light-colored dolomite; all set within a sparse microsucrosic dolomite matrix. Larger clasts appear to be dolomite mottles with both cross-cutting (fractured) and original smooth boundaries. The rounding of the smaller clasts strongly suggests the breccia has a depositional origin, however, the character of the larger clasts indicates the breccia is diagenetic (see text). Dolomite-filled molds of three-dimensional halite dendrites are present within former burrow fills at arrows. Brown well, 5745 ft.

    B. Dolomite breccia composed of the same clast types as in Fig. 9A (but finer grained) passing abruptly, but gradationally, into over- and underlying color-mottled dolomites. Both over- and underlying carbonates have been affected by pressure solution with overly-packed mottles that are locally compactively-fractured to generate clast-like idens similar to those in the breccia. Matrices are composed of skeletal grainstones, almost certainly diagenetically produced from skeletal wackestones, and "chips" or fragments of the lighter-colored marginal areas of mottles that appear to be wedged away from the mottles by microstylolitization and "rounded" by pressure solution in the matrix (best seen in the rock underlying the breccia). Brown well, 8740-41 ft.

have seen (elsewhere) evidence for shell distortion (implying the host sediment was plastic) and in many instances (Figs. 5A,5C; 6E; 9A) the shapes of dolomite mottles clearly reveal the former presence of skeletal material. This has subsequently been dissolved but there is no shell mold or cast; the matrix appears to have squeezed together.

The cores from South Dakota have yielded no further evidence which could be used to discriminate between the two scenarios for mottle dolomitization that were discussed in Kendall (1977). Preferential dolomitization of burrow fills and formerly-aragonitic shell casts can be explained either by reflux dolomitization during the deposition of the Upper Red River (Lake Alma) evaporites (reflux occurring preferentially through the more permeable burrow fills), or dolomitization occurred by a process of unmixing or cannibalistic dolomitization whereby magnesium from the highly magnesian sediments migrates to the more permeable parts of the sediment, there to cause local dolomitization. Longman and others (1984) have presented arguments against the second explanation and in favor of the first. However, it cannot be denied that seemingly identical dolomite-mottled limestones of the same age extend from Greenland, fully across the North American craton, to New Mexico and (except for the Williston Basin) are not associated with evaporites.

LITHOLOGIC VARIABILITY IN THE YEOMAN

The most obvious lithologic variations in Yeoman sediments are products of later diagenetic overprints - matrix dolomitization and chemical compaction. Variations in these effects, which can produce strikingly different rocks from apparently very similar precursors, does appear to have been controlled, at least in part, by more subtle variations in depositional and/or early diagentic conditions. Two main variables may be identified: (1) the location and relative extent of sediment lithification and/or cohesiveness, and (2) the abundance of skeletal material.

Conspicuously dolomite-mottled limestones like those in Figs. 5A, 5C and 6E possess large and complex burrow networks that have been preferentially dolomitized during early diagenesis. The differences in behavior to the dolomitization event between burrow fills and their matrices are attributed to a preferential retention of sediment permeability by the burrow fills. This in turn must reflect, if not lithification of the host sediment, at least a greater degree of sediment consolidation or cohesiveness (whether produced by differential cementation or by other processes cannot be determined). This difference also explains the preferred reburrowing and growth of halite dendrites in the burrow fills. The burrow fills thus acted during early diagenesis as less lithified parts of the sediment. After dolomitization of the burrows, however, the dolomite mottles behaved as more resistant parts (or idens) of the rock, with chemical compaction largely confined to the formerly more-resistant matrix between mottles (see next section).

In other Yeoman carbonates, the burrows have apparently acted from the outset as the more resistant idens so that matrices are preferentially dolomitized and subject to compaction. These are nodular limestones and have formed as a result of the preferential lithification (or consolidation) of the burrow fills. This different behavior of burrow fills to early diagenesis, generating very different products, is not unexpected. Fursich (1973) has shown how burrow networks can be converted to nodular limestones and the very different types of preservation that can be found for the same trace fossil.

Nodular carbonates can only be recognized with certainty where the nodules have resisted dolomitization. Where the whole rock has been dolomitized it is usually impossible to determine the relative timing between dolomitization of the burrow fill and its matrix, as well as the relative timing of chemical compaction. Nodular-appearing dolomites similar to those in Figs. 7D and 8C may have arisen by the subsequent dolomitization of limestone nodules in a dolomitic limestone similar to that in Fig. 7A, or alternatively, began as burrow-mottled dolomitic limestones (like that in Fig. 5A) having passed through a color-mottled dolomite stage (Fig. 6E) by a progressive later-diagenetic removal of the matrix. Both alternatives appear equally probable and indeed it is believed that both types of nodular dolomite are present in different parts of the core. Discrimination between the two types, if this is at all possible, can only be made by examination of neighboring sediments. Nodular dolomites arising from dolomite-mottled carbonates pass vertically and gradationally into color-mottled dolomites in which matrix compaction has occurred to a lesser extent. Nodular dolomites that were originally nodular dolomitic limestones, on the other hand, appear to be most closely associated with yet another sediment type in the Yeoman Formation - one first described by Carroll (1978) as "dolomitic mudstones".

"Dolomitic mudstones" (Figs. 6C, 7B, 7C and 7e) are sparsely fossiliferous wackestones or unfossiliferous mudstones in which burrow structures are absent (Fig. 6C) or subdued. Some (Fig. 6C) closely resemble and intergrade with nodular dolomites but possess burrow-fills or nodules that are more diffuse. Other types (Fig. 7E) also have diffuse mottle boundaries but cannot have originated from nodular precursors because there is no evidence of differential matrix compaction. Cross-sections of secondary burrows within these "dolomitic mudstones" commonly exhibit a degree of compactive flattening that is not apparent within mottled carbonates or the preferentially lithified parts of nodular carbonates. From this slender line of evidence it is believed that the major differences in Yeoman carbonates arise from differences in the location of preferential sediment lithification or consolidation. Mottled carbonate matrices were more lithified than their burrow fills, whereas in nodular carbonates it was the burrow fills that become selectively more coherent and resistant to compaction and dolomitization. Dolomitic mudstones, which appear to grade into both of these other types, suffered little to no early diagenetic lithification, or if this occurred there is less contrast between burrow fills and their matrices. Finally, some thin-bedded dolomite mudstones lack burrow structures altogether.

The second major variation, other than those produced by later diagenesis, that can be found within Yeoman carbonates concerns the abundance of skeletal material. It is very evident that many fully dolomitized carbonates lack or have little skeletal debris. This could be the result of the dolomitization process obscuring or eliminating skeletal material but, in the two South Dakota wells, such fossil-poor intervals contain storm-sheet deposits composed of skeletal debris. These sheets are also dolomitized within the dolomite intervals yet skeletal grains are well preserved within them. Scattered skeletal debris, usually echinoderm grains, is also present within the dolomitized sparsely-fossiliferous wackestones and dolomitization has similarly not eliminated skeletal material from many color-mottled dolomites from the Sohio #11-8 Brown well that can be correlated with skeletal-rich dolomitic limestones in the neighboring Webb Cenex #7-9 Turbiville well (Fig. 3). It is believed that the distribution of skeletal debris in the Yeoman carbonates was depositionally controlled and that this distribution has not been obscured by the dolomitization overprint. Difficulty is found however in discriminating between skeletal-rich and poor wackestones in instances where the matrix between mottles or nodules has been reduced to generate a diagenetic grainstone texture (see next section).

A plot of the depositional textures present in the cores from the two wells (Fig. 3) reveals that boundaries between skeletal-rich and skeletal-poor sediments are abrupt so that relatively thick units of each sediment type are present. Furthermore, boundaries between rock units so distinguished are traceable between wells (and other wells in the vicinity) and parallel thin beds of bituminous carbonate. These organic rich intervals almost certainly are isochronous in that they record episodes when the seafloor became stagnant and euxinic (Kendall, 1976; Kohm and Louden, 1978). Changes in skeletal abundance are probably also time parallel and record episodes of different environmental conditions.

All nodular and "dolomite mudstone" intervals lack abundant skeletal debris but dolomite-mottled limestones and color-mottled dolomites may possess either skeletal-rich or skeletal-poor matrices.

Derby and Kilpatrick (1985) record a similar fauna improverishment within the dolomitized "D" porosity zone of the Killdeer Field of North Dakota. They attribute this to the existence of elevated or seasonally fluctuating salinities and further suggest that such environmental conditions could have been responsible for dolomitization of burrow fills. Given that thin intervals in the Yeoman are present that record euxinic bottom conditions and that these are commonly closely associated with beds of unfossiliferous dolomite mudstone (an association previously noted by Carroll, 1978, from North Dakota), then it seems more reasonable to conclude that sparcity of fauna results from marginally oxygenated (dysaerobic) conditions. Sediment types that lack skeletal material but which are bioturbated are typical of dysaerobic environments (Byers, 1977) and typify much of the Yeoman Formation, particularly that part that is particularly prone to complete dolomitization and develops into the "D" porosity zone.

LATER DIAGENETIC CHANGES

Dolomitization of matrix carbonate within color-mottled dolomites is unrelated to depositional or early diagenetic environments since this type of dolomite is laterally discontinuous (Kohm and Louden, 1978; Longman and others, 1983; this study Fig. 3). As noted earlier I do not wish to discuss this dolomitization event at this time and have found nothing in this new core study to make me alter my views presented in 1984 in discussion of Longman and others (1983) paper. However, comments concerning its relations to other diagenetic events will be made where appropriate.

Chemical compaction has had a profound effect upon most Yeoman carbonates and sometimes has created rocks that bear little resemblance to what can be deduced about their sediment precursors. Some later diagenetic products of chemical compaction can also be confused with depositional or early diagenetic features - a confusion that causes incorrect determinations of environmental conditions.

Stylolites and pressure-solution seams containing concentrations of the less soluble sediment components are present throughout the Yeoman successions. Where cross-cutting relationships can be observed stylolites always cut solution seams and so post-date them. Solution has commonly occurred between dolomite mottles and their limestone matrices and at the same mottle boundaries in color-mottled dolomites. This similarity of solution contacts may indicate that pressure solution occurred prior to matrix dolomitization (an interpretation favored by the author), however, other explanations are possible.

Closely associated with both stylolites and solution seams are fractures (Figs. 6D, 6E, 7B, 7C, 8A, 8B, 8C, 9B, 10A, 10B, and 10D). They extend at right angles from the solution plane and may have uniform thickness or are conspicuously wedge shaped - always thinning away from the solution plane. Their association with pressure solution features was first reported from the Yeoman of Saskatchewan (Kendall, 1976) but was later explained by Nelson (1981) as resulting from the application of triaxial strain conditions. The fractures represent tension gashes. One feature not mentioned by Nelson is that wedge-shaped fractures can sometimes be partially filled by the less competent rock type involved in the pressure solution. Fractures in the Yeoman are commonly open (in which case they may significantly contribute to the rock's permeability) or are plugged by anhydrite (and saddle dolomite or calcite in other cores). Some, however, are partially filled with sediment which appears to have been injected into the opening (Fig. 7C). Carroll (1978) noted this feature and concluded from it that fracturing formed as a result of a volume reduction that accompanied dolomitization of the mottles. The fractures figured and interpreted by Carroll (1978) are identical with those here identified as being associated with pressure solution. Carroll's Fig. 9 also shows what appears to be a solution residue along the mottle-matrix boundary. It is suggested that sediment in such fractures must post-date matrix lithification otherwise there would not have been pressure solution between mottle and matrix. Matrix enters the fractures either because a stress acting over a long period of time will cause even a lithified limestone to flow plastically, or because after fracture of the mottle or nodule no further solution occurs opposite the fracture opening. As matrix continues to be removed at the mottle-matrix boundary a column of matrix enters the fracture.

Continued pressure solution at mottle/nodule and matrix boundaries may almost entirely eliminate the matrix so that mottles or nodules come into close contact. Pressure solution at nodule-nodule or mottle-mottle contacts results in an accommodation between these idens and their progressive change in shape. Matrix may be almost eliminated; it being confined to isolated areas or within solution seams (Figs. 8C, base of 8D). More commonly, and when wackestone matrices are affected, the skeletal material is less soluble than the carbonate mud and is concentrated to form a diagenetic packstone or grainstone between the dolomite mottles (Figs. 8B, 8D and 10D). The diagenetic origin of these textures is clearly demonstrated by the numerous transitions that occur between uncompacted and compacted rocks and by the association of diagenetic grainstones with overly-packed dolomite mottles. Non-recognition of the diagenetic origin of these rock textures leads to an incorrect assessment of depositional conditions.

Chemical compaction has also lead to the formation of diagenetic breccias which have not previously been reported from the Yeoman or Upper Red River but which may be represented by the brecciated hardgrounds of Derby and Kilpatrick (1985). Two such breccias are present in the Sohio #11-8 Brown core (Fig. 9). Both can easily be misidentified as being of depositional origin because they contain clasts of different lithology, some of which are rounded. Larger clasts contain burrow structures and are clearly reworked dolomite mottles. Breccias are clast-supported but fine-grained carbonate (now dolomite) occurs as a matrix.

The depositional origin of breccias can be established by examining the vertical contacts of the bed (Fig. 9B). These are abrupt but gradational and color-mottled dolomites above and below the breccia have been markedly affected by chemical compaction with the formation of diagenetic grainstone textures. Removal of most of the matrix from between mottles has caused their fracture and displacement. Breccias represent thin zones where this fracture and rearrangement have been developed to an extreme degree. Rounded clasts in the breccia appear to have formed by the spalling off of fragments of the mottle boundaries and their further modification within the diagenetic matrix by further pressure solution. No evidence remains which could shed light upon why the effects of pressure solution should have been concentrated to generate the thin breccia units. One possibility is that they mark the position of small faults, in which case the breccias could be regarded as a type of "diagenetic mylonite".

A somewhat similar brecciated appearance is found in some dolomite-mottled limestones from the Sohio #11-8 Brown core (Fig. 10B). Here stylolitization occurs at mottle-matrix boundaries and in the matrix between mottles. Preferential removal of the limestone matrix has caused the more rigid dolomitized burrow network to be fractured and the fragments to become rearranged. This process has not proceeded as far as that in the dolomitized breccias present lower in the core.

Other late diagenetic features present in the South Dakota cores are particularly significant to porosity formation and destruction but can only be briefly mentioned here. Much additional work needs to be done and the cores have yet to reveal much of interest about them. Much of the effective porosity within the "D" zone occurs in the form illustrated in Figure 10A and 10C. It is an intercrystalline porosity associated with the more coarsely microcrystalline dolomite that appears to have overprinted earlier dolomite. It may be a replacement or result from selective leaching of the earlier dolomite. Commonly it

Figure 10. Late diagenetic features.

A. Nodular dolomite with porosity (oil-stained) preferentially developed within nodules adjacent to nodule-matrix boundaries. Porosity resides within slightly more coarsely microsucrosic dolomite (compared with that of the matrix) and this, from its distribution may have originated along microstylolites (now obliterated). Turbiville well, 8510 ft.

B. Mottled dolomitic limestone with dolomite mottles (slightly lighter gray areas containing lighter-colored secondary burrows) separated from sparsely fossiliferous wackestone matrix by stylolites. Stylolites and minor pressure solution seams also cross matrix between dolomite mottles. Chemical compaction of matrix and boundaries with mottles has caused compactive fracture of the three-dimensional burrow network (new dolomite mottles) and profound shape changes to the original mottles. Many of the mottles have been displaced relative to each other so as to generate a rock type that approaches a breccia in appearance. Turbiville well, 8471 ft.

C. Color-mottled dolomite transected by graded storm-sheet layer (st s). Matrix and interior portions of some mottles (parts affected by secondary burrowing?) replaced by friable microsucrosic dolomite. Locally this friable dolomite has collapsed to produce a geopetal structure; the void so formed being filled with coarsely crystalline anhydrite. Brown well, 8753½ ft.

D. Color-mottled dolomite. Mottles, containing secondary burrow fills, exhibit fractured boundaries and are set within an abundant skeletal grainstone dolomite matrix of diagenetic origin. This rock bears little resemblance to the original sediment. The large amount of matrix indicates that a relatively large amount of wackestone host sediment has been pressure-solved and condensed to form the grainstone texture and this implies mottles were originally more widely spaced. The size, shape and relative position of mottles has also been profoundly modified by chemical compaction. Brown well, 8726 ft.

appears to be located along mottle or nodule and matrix boundaries (Fig. 10A) or selectively replace matrix and the central parts of mottles (Fig. 10C). Locally, this dolomite type is so porous as to be friable. The not uncommon presence of this dolomite within geopetal structures (Fig. 10C) suggests that the leaching or replacive event responsible for its formation was also capable of forming large voids. In almost all cases such voids are filled with coarsely crystalline anhydrite -- a cement that does not plug the intercrystalline porosity.

CONCLUSIONS AND DISCUSSION

Examination of the Yeoman carbonates suggests they possess subtle, yet significant, depositional and early diagenetic variations and that these have markedly affected later diagenetic changes.

Yeoman carbonates were deposited in anaerobic, dysaerobic and aerobic environments. Whereas it is tempting to use the water depths suggested by Byers (1977) for these environments and conclude that many Yeoman sediments (if not all of them) were deposited in a deep shelfal environment (dysaerobic environments occurring in 50 to 150 m of water) it must be emphasized that stratification in the Red River sea (responsible for the development of minimally oxygenated environment) may have been caused as much by variations in water salinity as by temperature differences. Boundaries between differently oxygenated water layers may have occurred at much shallower depths. Nevertheless the occurrence of storm-sheet deposits coupled with a total lack of fair-weather wave and current deposits; the general absence of skeletal algal material together with the craton-wide lateral

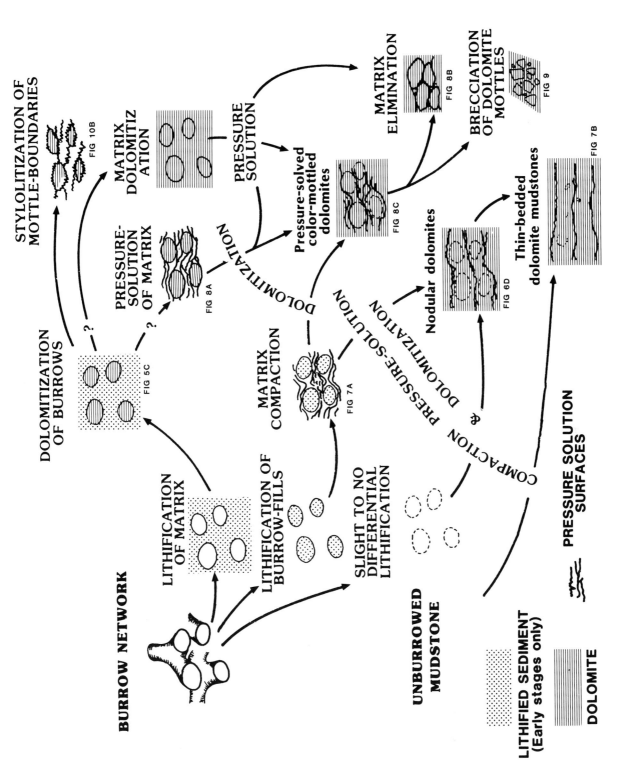

Figure 11. Diagenetic pathways taken by Yeoman carbonates.

distribution and absence of interbedding with other facies all strongly suggest that the Yeoman was deposited within a deep shelfal environment. This conclusion also helps explain why the early-diagenetic history of the Yeoman Formation, particularly the occurrence of aragonite dissolution, so closely resembles that of the northwest European Chalk - another deep shelf or epeiric shelf carbonate (Kendall, 1977).

Regardless of the correctness (or otherwise) of the deep shelf interpretation the recognition of dysaerobic conditions during formation of "D" porosity zone creates a problem of considerable interest - what is the origin of the carbonate mud in the Yeoman Formation? The "D" zone covers thousands of square miles in the central parts of the Williston Basin and extends through several tens of feet. This thick and widespread unit is unlikely to have been formed by reworking earlier deposits and contributions from skeletal disaggradation or algal precipitation are equally improbable because of the adverse environmental conditions. In the absence of pelagic carbonate skeletons only an inorganic (?) precipitate origin appears to remain as a possibility.

Dissolution of skeletal aragonite during early diagenesis indicates that Yeoman sediments and any early-diagenetic cements added to them were calcitic (Kendall, 1977). Whether or not these sediments experienced early diagenetic cementation is uncertain because the presence of undeformed shell molds and of hardgrounds (present elsewhere in the Yeoman Formation, Kendall, 1977) merely indicates that the sediments were coherent or stiff such they they resisted compaction and were capable of being eroded. Partial cementation, however, is still the most likely explanation for sediment rigidity. Variations in the location and extent of this sediment stiffening are responsible for much of the lithologic variability now seen in Yeoman carbonates because it was this sediment attribute that has largely controlled dolomitization and chemical compaction. Dysaerobic sediments appear to have been less lithified (or were more homogeneous) after early diagenesis and are consequently more prone to complete dolomitization. It is also possible that these sediments, having been formed in more restricted conditions, were composed of more magnesian calcites and thus may be dolomitized more readily. The existence of an apparently basin-wide unit of dysaerobic sediments has lead to the development of an equally widespread unit that commonly has been completely dolomitized (the "D" porosity zone). Other fully dolomitized carbonates are not stratigraphically controlled so that within the Yeoman Formation dolomitization is both stratigraphically and facies controlled as well as being independent of these factors.

Chemical compaction in Yeoman carbonates has 1) caused nodular and mottled dolomitic limestones to resemble each other after complete dolomitization, 2) preferentially removed matrix so that nodules or mottles have become closer packed and/or brecciated, and 3) created diagenetic grainstone textures. The result of these changes is commonly the creation of rocks, like that illustrated in Figure 10D, in which no component resembles that inferred to have been present in the sediment precursor.

There appears to be a relationship between complete dolomitization and the degree of chemical compaction within dolomite-mottled carbonates. Those that are fully dolomitized are characteristically those that are more affected by chemical compaction. This may be explained in several different ways. In the first, dolomitization creates sediments that are particularly prone to compaction. Compaction may occur subsequently, or may be more or less contemporaneous with dolomitization. It is even possible that the very porous and friable dolomites (Figs. 10A, 10C) are not later replacements but are instead locally preserved remnants of the rock type that elsewhere suffered extreme compaction and removal.

A contrasting explanation is that dolomitization may have preferentially affected rocks that had suffered chemical compaction. Perhaps the planes of pressure solution acted as pathways for dolomitizing fluids. This interpretation implies that dolomitization post-dates the products of pressure solution and this should be capable of verification petrographically. In fact this assessment of relative timing has been extraordinarily difficult to make. Many solution seams in dolomites are diffuse and may have been overprinted by the dolomitization. This assessment is, however, subjective and furthermore is itself capable of other interpretation. The type of pressure solution and its products are similar in both dolomitic limestones and fully dolomitized equivalents. I find it difficult to believe that pressure solution would have generated similar products in these very different materials. It is easier to believe that pressure solution affected dolomitic limestones and later some of them were more completely dolomitized, with dolomitization overprinting the earlier pressure solution effects. Once again, however, this is a subjective assessment.

Since fully dolomitized rocks appear in general to have suffered more compaction than other carbonates, it would be expected that dolomite rock units will be thinner (more compacted) than equivalent rock units that have remained dolomitic limestones. Unfortunately the present study is ambiguous. Using depths marked on the cores, stratigraphic units have the same thickness or are thinner in the more dolomitic Sohio #11-8 Brown well (Fig. 3). Log correlations, however, suggest there is no thickness changes between the two wells.

The present study of two South Dakota wells has probably raised more questions about the Yeoman (Lower Red River) than it has solved. Much remains to be done and many of the problems will only be solved by detailed examination of a much larger population of cores from a larger geographical area. In particular, the deep shelf interpretation of depositional environment focuses attention upon the contact of the Yeoman with the overlying evaporative carbonate and anhydrite deposits of the Upper Red River (= Herald Formation, Lake Alma Member). A regional study of the type and frequency of storm sheet deposits in the upper Yeoman carbonates may also produce results significant to assessing the depositional environment. However, over-riding these are studies that need to be made upon dolomitization in the Yeoman - how many dolomitization events, their timing and the nature of the dolomitizing processes. Much work also is required to fully assess the effects of chemical compaction in these rocks. I have the impression that those in the South Dakota cores have suffered more from the effects than those of Saskatchewan and Manitoba; this despite comparable present-day depths of burial. A quantitative regional study of chemical compaction in the Yeoman would probably produce interesting results.

REFERENCES

Aigner, T., 1982, Calcareous tempestites: storm-dominated stratification in Upper Muschelkalk limestones (Middle Trias, SW-Germany): in Einsele, G., and Seilacher, A. (eds.), Cyclic and Event Stratification, Springer-Verlag, Berlin, p. 180-198.

Baillie, A. D., 1952, Ordovician geology of Lake Winnipeg and adjacent area, Manitoba: Manitoba Dept. Mineral. & Nat. Resourc. Mines Br. Publ. 51-6.

Byers, C. W., 1977, Biofacies patterns in euxinic basins: a general model: in H. E. Cook and P. Enos (eds.), Deep-Water Carbonate Environments, Soc. Econ. Paleont. Mineral. Spec. Publ. 25, p. 5-17.

Carroll, W. K., 1978, Depositional and paragenetic controls on porosity development, Upper Red River Formation, North Dakota: 1978 Williston Basin Symposium, Montana Geol. Soc. Guidebook, p. 79-94.

Cowan, J., 1971, Ordovician and Silurian stratigraphy of the Interlake area, Manitoba: in Turnock, A. C. (eds.), Geoscience Studies in Manitoba, Geol. Assoc. Canada Spec. Paper 9, p. 235-241.

Derby, J. R., and Kilpatrick, J. T., 1985, Red River dolomite reservoirs, Killdeer Field, North Dakota: in Roehl, P. O., and Choquette, P. W. (eds.), Carbonate Petroleum Reservoirs, Springer-Verlag, New York.

Dowling, D. B., 1900, Report on the geology of the west shore and islands of Lake Winnipeg: Geol. Surv. Canada, Ann. Rept. 1898, pt. F.

Fursich, F. T., 1973, Thalassinoides and origin of nodular limestones in the Corallian Beds (Upper Jurassic) of southern England: Neues Jahrb. Geol. Palaont. Mh., p. 136-156.

Kendall, A. C., 1976, The Ordovician carbonate succession (Bighorn Group) of southeastern Saskatchewan: Saskatchewan Dept. Miner. Resourc. Rept. 180, 185 p.

Kendall, A. C., 1977, Origin of dolomite mottling in Ordovician limestones from Saskatchewan and Manitoba: Bull. Canadian Petrol. Geology, v. 25, p. 480-504.

Kendall, A. C., 1984, Origin and geometry of Red River dolomite reservoirs, western Williston Basin: Discussion: Am. Assoc. Petrol. Geol. Bull., v. 68, p. 776-779.

Kohm, J. A., and Louden, R. O., 1978, Ordovician Red River of eastern Montana - western North Dakota: Relationships between lithofacies and production: 1978 Williston Basin Symposium, Montana Geol. Soc. Guidebook, p. 99-117.

Longman, M. W., Fertal, T. G., and Glennie, J. S., 1983, Origin and geometry of Red River dolomite reservoirs, western Williston Basin: Am. Assoc. Petrol. Geol. Bull., v. 67, p. 744-771.

Longman, M. W., Fertel, T. G., and Glennie, J. S., 1984, Origin and geometry of Red River dolomite reservoirs, western Williston Basin: Reply: Am. Assoc. Petrol. Geol. Bull., v. 68, p. 780-784.

Nelson, R. A., 1981, Significance of fracture sets associated with stylolitic zones: Am. Assoc. Petrol. Geol. Bull., v. 65, p. 2417-2425.

DEPOSITIONAL ENVIRONMENT AND DIAGENESIS OF THE RED RIVER FORMATION, "C" INTERVAL, DIVIDE COUNTY, NORTH DAKOTA AND SHERIDAN COUNTY, MONTANA

DOUGLAS G. NEESE
Conoco Inc.
12600 West Colfax Avenue A500
Lakewood, Colorado 80215

ABSTRACT

Data from the "C" interval of the Ordovician Red River Formation in Divide and Sheridan counties show depositional environments of the "C" burrowed carbonate, "C" laminated carbonate, and "C" anhydrite are remarkably persistent. Diagenetic events, however, are highly localized, and dolomitic porosity is discontinuous.

Environmental conditions changed during deposition of the Red River "C" interval. At the base of the interval is a burrowed carbonate deposited in a well-circulated, subtidal, normal marine environment. Gradually a restricted subtidal, relatively quiet water environment developed and the "C" laminated carbonate was deposited. Intertidal conditions may have existed intermittently. The sequence is capped by the "C" anhydrite deposited in a hypersaline, subtidal restricted environment. Red River paleotopographic highs coincide with "C" laminated carbonate thins and contain vertical microfractures healed with anhydrite. In contrast to laterally persistent dolomite of the "C" laminated mudstone, dolomite of the "C" burrowed carbonate is discontinuous. Anhydrite crystals within the "C" burrowed carbonate are always associated with dolomite and may have been derived from downward percolating sulphate-rich brines from the overlying "C" laminated dolomite, or "C" anhydrite.

INTRODUCTION

The Red River Formation in the Williston Basin has been the focus of much research and petroleum exploration. While terminology and overall depositional interpretations have caused little argument, the diagenetic mechanisms and sequence of events have provided many points of contention. Of principal interest are the mechanisms and timing of dolomitization.

Core data from the Conoco State #16-7, located in the Skjermo field area (Fig. 1), show a typical depositional sequence recognized from core studies. Although depositional fabrics show remarkable persistence, diagenetic events, including dolomitization, are extremely variable between adjacent wells.

Within the Red River "C" interval (Fig. 2) are three major units which persist across Divide and Sheridan counties. At the base is a burrowed, fossiliferous dolomitic limestone. This is overlain by a mud-dominated laminated dolomite 23 to 28 feet (7-8.5 m) thick, and the sequence is capped by an anhydrite 22 to 26 feet (7-7.5 m) thick.

Evidence for relatively stable conditions during the Ordovician in the Williston Basin can be seen in the remarkable continuity of the three units of the Red River "C" interval. Discussions by Kendall (1984) and Longman and others (1984) indicate that detailed studies of the Red River in one portion of the Williston Basin may not necessarily provide an interpretation applicable for the entire basin. Local variations in depositional and diagenetic fabrics account for some differences in the conclusions drawn, and such is the case in the Skjermo field area. Patterns of dolomitization in the Red River from

Figure 1. Generalized map of the Williston Basin showing the outline of many of the major producing fields within the basin. Skjermo Field is shown located in Divide County in northwestern North Dakota. Production from Skjermo Field is from the "C" laminated dolomite. Other nearby Red River production is found in Big Dipper, Writing Rock, and Fortuna fields.

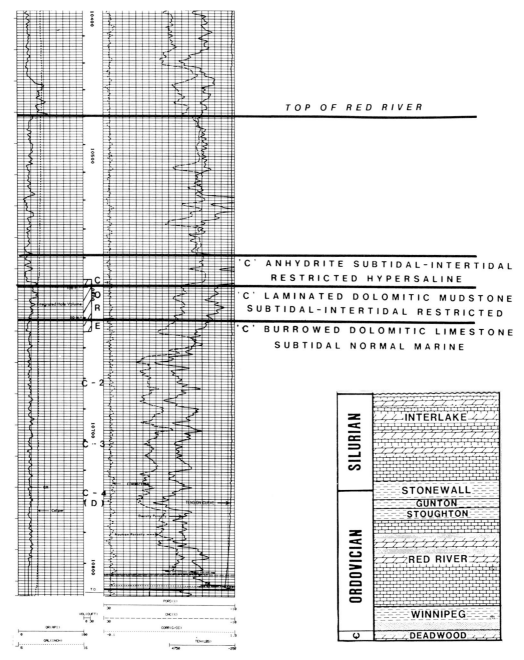

Figure 2. Neutron-density log showing the major units of the Red River Formation. The cored interval contains rocks from the "C" anhydrite, "C" laminated dolomite, and "C" burrowed dolomitic limestone. Lithofacies are based upon the cored interval from the Conoco 16-7 State, and other conventional cores penetrating the Red River "C" interval from Divide and Sheridan counties. The Ordovician Red River lies stratigraphically above the Ordovician Winnipeg Formation and below the Ordovician Stoughten member.

Divide County are similar to those described by Kendall (1976, 1977) in Saskatchewan, and Longman and others (1983) in Richland County, Montana. Timing and mechanisms of dolomitization in the Red River carbonates appear to be varied. The "C" laminated carbonate is dolomitized throughout Sheridan and Divide counties, whereas the burrowed carbonate contains dolomite correlative for at most only a few miles.

Kendall (1976, 1977) has proposed that there were two episodes of dolomitization in the "C" burrowed carbonate: one responsible for dolomitization of the burrows, the other for dolomitization of the host matrix. In the Skjermo field area there is variability of dolomite crystal size between the burrows and matrix, similar to the fabric discussed by Kendall (1976). The spatial distribution of dolomites within the "C" burrowed carbonate is not as extensive as in the dolomites in Saskatchewan, but more closely resembles patterns present in Richland County, Montana (Longman and others, 1983). Dolomitic porosity may be present in the laminated carbonate, the burrowed carbonate, or both. The "C" laminated dolomite may be either porous or tight, in contrast to the burrowed carbonate which, where dolomitized, always has at least some porosity.

The "C" laminated dolomite is present everywhere across Sheridan and Divide counties. This is in contrast to Richland County where the "C" laminated carbonate is locally limestone. Best hydrocarbon production is usually from the "C" laminated dolomite in Divide County and the distribution of this porosity and its predictability is of major concern to explorationists.

Well log suites and available core data throughout Sheridan and Divide counties were analyzed to establish a depositional and diagenetic sequence of events.

Stratigraphic terminology (Porter and Fuller, 1959; Kendall 1976), is consistent with correlations derived from well log data in Divide and Sheridan counties. Many workers have interpreted the Red River "C" interval as a brining-upward subtidal sequence (Fuller, 1961; Kohm and Louden, 1978; Longman and others, 1983), and similarly in Divide and Sheridan counties there is a lack of evidence for long term subaerial exposure. Others have suggested shallowing or shoaling upward sequences for the Red River (Asquith, 1978; Carroll, 1978; Kendall, 1984) which is consistent with some depositional textures present in cores from Skjermo Field. Core data from the Skjermo Field indicate the Red River "C" interval to be a shoaling and brining-upward sequence from subtidal normal marine limestones through a low intertidal restricted facies into a subtidal hypersaline depositional environment.

LITHOLOGIC UNITS

"C" Burrowed Carbonate (10,623-10,628 in the Conoco State #16-7 Core)

The "C" burrowed carbonate is predominately mixed skeletal, non-skeletal packstones and wackestones. Burrows are irregular and occur throughout the lower Red River Formation. The consistent gamma ray character on well logs suggest that this unit persists to the underlying Winnipeg Formation. Gamma ray character for this unit is easily recognizable on logs (Fig. 2) and persists across Sheridan and Divide counties. In cross-section, burrows range from 0.5 to 4 inches (1 to 10 cm) and are distributed nonuniformly through the interval. Core data (Fig. 3) show a light to dark brown mottled texture, commonly with distinct color changes between the burrow fill and matrix. Organisms most likely responsible for the network of burrows include mollusks, polychaetes, and crustaceans. The carbonate matrix is generally mud-supported, whereas some burrow fills are grain-supported. However, many burrows show no difference in carbonate mud percentage between the burrow fill and matrix. In such cases a distinction may be made from color and grain type. Carbonate grains include skeletal fragments of echinoderms, brachiopods, trilobites, corals, and bryozoans (Fig. 4). Non-skeletal grains include peloids, intraclasts and micrite lumps.

Extent of dolomitization of the burrowed carbonate in the Conoco 16-7 State can be seen in separation and cross plot of the compensated neutron density log. Dolomite of the "C" burrowed carbonate is present in the "C-2", "C-3", and "D" porosity zones of the Conoco 16-7 State. The lowermost dolomitic porosity of the burrowed carbonate has been termed the "D" zone by Carroll (1979). In Skjermo field and across Divide County lithologic characteristics are similar to the rest of the burrowed carbonate, and indistinguishable by gamma ray character. In many wells the "D" dolomitic porosity is absent, but burrow and grain types are consistent with other "C" burrowed porosity zones. To maintain consistency with the other burrowed dolomitic porosity horizons, the label "C-4", instead of "D" is applied here.

Figure 3. Slabbed core samples showing typical fabric from the Red River "C" interval including anhydrite, laminated dolomite, and burrowed dolomitic limestone.

A. "C" burrowed dolomitic limestone shows anastomosing burrows. Darker burrow fills contain skeletal and non-skeletal grains including crinoids, brachiopods, bryozoans, corals and trilobites.

B. Burrow fills in the "C" dolomitic limestone show grain-supported textures whereas the lighter matrix may be either mud-supported or grain-supported. Vertical microfractures are common and are healed with calcite. In the vicinity of some microfractures, anhydrite crystals locally occur.

C. Typical bedded to nodular anhydrite showing minor wispy laminations. Within the "C" anhydrite enterolithic structures occur.

D. The "C" laminated dolomite shows light to dark gray and brown laminations. Organic material material within laminations is variable. Stylolites may occur parallel to bedding and almost perpendicular, (Conoco Gjesdal #28-1). Orientation of stylolites other than parallel or subparallel to bedding is rare.

E. Local current activity has produced scour and fill structures in the "C" laminated dolomite, (Conoco Drawbond #27-6).

F. Laminations shown in the "C" laminated dolomite are irregular and pronounced from organic stain. Silt-sized carbonate grains are common in this interval, and irregular laminae suggest current transport, (Conoco Drawbond #27-6).

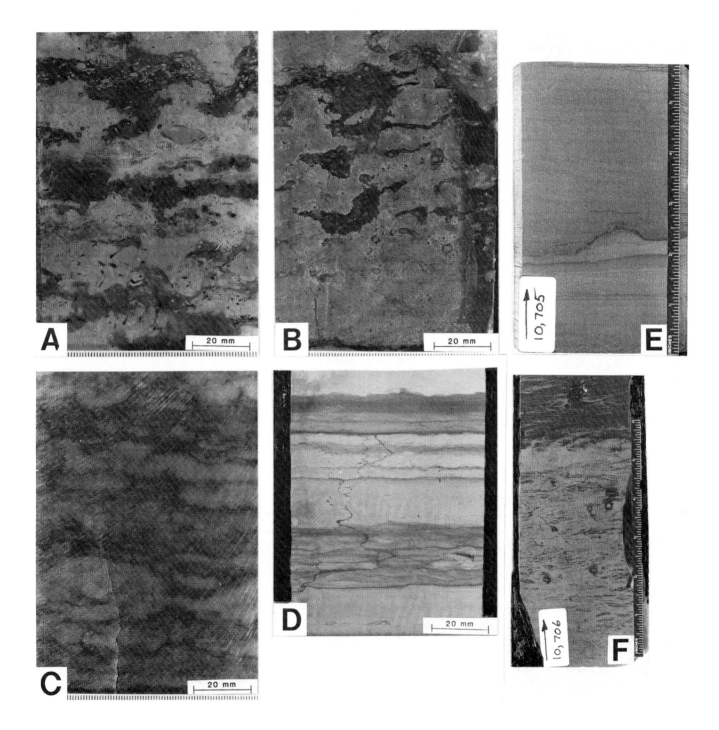

Figure 4. Carbonate textures and diagenetic fabrics which occur in the "C" burrowed dolomitic limestone.

A. Vugs are locally common in the "C" burrowed carbonate. This sample contains non-skeletal grains and brachiopod fragments. Baroque dolomite occurs as a coarsely crystalline pore-filling cement formed during later burial.

B. Mixed skeletal, non-skeletal packstone from a burrow fill containing crinoid and brachiopod fragments. Dolomite crystals range in size from 10 to 25 microns.

C. Anhydrite nodules occur as a secondary pore filling in burrows or matrix associated with dolomite. Skeletal fragments are crinoids and brachiopods.

D. Secondary calcite commonly occurs with stylolites in the burrowed carbonate. The Stylolite is filled with organic material, an calcite occurs both above and below the sutured surface.

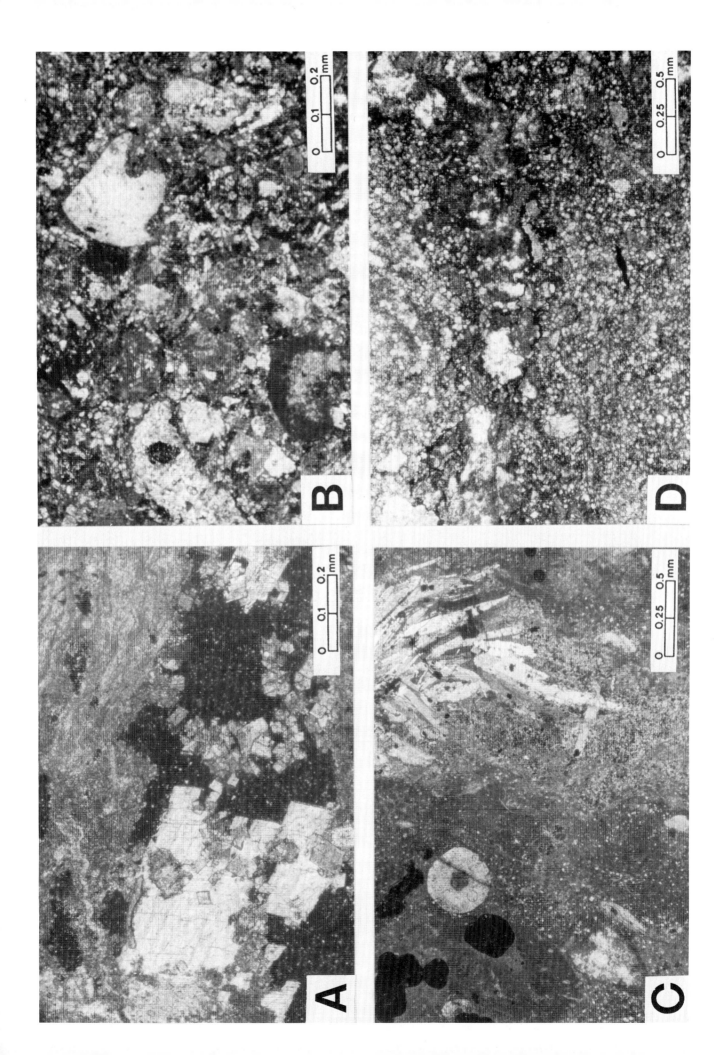

Figure 5. Carbonate textures and diagenetic fabrics of the "C" laminated dolomite.

A. The dolomitic mudstone contains minor coarse silt size non-skeletal carbonate grains. Anhydrite crystals are common in this sample from the upper "C" laminated dolomite. Stylolites occur parallel to bedding and contain organic material.

B. Elongate micrite intraclasts parallel to bedding. Calcite occurs as void fill and stylolites are filled with black organic material. Some intraclasts show scalloped surfaces.

C. High magnification view of an intraclast shown in the lower center of the previous photograph. This clast has an irregular surface. The scalloped surface may be the result of algal activity.

D. Vertical fractures in the dolomitic mudstone healed with anhydrite. Crystals of anhydrite are commonly associated with the anhydrite-filled fractures.

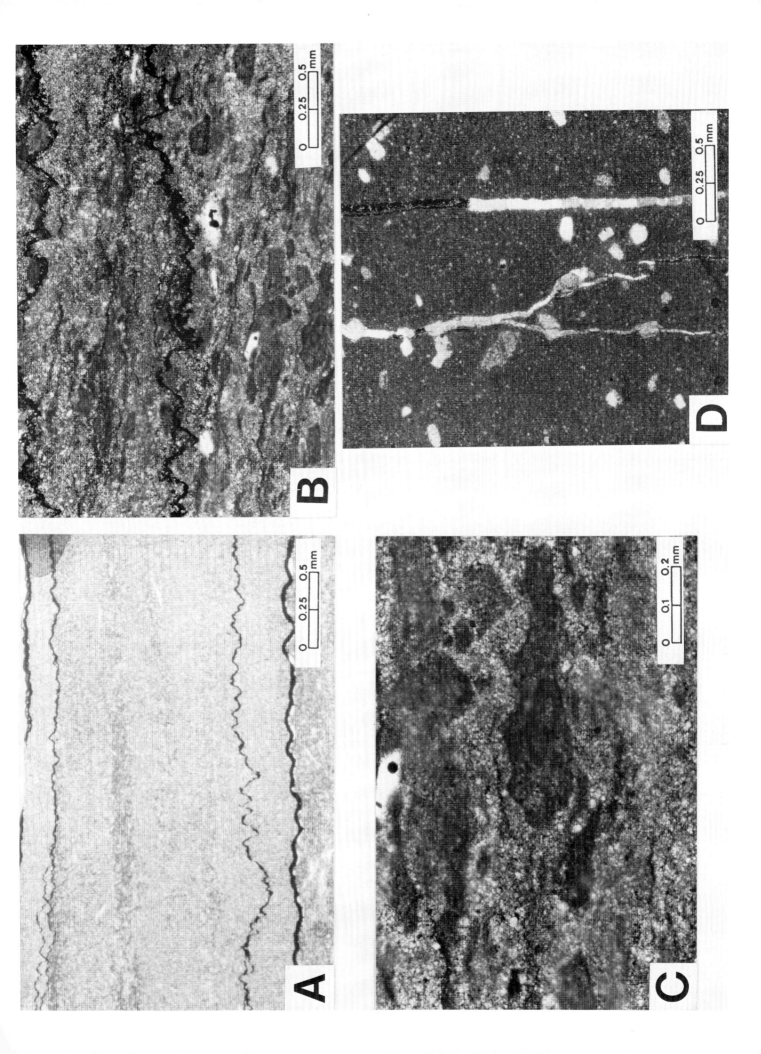

Associated with the dolomite of the "C" burrowed carbonate are anhydrite nodules 1-3 inches (2.5 to 8 cm) in diameter, and anhydrite crystals (Fig. 5). Anhydrite is not abundant in dolomite of the "C" burrowed carbonate, but occurs in the "C-2", "C-3, and "C-4" ("D") porous dolomite horizons. Baroque dolomite is locally present in vugs within the burrowed carbonate as a late stage burial pore filling (Fig. 4), but is not common.

Core samples from depths of 10,618.5 to 10,623 feet (3236.5 to 3237.8 m) contain depositional textures transitional between the majority of textures and burrow types in the "C" burrowed carbonate, and the overlying "C" laminated dolomite. Rock types within this interval include dolomitic packstones, wackestones, and mudstones. Carbonate grain types are predominantly non-skeletal and include peloids and intraclasts 0.1 to 0.8 inches (0.2 to 2 cm) in length. Skeletal grains include ostracod fragments, and rare brachiopod and crinoid debris. Minor vertical fractures are filled with calcite, or anhydrite. Burrows are variable, 0.5 to 4 inches (1 to 10 cm) in cross section, and color mottling is poorly developed. Fine wispy "algal" laminations, 0.1 to 0.5 inches (0.2 to 1.25 cm) thick, are common within this transitional interval. Across Sheridan and Divide counties the transitional zone varies in thickness from 1 to 5 feet (0.3 to 1.5 m).

"C" Laminated Carbonate (Core depth, 10,593.5-10,618.5 Feet)

The "C" laminated unit is predominantly a light tan dolomitic mudstone. All cores and logs penetrating the "C" carbonate of the Red River in Sheridan and Divide counties contain a "C" laminated dolomite from 24 to 30 feet (8 to 10 m) thick. Within the dolomitic mudstone, light tan laminations (0.5 to 2 inches) are laterally persistent. Laminations are present throughout the interval and are often separated by non-laminated dolomudstone (Fig. 3). Contact with the underlying burrowed interval is gradational, with dolomitic wackestones abundant at the base of the "C" laminated member. Carbonate grains include minor ostracods, peloids, micrite lumps, and elongate micrite intraclasts, some with scalloped surfaces (Fig. 5). In the upper five feet, anhydrite nodules 1 to 3 inches (2.5 to 7.5 cm) in diameter may occur, but are rare. Anhydrite crystal laths are common in the upper 1-3 feet of the "C" laminated dolomite and are present in all cores of the Skjermo field area (Fig. 5). Within the "C" laminated dolomite, primary depositional structures suggestive of traction transport may occur in units from 2-10 inches (5 to 25 cm) thick (Fig. 3). Porosity is absent in the "C-1" laminated dolomite from the #16-7 State, but may occur as microcrystalline porosity where the "C-1" laminated dolomite is thicker. Locally, thin laminations (0.1 to 0.6 inches) are present as wispy dark gray bands, and are suggestive of algal origin.

"C" Anhydrite (Core Depth, 10,590-10593.5 Feet)

The "C" anhydrite is approximately 20 feet (6 m) thick, and is ubiquitous across Sheridan and Divide counties. The light tan to gray anhydrite exhibits both nodular and bedded textures. Bedded and local enterolithic textures are common throughout the "C" anhydrite. Minor wispy "algal"(?) laminations occur throughout the entire 20-foot

(6 m) anhydrite interval. Fine-grained lath-like crystals are abundant Fig. 5), and are identical to those present locally in the upper few inches of the "C" laminated dolomite. In all cores, evidence for long term subaerial exposure is absent.

DEPOSITIONAL SUMMARY

The "C" interval (burrowed dolomitic limestone, laminated dolomite, and anhydrite), represents a shallow, brining and shoaling upward, subtidal to intertidal sequence (Fig. 6). Regional and blanket-like extent in each of the three units indicates relatively stable conditions across much of the Williston Basin area during Ordovician time. A diverse faunal assemblage and abundant bioturbation indicate that the "C" burrowed carbonate was deposited in a subtidal normal marine environment. Depositional conditions changed from a open marine environment to restricted relatively quiet water conditions during deposition of the "C" laminated carbonate. Mud supported textures predominate, and the faunal assemblage is restricted and sparse. Elongate micritic intraclasts, some with scalloped edges (algal?), mat laminations, minor current laminations and local scour features may occur in a shallow subtidal to low intertidal regime. Evidence for deposition of units of the "C" laminated carbonate within the subtidal-intertidal regime is not absolute. However, some features present in cores from the Skjermo area are similar to intertidal deposits in other carbonate systems, (cf. Logan and others, 1974; Ginsburg, 1975). Isopach maps of the "C" laminated dolomite show thins corresponding closely with non-porous laminated dolomite. This suggests that these areas were high during the Ordovician. Such Red River paleohighs are sometimes coincident with present day structure (Fig. 7).

Like the "C" laminated dolomite, the "C" anhydrite shows no evidence of longterm subaerial exposure. Minor wispy and mat lamination are present in varied portions of the "C" anhydrite. Rare poorly developed fenestral fabric is present in mat-laminated dolomite within the "C" anhydrite interval. Textures of the "C" anhydrite suggest deposition in a subtidal, restricted hypersaline environment. During deposition of the "C" anhydrite, local conditions may have intermittently existed in which mat-laminated dolomite was deposited in an intertidal environment.

DOLOMITIZATION

The distribution of dolomite and its associated porosity is complex. Dolomite in the "C" laminated carbonate is everywhere present, while portions of the "C" burrowed carbonate may or may not be dolomitized. The "C-2", "C-3", and basal "C-4" ("D") dolomitic porosity intervals are 10 to 20 feet thick (3 to 6 m), often vertically separated by burrowed limestone, and may be correlative over several miles. Dolomitic porosity in the burrowed carbonate is discontinuous, and pinches out laterally into tight limestone. In some areas dolomitic burrowed carbonate is completely absent. The gross vertical distribution of dolomite is at most continuous for 175 feet (50 m) beneath the "C" anhydrite. Underlying the dolomite of the burrowed

Figure 6. This generalized depositional model shows the sequential development of facies in the Red River "C" interval. During deposition of the burrowed carbonate (time A) conditions were typical of a well-circulated normal marine environment. Light and dark areas indicate bottom irregularities. A diverse faunal assemblage flourished in the present day area of Divide and Sheridan counties. Environmental conditions became restricted during deposition of the "C" laminated carbonate (time B). Water depths decreased, and locally a change from subtidal to intertidal conditions may have existed. Coincident with restricted shoal water conditions, early dolomitization of the "C" laminated carbonate took place regionally. Paleotopographic highs were the site of increased anhydrite precipitation, and downward percolation of Mg-rich brines may have begun during the latest stages of deposition of the "C" laminated dolomite. The white areas in the burrowed carbonate in stage B are areas in which some dolomitization had begun. During deposition of the "C" anhydrite (time C) conditions across the Divide and Sheridan County areas were restricted and hypersaline. Locally intertidal conditions may have existed. Brines present during precipitation of the "C" anhydrite were a likely source for anhydrite within the fractures of the "C" laminated dolomite, anhydrite crystals within the "C" burrowed dolomitic limestone, and a source of Mg-rich fluids for continued dolomitization of the underlying "C" laminated carbonate and "C" burrowed carbonate.

DEPOSITIONAL MODEL - RED RIVER 'C' INTERVAL

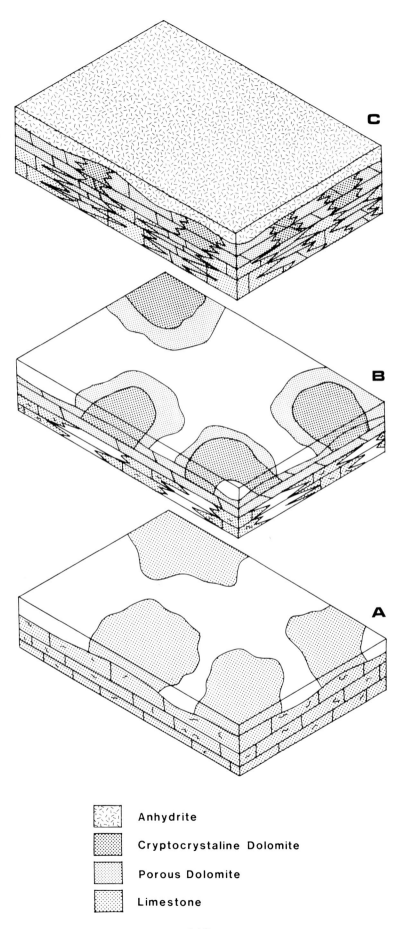

Anhydrite
Cryptocrystaline Dolomite
Porous Dolomite
Limestone

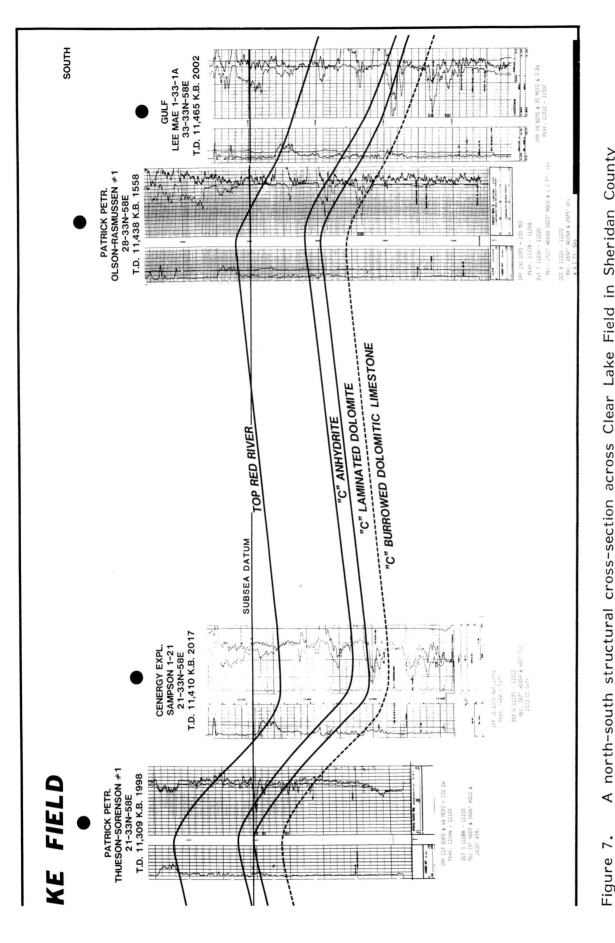

Figure 7. A north-south structural cross-section across Clear Lake Field in Sheridan County shows how present day structure may coincide with paleostructure. The Patrick Guenther-Anderson lies structurally downdip from the Patrick State #2. Well-developed "C" laminated porosity occurrs in the Guenther-Anderson well and coincides with a thicker "C" laminated dolomite. With adequate well control, the stratigraphic and diagenetic relationships may be better predicted, but with significant structural closure production may occur from any one or a number of dolomitic porosity zones within the "C" burrowed carbonate.

Figure 7. Continued

"C-4" ("D") dolomitic porosity zone is 150 to 200 feet (45 to 60 m) of tight burrowed limestone, continuous to the Winnipeg Formation.

Dolomite of the "C" Burrowed Carbonate

Dolomite distribution within the burrowed carbonate is irregular. However, in most cases where dolomitic porosity is abundant in the burrowed carbonate, porosity in the overlying "C" laminated dolomite is absent. The converse is also true. Dolomite crystal sizes vary from 5 to 35 microns, with most ranging from 15-30 microns. The largest crystals are generally present within burrows. Anhydrite crystals and nodules present in the burrowed zone are less than 2 inches (5 cm) in diameter and are always associated with dolomite (Fig. 4). Locally the "C-2", "C-3", and "C-4" ("D") porosity zones may be separated vertically by tight beds of burrowed limestone 1 to 5 feet (0.3 to 1.6 m) thick.

Anhydrite crystals indicate that at some time the normal marine burrowed carbonate was subjected to fluids rich in sulphate. Sources for such waters could include brines present during latest deposition of the "C" laminated dolomite, or sulphate-rich brines present during precipitation of the "C" anhydrite. Dolomitization through downward percolation of shallow subsurface waters has been documented in other rocks by Dunham and Olson (1980) through use of trace element and stable isotope data. Such a model supports petrographic evidence at Skjermo field within the Red River burrowed carbonate for dolomitization by downward percolating brines.

Dolomite of the "C" Laminated Carbonate

The "C" laminated carbonate is dolomite across Sheridan and Divide counties. Porous "C" laminated dolomite occurs as local discontinuous lenses up to 18 feet (6 m) thick. Dolomite crystals of the porous "C" laminated carbonate range from less than 5 to 18 microns while non-porous "C" laminated carbonate contains dolomite crystals less than 5 microns. Commonly associated with the non-porous "C" laminated dolomite are euhedral anhydrite crystals (Fig. 4). Vertical fractures may be filled with calcite or anhydrite and are most commonly associated with non-porous "C" laminated dolomite. Dolomite crystals of the "C" laminated carbonate are generally smaller and more homogeneous than the underlying dolomites of the "C" burrowed carbonate. The very fine crystalline texture of the "C" laminated dolomite is similar to textures present in early dolomitized rocks (cf. Shinn and others, 1965; Illing and others, 1965; Kendall and Skipwith, 1969; Behrens and Land, 1972; and Friedman, 1980). Textures of the "C" laminated dolomite indicate penecontemporaneous dolomitization in a shallow penesaline environment.

Alternative Dolomite Hypothesis

Burial dolomitization of the Red River by fluids expelled from the Winnipeg Formation, proposed by some industry workers, provides an alternative to the interpretations already discussed. However, several problems must be considered when invoking this hypothesis. First, it should be noted that the carbonates immediately above the Winnipeg are not dolomitized. Second, sulphates present in the burrowed dolomite are unlikely to be precipited from fluids generated from the Winnipeg. Third, whole rock burial dolomitization would require an exceedingly large volume of Mg-rich fluids to be generated in the Winnipeg.

Subsurface dolomitization by fluids generated from buried shales may occur under certain conditions (Mattes and Mountjoy, 1980). Dolomites of the Miette buildup (Upper Devonian) have been interpreted by Mattes and Mountjoy (1980) as being dolomitized by vertical movement of deep burial brines along fracture-controlled conduits. As pointed out by Mattes and Mountjoy (1980), a major barrier to incorporation of Mg^{++} into a carbonate lattice is the size of the hydrated ion, (4.23A hydrated, 0.66A unhydrated). Only with increased temperatures, or high Mg:Ca ratios, is dolomitization favored in solutions with high ionic strengths, typical of many subsurface waters. At elevated temperatures around 140°F., solutions with a Mg:Ca ratio of 0.33 to 2.00 are at equilibrium with dolomite and dolomitization may occur (Murata and others, 1972). Using a thermal gradient of 1°F. per 100 feet, a depth of burial of at least 4,000 feet would be necessary before dolomitization of limestone would be favored, and in the case of the Red River Formation, this burial depth would occur sometime during the Devonian.

The mechanism for precipitation of sulphates in the burrowed carbonate is difficult to explain with fluids generated from shales during burial. Even though subsurface waters associated with oil fields contain 50,000 to 350,000 ppm dissolved solids (Dickey, 1966), deep subsurface brines commonly contain little or no sulfate or bicarbonate, as chloride is the major anion present.

Formation waters often have an increase in Ca^{++} and a decrease in Mg^{++} relative to present day sea water and many authors have attributed this to dolomitization of limestones (e.g., Collins, 1967). Hitchon and others (1981) have studied formation waters in western Canada sedimentary basins and have shown a gain and loss of Ca and Mg, respectively, but according to their calculations, if all the Mg lost from the volume-weighted mean formation water was used in dolomitization, 190 cubic miles of limestone could be converted to dolomite. However, the volume of dolomite in the western Canada (Alberta) basin is 62,950 cubic miles. These subsurface waters may be similar to those of the Williston Basin and the conclusion drawn by Hitchon and others (1981) may be equally valid. They conclude "if the loss of Mg^{++} and gain of Ca^{++} is due to dolomitization, then the amount of dolomite formed is an insignificant portion of the dolomite now present in the basin... and the process of dolomitization (of most dolomites) was essentially a penecontemporaneous reaction...".

Stylolitization

Throughout the carbonate sequence of the Red River, stylolites are present in varied abundance. Pressure solution from compaction has resulted in stylolitization in both the "C" laminated dolomite, and "C" burrowed dolomitic limestone. Stylolites are more common in the burrowed dolomitic limestone than in the "C" laminated dolomite, and this can in part be attributed to the greater solubility of calcite

($10^{-8.2}$) compared to dolomite ($10^{-16.7}$). In the burrowed interval, stylolites are most common where porosity and permeability are low. Secondary calcite (Figure 4) is responsible for some pore occlusion, and much of the diagenetic calcite cement may have been generated during stylolitization. Burial cementation by calcite generated through stylolitization has been shown by Wong and Oldershaw (1981) to have a significant effect on pore occlusion in carbonates. Localized variations in the vertical stress may account for increases in the intensity of stylolitization. Stylolites are most commonly parallel or sub-parallel to bedding, but some are oriented almost perpendicular to bedding (Fig. 3), indicating horizontal compression. The resulting relatively "tight" burrowed carbonate may have routed hydrocarbon migration to overlying reservoirs within the laminated dolomite.

CONCLUSIONS

1. The depositional sequence of the Red River "C" interval represents a brining and shoaling upward unit. The base of the unit is a subtidal, normal marine burrowed carbonate. This grades upward into a penesaline restricted subtidal-intertidal, mud-supported dolomite, and is capped by a subtidal, restricted, hypersaline anhydrite.

2. Dolomitization of the "C" laminated carbonate was penecontemporaneous with deposition, and occurs everywhere across Sheridan and Divide counties. Dolomitization of the underlying "C" burrowed limestone is variable and occurred after "C"

laminated dolomitization. The diagenetic fabric and the geometry of the "C" burrowed dolomitic porosity suggest downward percolation of Mg-rich brines. Microfractures healed with anhydrite in the overlying "C" laminated dolomite indicate cohesiveness and a pathway for downward percolation of sulphate-rich brines during latest deposition of "C" laminated dolomite and/or precipitation of the "C" anhydrite.

3. Dolomitic porosity in the "C" laminated carbonate developed in depositional or tectonic low flank areas. This is suggested by isopach thicks in the Red River which coincide with the best "C" laminated porosity development. Porosity is discontinuous in both the "C" laminated and "C" burrowed horizons. Porosity in the "C" burrowed carbonate is correlative at most for a few miles, before grading into tight limestone.

4. Local variation in vertical tectonics has resulted in a greater intensity of stylolitization in the "C" burrowed interval in some areas, and the secondary calcite generated by pressure solution has destroyed some of the dolomitic porosity by pore occlusion.

5. Non-porous "C" laminated dolomite is usually thinner than porous "C" laminated dolomite, and suggests that the cryptocrystalline, anhydrite-plugged, laminated dolomites were probably paleohighs. Where present day structure coincides with Red River paleostructure (e.g., in Skjermo, Comertown, Raymond, and Clear Lake

fields), prediction of porosity and hydrocarbon production is less complicated.

REFERENCES

Asquith, G.B., Parker, R.L., Gibson, C.R. and Root, J.R., 1978, Depositional history of the Ordovician Red River C and D zones, Big Muddy Creek field, Roosevelt County, Montana; Montana Geological Society Guidebook, 1978, Williston Basin Symposium, p. 71-76.

Carroll, W.K., 1978, Depositional and paragenetic controls on porosity development, Upper Red River Formation, North Dakota: Montana Geological Society Guidebook, 1978 Williston Basin Symposium, p. 79-94.

Carroll, W.K., 1979, Depositional environments and paragenetic porosity controls, upper Red River Formation, North Dakota: North Dakota Geological Survey, Report of Investigations 66, 51 p.

Collins, A.G., 1967, Geochemistry of some Tertiary and Cretaceous age of oil-bearing formation water: Environ. Sci. Technol., v. 1, p. 79-94.

Dickey, P.A., 1966, Patterns of chemical composition in deep subsurface brines: Am. Assoc. Petrol. Geol., Bull., v. 50, p. 2472-2478.

Dunham, J.B., and Olson, E.R., 1980, Shallow subsurface dolomitization of subtidally deposited carbonate sediments in the Hanson Creek Formation (Ordovician-Silurian) of central Nevada: Soc. Econ. Paleont. Mineral. Spec. Publ. 28, p. 139-161.

Folk, R.L., and Land, L.S., 1975, Mg/Ca ratio and salinity: Two controls over crystallization of dolomite: Am. Assoc. Petrol. Geol. Bull., v. 59, p. 60-68.

Friedman, G.M., 1980, Dolomite is an evaporite mineral: Evidence from the rock record and form sea-margin ponds of the Red Sea: Soc. Econ. Paleont. Mineral. Spec. Publ. 28, p. 69-80.

Fuller, J.G.C.M., 1961, Ordovician and continuous formations in North Dakota, South Dakota, Montana, and adjoining areas of Canada and United States: Am. Assoc. Petrol. Geol. Bull., v. 45, p. 1334-1363.

Ginsburg, R.N., 1975, Tidal deposits: A Casebook of Recent Examples and Fossil Counterparts: Springer-Verlag, New York, 428 p.

Hitchon, B., Bullings, G.K., and Klovan, J.E., 1971, Geochemistry and origin of formation waters in the western Canada sedimentary basin - III. Factors controlling chemical composition: Geochimica et Cosmochimica Acta, v. 35, p. 567-598.

Illing, L.V., Wells, A.J., and Taylor, J.C.M., 1965, Penecontemporary dolomite in the Persian Gulf: in Pray, L.C. and Murray, R.C. (eds), Dolomitization and Limestone Diagenesis: Soc. Econ. Paleont. Mineral. Spec. Pub. 13, p. 89-111.

James, N.P., 1977, Shallowing-upward sequences in carbonates; facies models 8: Geoscience Canada, v. 4, p. 126-136.

Kendall, A.C., 1976, The Ordovician carbonate succession (Big Horn Group) of southeastern Saskatchewan: Department of Mineral Resources, Saskatchewan Geol. Surv. Report 180, 185 p.

Kendall, A.C., 1977, Origin of dolomite mottling in Ordovician limestones from Saskatchewan and Manitoba: Bull. Canadian Petrol. Geol., v. 25, p. 480-504.

Kendall, A.C., 1984, Origin and geometry of Red River dolomite reservoirs, western Williston Basin: Discussion: Am. Assoc. Petrol. Geol. Bull., v. 68, p. 776-779.

Kendall, C.G.ST.C., and Skipwith, P.A., 1969, Holocene shallow-water carbonate and evaporite sediments of Khol al Bazam, Abu Dhabi, southwest Persian Gulf: Am. Assoc. Petrol. Geol. Bull., v. 53, p. 841-869.

Kohm, J.A., and Louden, R.O., 1978, Ordovician Red River of eastern Montana-western North Dakota: Relationships between lithofacies and production: Montana Geological Society Guidebook, 1978 Williston Basin Symposium, p. 99-117.

Logan, B.W., Read, J.F., Hagan, G.M., Hoffman, P., Brown, R.G., Woods, P.J., and Gebelein, C.D., 1974, Evolution and diagenesis of Quaternary carbonate sequences, Shark Bay, western Australia: Am. Assoc. Petrol. Geol. Memoir 22, 358 p.

Longman, M.W., Fertal, T.G., and Glennie, J.S., 1983, Origin and geometry of Red River dolomite reservoirs, western Williston Basin: Am. Assoc. Petrol. Geol. Bull., v. 67, p. 744-771.

Longman, M.W., Fertal, T.G., and Glennie, J.S., 1984, Origin and geometry of Red River dolomite reservoirs, western Williston Basin: Reply: Am. Assoc. Petrol. Geol. Bull., v. 68, p. 779-784.

Mattes, B.W., and Mountjoy, E.W., 1980, Burial dolomitization of the upper Devonian Miette buildup, Jasper National Park, Alberta: Soc. Econ. Paleont. Mineral. Spec. Publ. 28, p. 259-297.

Murata, K.S., Friedman, I., and Cremer, M., 1972, Geochemistry of diagenetic dolomites in Miocene marine formations of California and Oregon: U.S. Geol. Surv. Prof. Paper 724-C, p. C1-C12.

Porter, J.W., and Fuller, J.G.C.M., 1959, Lower Paleozoic rocks of northern Williston Basin and adjacent areas: Am. Assoc. Petrol. Geol. Bull., v. 43, p. 124-189.

Shinn, E.A., Ginsburg, R.N., and Lloyd, R.M., 1965, Recent supratidal dolomite from Andros Island, Bahamas: in Pray, L.C. and Murray, R.C., (eds.), Dolomitization and Limestone Diagenesis: A Symposium: Soc. Econ. Paleont. Mineral. Spec. Publ. 13, p. 112-123.

Wong, P.K., and Oldershaw, A., 1981, Burial cementation, Kaybob Reef: Jour. Sed. Petrol., v. 51, p. 507-520.

DEPOSITIONAL ENVIRONMENTS, PALEOECOLOGY AND DIAGENESIS OF SELECTED WINNIPEGOSIS FORMATION (MIDDLE DEVONIAN) REEF CORES, WILLISTON BASIN, NORTH DAKOTA

NANCY A. PERRIN
2108 W. Silver Fox Dr.
Edmond, Oklahoma 73034

WILLIAM F. PRECHT
Champlin Petroleum Co.
P. O. Box 1257
Englewood, Colorado 80150

ABSTRACT

The Winnipegosis carbonates were deposited in three episodes during the initial transgressive-regressive cycle of the Kaskaskia sequence. The first episode is represented by a normal, open marine environment. During the second episode, the basin differentiated into two regions: the shallower shelves and a bathymetrically "deeper" basin. This deeper basin was comprised of a restricted environment with large pinnacle reefs. Shallow marine, patch reef, lagoon, and tidal flat environments became established in the shelf region. During the third episode, supratidal stromatolites and lime mudstones were deposited on top of the shelf and pinnacle reefs while stromotolites, dolomites, and starved-basin anhydrites accumulated in the inter-reef areas. The pinnacle reefs contain four lithofacies: the Stromatoporoid, Tabulate Coral Boundstone; the Codiacean Algae, Calcisphere, Peloid Packstone; the Porous Dolomite; and the Pisolite Dolomite Lithofacies. The patch reef is composed of the Stromatoporoid, Tabulate Coral Boundstone Lithofacies.

The pinnacle reefs exhibit a three-stage successional biofacies development: (I) pelmatozoan packstones which represent an intrinsic preparation of the substratum for reef growth; (II) establishment of reef-building domal and dendroidal tabulate corals; and (III) alternating zones of more massive thamnoporid corals and stromatoporoids. Stromatoporoid-algal boundstones are also common. During Stage III a lateral zonation also developed with higher energy environments on an outer reef rim and a lower energy backreef lagoon. The backreef deposits are composed of codiacean algae, calcispheres, peloids and Amphipora.

Patch reefs show a four-stage, shallowing-upward succession: (I) pelmatozoan skeletal sand shoals; (II) growth of domal and branching tabulate corals; (III) increasing size, number of genera, and growth forms of the reef building taxa, with growth shapes indicative of a higher energy, above mean wave-base environment predominating; and (IV) large massive, hemispherical stromatoporoids which dominate in turbulent high energy environments.

Diagenetic fabrics in the pinnacle reefs include partial to extensive dolomitization. The extent of dolomitization and the destruction of the allochems and textures decreases downward within the reef core. The diagenesis of the patch reef is minor in comparison with that of the pinnacle reef. Diagenetic events include isopachous calcite and equant calcite spar cements, stylolites, local dolomitization of allochems and the mudstone matrix and rare bladed or blocky anhydrite.

INTRODUCTION

The Middle Devonian (Givetian) Winnipegosis Formation is the first major carbonate unit of the Kaskaskia sequence (Sloss, 1963; Gerhard, and others, 1982) (Fig. 1). Microfacies analyses performed on 39 cores from the Winnipegosis in northwestern North Dakota form the basis for the interpretation of lithofacies, environments of deposition and diagenesis (Perrin, 1982, 1985) (Fig. 2).

This paper deals with the Winnipegosis reefs and focuses on the reef lithofacies (depositional environments), biofacies (paleoecology), and diagenesis as shown in three representative cores. The cores selected for this paper (arrows on Fig. 2) include a patch reef, McMoran Oil--Deraas #1 (SW SW 24-161-76, North Dakota Geological Survey well #5280, Bottineau County) and two pinnacle reefs, Shell Oil--Greek #41-2 (NE NE 2-161-83, N.D.G.S. well #6535, Bottineau County) and Shell Oil--Osterberg #22x-1 (SE NW 1-161-85, N.D.G.S. well #6624, Renville County). For comparison purposes, a basinal core, Shell Oil--Osterberg #21-2 (NE NE 2-161-85, N.D.G.S. well #6684, Renville County), is included. Thin section and slab photographs from other Winnipegosis cores have also been used.

PALEOGEOGRAPHIC SETTING

In the Late Silurian much of the North American craton was exposed. During the Early Devonian, marine seas transgressed southeastward from the Cordilleran Basin and inundated a subsiding

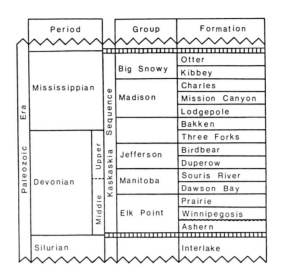

Figure 1. Stratigraphic column for the Kaskaskia sequence in the study area.

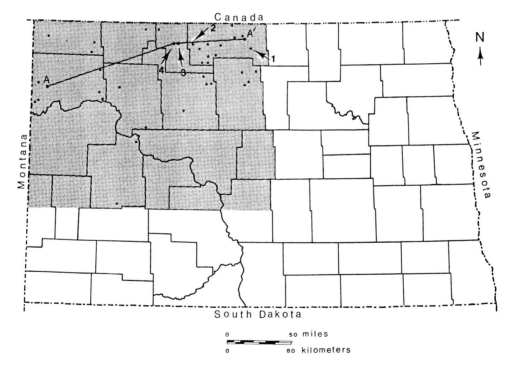

Figure 2. Study area (shaded), with locations of wells with Winnipegosis cores, cores selected for this study indicated by arrows, and line of cross section, A – A' (Fig. 5). 1 – McMoran Oil-Deraas #1, 2 – Shell Oil Greek #41-2, 3 – Shell Oil – Osterberg #22x – 1, and 4 – Shell Oil – Osterberg #22 – 2.

portion of the craton to form the Elk Point Basin (Fig. 3). Southeastward transgression occurred during the Early and Middle Devonian and marine seas ultimately covered parts of Montana and North Dakota (Grayston and others, 1964).

As the seas advanced into the southeastern portion of the Elk Point Basin (Williston Basin), the Ashern Formation, a fine-grained lime mudstone (now dolomitized), was deposited (Fig. 4). Following a breif erosional hiatus, three episodes of marine deposition produced the Winnipegosis Formation. The first, and most of the second episode comprise the transgressive phase of the Winnipegosis. Regression began near the end of the second episode and continued with minor transgressive pulses through the rest of the Winnipegosis and overlying Prairie Evaporite depositional periods. Cross section A - A' (Fig. 5) extends from the western shelf (Williams County), across the deep basin (Renville County) and on to the eastern shelf (Bottineau County). The section exhibits typical log profiles and thicknesses in these settings.

REEF LITHOFACIES

Two different reef environments are represented in the Winnipegosis Formation: patch reef and pinnacle reef. Patch reefs developed either at the shelf margin or upon the shelf platform. The pinnacle reefs developed in the "deeper" basin environment. Because the reef organisms produced sufficient carbonate sediment to maintain continued reef growth, this environment was able to keep pace with

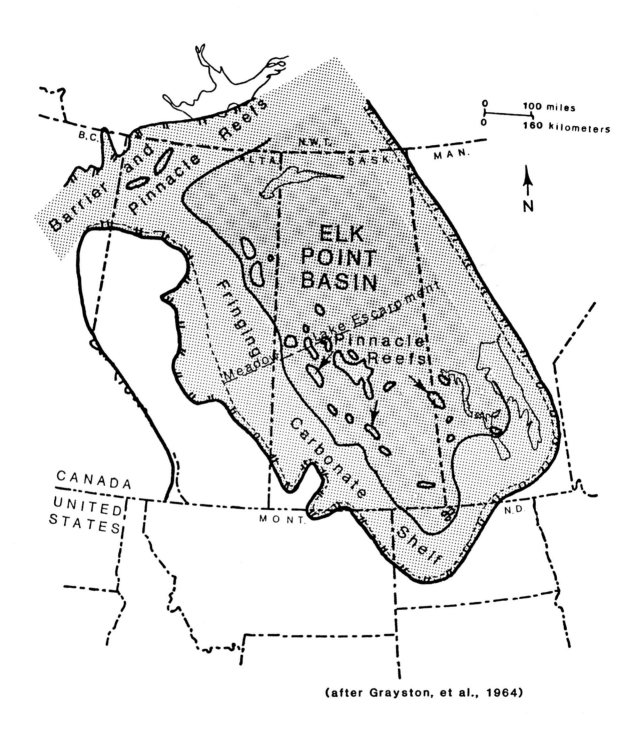

Figure 3. Elk Point Basin at the beginning of the Kaskaskia sequence (after Grayston and others, 1964).

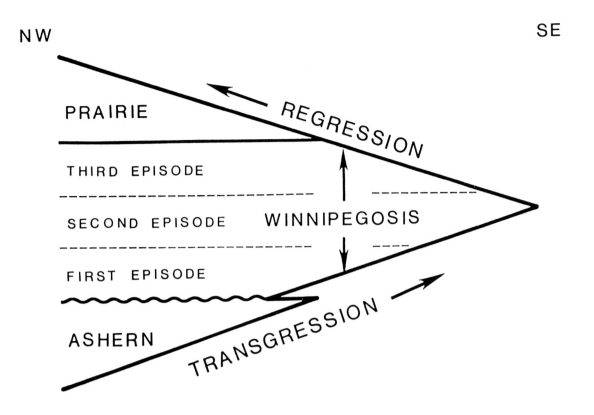

Figure 4. Schematic representation of the first transgressive-regressive cycle of the Kaskaskia sequence.

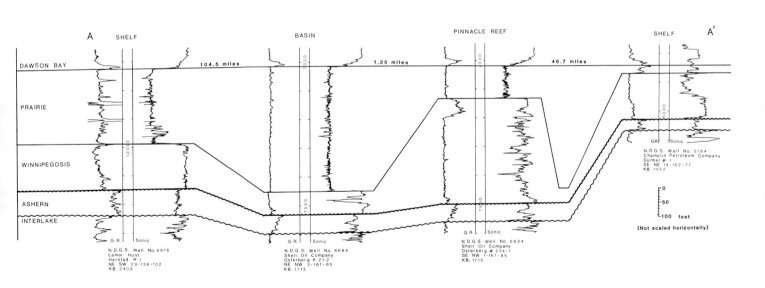

Figure 5. Cross section, A - A' (Indexed on Fig. 2), from the western shelf region in Williams County to the basin region in Renville County (showing both basinal and pinnacle reef logs approximately 1 1/4 miles apart) and to the eastern shelf region in Bottineau County.

the rising sea level. The pinnacle reefs developed as "pseudo-atolls" with an outer margin and an inner backreef lagoon.

In the Middle Devonian, corals and stromatoporoids were common and diverse. In the Winnipegosis, they occur as framebuilders and encrusters in the reef environment and as isolated, large fossils in the shallow shelf environment. The corals include <u>Thamnopora</u>, <u>Coenites</u>, <u>Syringopora</u>, <u>Hexagonaria</u>, <u>Favosites</u>, <u>Alveolites</u>, solitary rugosans, and chaetitids. Stromatoporoids include common tabular and bulbous to hemispherical growth forms, as well as rare dendroid forms. These fossils locally occur associated with much smaller allochems and fragmented shells within wackestones or packstones (Fig. 6A). These reefs are true ecologic reefs as defined by Dunham (1970) and Heckel (1974).

Winnipegosis patch reefs are composed of the Stromatoporoid, Tabulate Coral Boundstone Lithofacies. The pinnacle reefs contain four lithofacies: the Stromatoporoid, Tabulate Coral Boundstone; the Codiacean Algae, Calcisphere, Peloid Packstone; the Porous dolomite; and the Pisolite Dolomite Lithofacies. The pinnacle reefs are capped by the Crinkly Stromatolite Lithofacies which was deposited as sea level fell.

Stromatoporoid, Tabulate Coral Boundstone Lithofacies

Stromatoporoids and tabulate corals are the dominant megascopic allochems of this lithofacies. Bulbous to almost hemispherical stromatoporoids and tabular forms, sometimes only a few laminae thick, are common. Laminae may encrust or surround other organisms.

Although the most abundant tabulate coral is Thamnopora, chaetitids are also abundant. Of special importance, but identified only in microscopic views, are blue-green algae represented by tubular filaments. These algae seem to encrust almost everything (Fig. 6B), and may have formed a symbiotic association with stromatoporoids (N. P. James, pers. comm., 1985), because they are commonly found within the stromatoporoid colony between the laminae (Fig. 6C).

Stromatoporoids encrust Thamnopora (Fig. 6D), chaetitids (Fig. 6E), and solitary and colonial rugose corals, as well as the matrix. Accessory organisms include bryozoans and foraminifers. Packstone and boundstone textures are common throughout this lithofacies, but they are generally impossible to differentiate in hand specimen.

This lithofacies comprises the outer margin of the pinnacle reef and the entire patch reef. The major difference between the development of this lithofacies in the pinnacle reef and the patch reef is that the pinnacle reefs are extensively dolomitized whereas the patch reefs are largely limestone.

The organisms which are most useful in identification of the reef lithofacies include the corals (especially Thamnopora), stromatoporoids, and codicean algae. As dolomitization proceeds, the fossil skeleton becomes progressively replaced and loses the characteristics necessary for identification. By noting the size, shape, and orientation of pores (intraparticle porosity) in well preserved specimens, it is possible to trace each type of organism through the dolomitization process, until all that remains for identification is its characteristic "ghost".

Figure 6. Core slab photographs and thin section photomicrographs from the reef lithofacies.

    A. McMoran Oil - Deraas #1 4734/35 feet. Core slab photograph. Large fossils (stromatoporoid and pelmatozoans) associated with much smaller allochems within packstone.

    B. Hunt Oil - Harstad #1 12046.4 feet. Thin section photomicrograph (crossed polars). Blue-green algae located in "muddy" area between stromatoporoid laminae. See Figure 6-C for enlargement of this area.

    C. Hunt Oil - Harstad #1 12046.4 feet. Thin section photomicrograph (crossed polars). Blue-green algal filaments found in "muddy" area between stromatoporoid laminae (part of laminae in upper right corner).

    D. McMoran Oil - Deraas #1 4739.9 feet. Thin section photomicrograph *Thamnopora* encrusted by blue-green algae which, in turn was encrusted by a stromatoporoid.

    E. Marathon Oil - Adams #1 6650 feet. Core slab photograph. Stromatoporoid (light) encrusting chaetitid. *Thamnopora* on right.

    F. Marathon Oil - Adams #1 6710 feet. Core slab photograph. Codiaean algae, *Litanaia*, thrived in quiet water lagoonal environment in a backreef setting.

    G. Shell Oil - Greek #41-2 6727.9 feet. Thin section photomicrograph Codiacean algae, *Litanaia*, in transverse view, showing six internal circular tubes.

    H. Shell Oil - Greek #41-2 6775.7 feet. Thin section photomicrograph Codiacean algae, *Litanaia*, in longitudinal view, with spiney, wave-like outline and long *en-echelon* tubes.

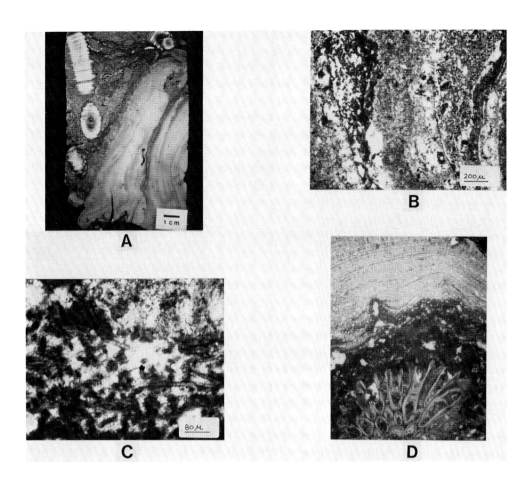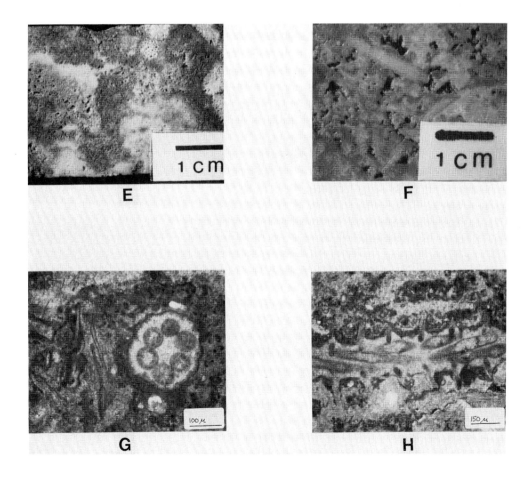

Figure 7. Core slab photographs and thin section photomicrographs from the reef lithofacies illustrating progressive loss of recognizable skeletal characteristics as dolomitization proceeds for chaetitids.

    A. Marathon Oil - Adams #1 6567 feet. Core slab photograph. Stromatoporoid encrusted chaetitid (lower left); Favosites (lower right), represented by pattern of internal pores.

    B. Marathon Oil - Adams #1 6617 feet. Core slab photograph. Stromatoporoid encrusting chaetitids recognized by pattern of porosity.

    C. McMoran Oil - Deraas #1 4746 feet. core slab photograph. Well-preserved chaetitid.

    D. Union Oil - Huber #1-A-2 5910 feet. Core slab photograph. Dolomitized chaetitid with characteristic pattern of porosity.

    E. McMoran Oil - Deraas #1 4746.6 feet. Thin section photomicrograph of well-preserved chaetitid, longitudinal view.

    F. California Oil - Henry #4 5631.4 feet. Thin section photomicrograph of dolomitized chaetitid, longitudinal view.

    G. McMoran Oil - Deraas #1 4746.6 feet. Thin section photomicrograph of well-preserved chaetitid, transverse view.

    H. California Oil - Henry #4 5631.4 feet. Thin section photomicrograph of dolomitized chaetitid, transverse view.

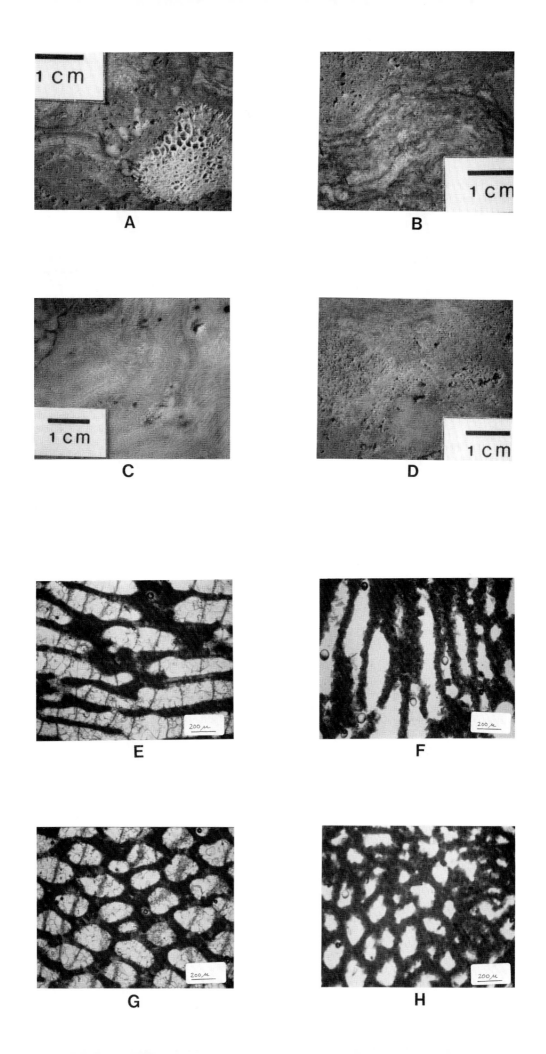

Figure 8. Thin section photomicrographs from the reef lithofacies illustrating progressive loss of recognizable skeletal characteristics as dolomitization proceeds for thamnoporid corals and stromatoporoids.

    A. Shell Oil - Greek #41-2 6514.7 feet. Thin section photomicrograph of well-preserved Coenites (coral), "longitudinal" view.

    B. Marathon Oil - Adams #1 6650.4 feet. Thin section photomicrograph of dolomitized Coenites (coral), "longitudinal" view.

    C. McMoran Oil - Deraas #1 4715.6 feet. Thin section photomicrograph of well-preserved Thamnopora (coral), transverse view.

    D. Marathon Oil - Adams #1 6650.4 feet. Thin section photomicrograph of dolomitized Thamnopora (coral), transverse view.

    E. Marathon Oil - Adams #1 6650.4 feet. Thin section photomicrograph of dolomitized coral, probably Thamnopora, but only porosity pattern remains.

    F. Marathon Oil - Adams #1 6571.9 feet. Thin section photomicrograph of dolomitized tabulate coral.

    G. McMoran Oil - Fairbrother #1 5280.3 feet. Thin section photomicrograph of well-preserved stromatoporoid illustrating wall and pillar structure.

    H. McMoran Oil - Deraas #1 4745.8 feet. Thin section photomicrograph of slightly dolomitized stromatoporoid, illustrating wall and pillar structure.

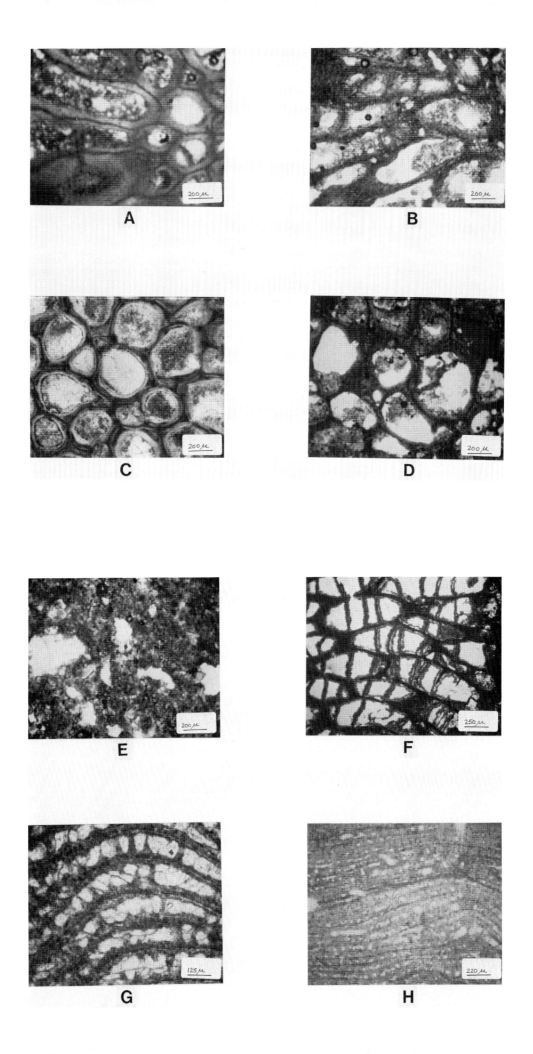

Figure 9. Thin section photomicrographs from the reef lithofacies illustrating progressive loss of recognizable skeletal characteristics as dolomitization proceeds for stromatoporoids and codiacean algae.

    A. McMoran Oil - Deraas #1 4736.1 feet. Thin section photomicrograph of well-preserved stromatoporoid, view parallels laminae.

    B. Union Oil - Huber #1-A-2 5921.3 feet. Thin section photomicrograph of dolomitized Stromatoporoid. View parallels laminae.

    C. Shell Oil - Greek #41-2 6713.5 feet. Thin section photomicrograph of fairly well-preserved codiacean algae, Litanaia in longitudinal view with dolomitized matrix.

    D. Shell Oil - Greek #41-2 6627.6 feet. Thin section photomicrograph of slightly dolomitized codiacean algae, Litanaia in longitudinal view.

    E. Shell Oil - Greek #41-2 6709.8 feet. Thin section photomicrograph of dolomitized codiacean algae, Litanaia in longitudinal view.

    F. Shell Oil - Greek #41-2 6643.7 feet. Thin section photomicrograph of slightly dolomitized codiacean algae, Litanaia in transverse view.

    G. Shell Oil - Greek #41-2 6643.7 feet. Thin section photomicrograph of dolomitized codiacean algae, Litanaia in transverse view.

    H. Marathon Oil - Adams #1 6663.1 feet. Thin section photomicrograph of dolomitized codiacean algae, Litanaia in transverse view.

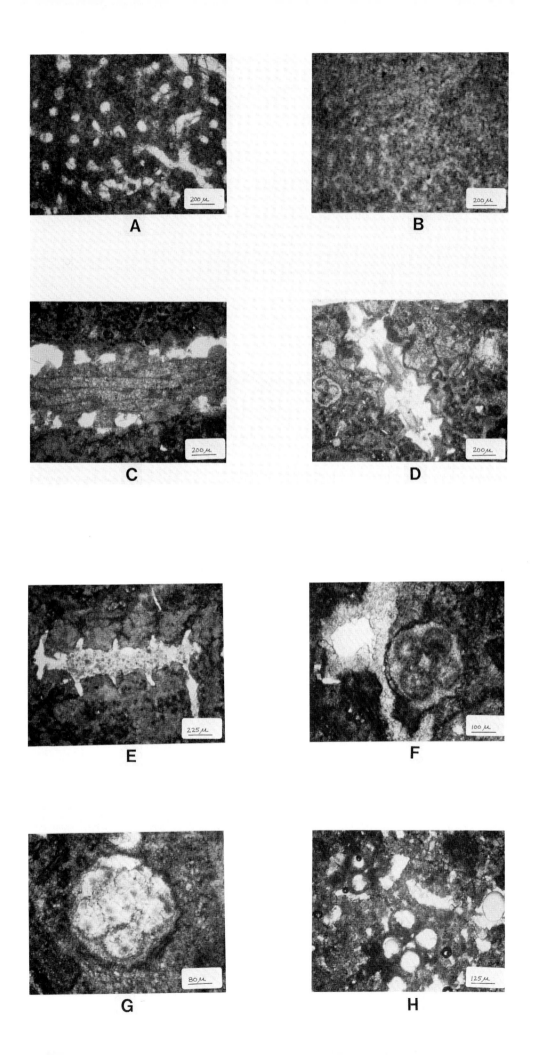

The four key organisms in the reef lithofacies in various stages of dolomitization are shown in Figures 7 to 9. The organization of the photographs for each organism is from well preserved (longitudinal and transverse views) to complete dolomitization. Most of the photographs are at the same magnification to facilitate comparison.

Codiacean Algae, Calcisphere, Peloid Packstone Lithofacies

In the center of the pinnacle reefs, the codiacean green alga, Litanaia, thrived in a backreef, lagoonal environment (Fig. 6F). The algal grains are approximately circular in transverse outline and generally contain from three to six internal circular tubes (Fig. 6G), although more tubes are present in a second possible species. In longitudinal section, the algae show a spiney, wave-like, or straight outline, and long en-echelon tubes begin near the margin of the wall and end near the center of the sheath. The elongate grains have locally been aligned by currents. Codiacean algae lived in warm, well-lit waters, at or close to low tide level (Johnson, 1961; Wray, 1977).

Calcispheres form a second abundant allochem in this lithofacies. They have a spar-filled center and a circular wall and some have spines which protrude into the matrix (Fig. 10A). Abundant peloids seem to float in pseudospar and there are a few samples in which blue-green algal filaments occur in the matrix. Blue-green algae may have been much more common than presently recognized, recrystallization of the matrix to microspar and pseudospar and its replacement by dolomite have obscured many textures (Fig. 10B).

Bioturbated packstones and wackestones dominate this lithofacies. There are common to abundant vugs and cavities in the matrix. (Fig. 10C). These voids parallel the edges of allochems such as codiacean algae and penetrate the matrix, but they do not seem to cut across allochems.

As relative sea-level began to fall, stromatolites developed in the high intertidal to supratidal zones on both the pinnacle and patch reefs.

Crinkly Stromatolite Lithofacies

This lithofacies is composed of porous, wavy and irregular, laminar dolomitic mudstone of probable stromatolitic origin (Fig. 10D). The stromatolites occur on top of the pinnacle reefs and on the eastern shelf and apparently grew in a high intertidal to supratidal environment under restricted conditions. Had they formed in an environment supporting a normal marine biota, the algal mats would have been destroyed by grazing organisms such as tribolites and gastropods (Kepper, 1974) and burrowers (Monty, 1973). Fenestral fabrics occur between and within laminations. Stylolites are common between the laminations and stromatolite heads. Emergence and reflux dolomitization of the upper part of the reef probably began at this time.

BASINAL LITHOFACIES

The basinal core (Shell Oil--Osterberg #21-2) includes rocks from all three episodes of Winnipegosis deposition. The Laminated Mudstone

Figure 10. Core slab photographs and thin section microphotographs from the reef lithofacies.

    A. Shell Oil - Greek #41-2 6716.1 feet. Thin section photomicrograph of calcispheres with spines.

    B. Shell Oil - Greek #41-2 6515.1 feet. Thin section photomicrograph (crossed polars) of matrix and blue-green algae being replaced by dolomite.

    C. Shell Oil - Greek #41-2 6607 feet. Core slab photograph of calcite-lined vugs in dolomitized codiacean packstone.

    D. Shell Oil - Greek #41-2 6513/14 feet. Core slab photograph of porous, wavy and irregular, laminar dolomite mudstone of probable stromatolitic origin.

    E. Shell Oil - Osterberg #21-2 7503 feet. Core slab photograph of laminated limestone deposited in deep basin coincident with reef development.

    F. Shell Oil - Osterberg #21-2 7497.2 feet. Thin section photomicrograph of laminations in lime mudstone emphasized by the developments of stylolite and microstylolites located between laminae.

    G. Shell Oil - Osterberg #21-2 7492.0 feet. Thin section photomicrograph of small burrows and clumps of small peloids occur within laminations; compaction occurs around these areas.

    H. McMoran Oil - Deraas #1 4746 feet. Core slab photograph of compaction indicated a broken and slightly offset fossil.

Lithofacies which formed during the second episode is composed of limestones which were deposited in the inter-reef deep basin coincident with the development of the pinnacle and patch reefs. A comparison of closely spaced pinnacle reef and basinal wells suggests that the depth of the basin in eastern Renville County was a maximum of approximately 230 feet (70 m) (Fig. 5).

Laminated Mudstone Lithofacies

While pinnacle reef and shelf deposition kept pace with rising sea level, laminated limestones accumulated slowly in the deepening basin between the pinnacle reefs (Fig. 10E). Similar laminated rocks have been studied in the Winnipegosis and Keg River Formations in Canada (Davies and Ludlam, 1971; 1973). In most samples, the lamination is emphasized by the development of microstylolites between laminae (Fig. 10F). There are abundant small burrows which occur within laminations but which do not seem to cross lamination boundaries, although microstylolitization might be masking the burrow-lamination relationship. There is compaction of the laminations around these burrows (Fig. 10G). Locally, concentrations of small peloids occur. The laminations resist compaction in these areas. In those few cases in which microstylolites did not form, there is organic-rich material present between the laminations.

Other diagenetic features include lath-shaped, generally isolated, crystals of anhydrite.

ENVIRONMENTS OF DEPOSITION

Following a brief hiatus after deposition of the Ashern carbonates, the Winnipegosis carbonates were deposited in three episodes. The first episode represents renewed transgression. As this transgression began, a broad, fairly uniform, normal marine environment became established (Fig. 11A). The deposited sediments include abundant brachiopods, crinoids, bryozoans, and mollusks suggesting good circulation in an open marine environment (Heckel, 1972).

During the second episode of Winnipegosis deposition, two depositional regions developed: the shallower shelves and a bathymetrically "deeper" basin. At this time, the Middle Devonian (Givetian) sea transgressed to its maximum extent in the Elk Point Basin, but shallowing may have begun late in the episode. Thick carbonates accumulated on the shallow shelves and on pinnacle reefs, but only thin sediments accumulated in the deeper basinal environment. In the shelf region, four separate environments of deposition became established: shallow marine, patch reef, lagoon, and tidal flat (Fig. 11B). In the basin region, two environments existed: pinnacle reefs and a sediment-starved basin. Pinnacle reef growth was rapid and eventually these reef buildups towered above the surrounding basin floor.

During the third episode of Winnipegosis deposition, the overall regression continued although there were several minor fluctuations in relative sea level (Fig. 11C). Supratidal deposition occurred both on the shelf and on the pinnacle reefs. Finally, restricted, shallower

Figure 11. Schematic view of environments of deposition during the three episodes of deposition of the Winnipegosis Formation.

    A. First Episode: a shallow marine environment baceme widely established in the North Dakota portion of the Elk Point Basin.

    B. Second Episode: the basin differentiated into two regions. In the self region, four separate environments became established: shallow marine, patch reef, lagoon, and tidal flat. In the basin region, pinnacle reefs and inter-reef "deeper" starved basin sediments accumulated.

    C. Third Episode: early in the regression, supratidal deposits occurred both on the shelf region and on top of the pinnacle reefs of the basin region; then following a significant fall of sea level, restricted shallower water deposits accumulated in the basin region.

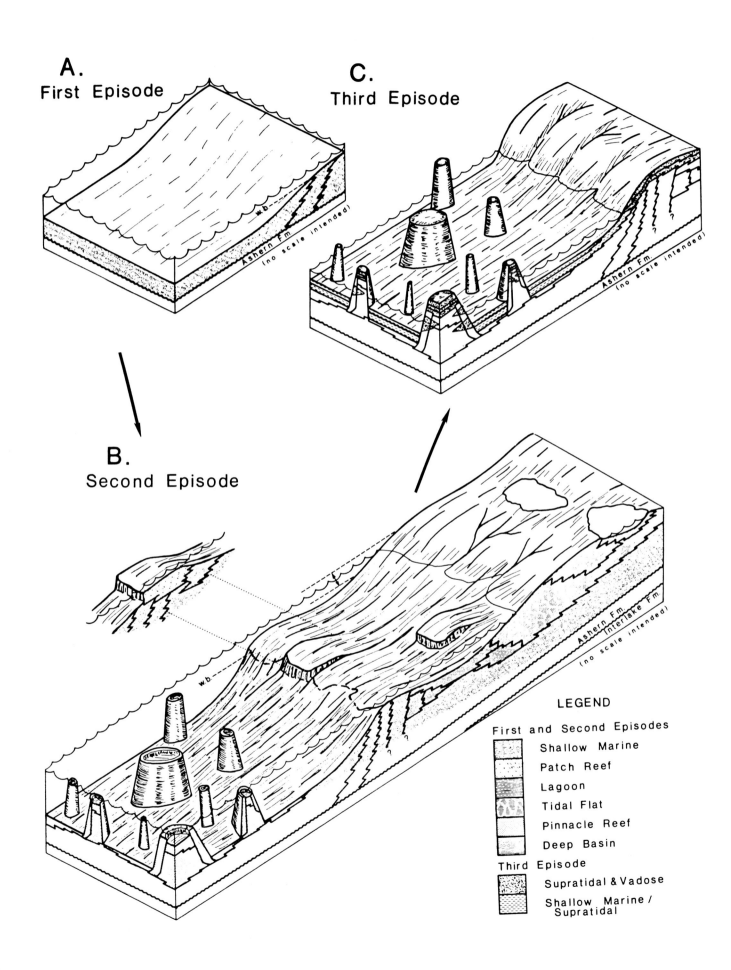

water deposits accumulated in the basin region. These events are summarized in Figure 12.

A schematic diagram of the pinnacle reef and basin at the end of the second episode when the sea covered the reefs is shown in Figure 12A. Subsequently, as sea level fell (Fig. 12B) stromatolites developed in a high intertidal to supratidal environment above the pinnacle reefs. Then fine-grained supratidal lime mudstones which were penecontemporaneously dolomitized were deposited on top of the stromatolites (Fig. 12C).

As the regression continued, sea level fell substantially. The pinnacle reefs were extensively exposed and subjected to subaerial conditions (Fig. 12D). Terra rosa formed on top of the reefs and the exposed pinnacles experienced dolomitization and vadose diagenesis. In the area between emergent reefs, a shallower and restricted marine environment formed, and a succession of different rock types was produced including stromatolites, dolomite, and starved-basin anhydrite. These rocks, in turn, are overlain by the Prairie Formation.

PALEOECOLOGY

The pinnacle and patch reefs of this study illustrate a distinct vertical zonation of biofacies. This zonation can be directly correlated with a successional buildup model which shows a complex relationship between species diversity and environment (Fig. 13). For many years, workers in Paleozoic reef paleoecology (i.e., Lowenstam, 1950; 1957; Nicol, 1962; Lecompte, 1970; Embry and Klovan, 1972; Copper,

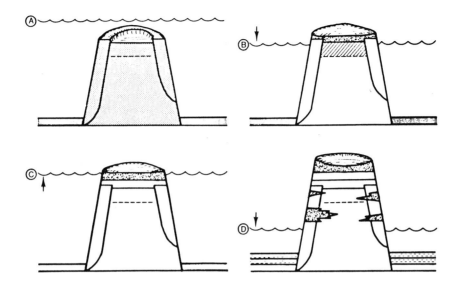

Figure 12. Stages of deposition during the third episode. Dashed horizontal lines indicate the general level of severely dolomitized reef; cross-hatching in 12-B indicates the areas completely dolomitized to form the Porous Dolomite Lithofacies. Other symbols in legend of Figure 11.

    A. At the end of the second episode, sea level covered the top of the pinnacle reefs.

    B. As the third episode began, sea level fell and stromatolites were deposited on top of the pinnacle reefs; the uppermost part of the reefs were periodically exposed and dolomitization of the upper part of the pinnacle reefs probably began at this time.

    C. Following this sea-level drop, relative sea-level maintained the same apparent level, and supratidal dolomites were deposited over the pinnacle reefs.

    D. Subsequently, sea level fell substantially. Terra rosa formed on top of the pinnacle reefs, dolomitization of the upper part of reefs continued, vadose diagenesis altered the exposed reef mass, and in the area between the emergent reefs a restricted basin shallower than before continued. Mottled and laminated dolomites, stromatolites, and starved basin anhydrites formed in the inter-reef areas surrounding the partly emergent reefs.

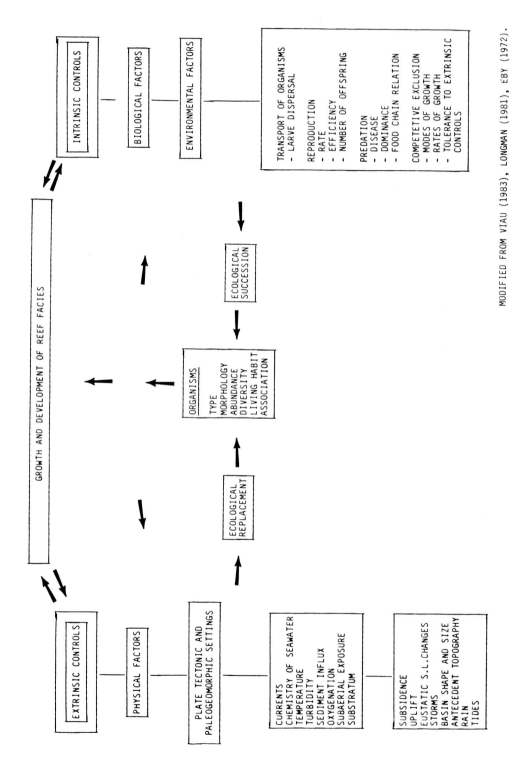

Figure 13. Relationship between intrinsic and extrinsic controls on reef facies.

1974; Tsien, 1974; Hoffman and Narkiewicz, 1977) have recognized successions of biofacies which, in most cases, represent a shallowing upward motif. These authors broadly relate the successional changes of the reef facies to the extrinsic control of water turbulence.

Alberstadt and Walker (1973) and Walker and Alberstadt (1975) discuss a four-stage successional model for Phanerozoic reef communities combining community controlled (autogenic-intrinsic) succession and environmentally controlled (allogenic-extrinsic) succession. The stages in vertical sequence, are termed: (I) stabilization; (II) colonization; (III) diversification; and (IV) domination. The first three stages are considered autogenic and intrinsic and the last stage, allogenic and extrinsic.

James (1979; 1983) has modified and expanded the model of Walker and Alberstadt (1975) into a four-stage model relating lithologic texture, species diversity, growth forms of the reef-building organisms, and the energy regime with each successive development stage. This classification is used in this paper.

Pinnacle Reef Biofacies

The pinnacle reef morphology suggests development during a rapid rise in sea-level from a "deeper" basin environment. These pinnacle reefs probably built upon small antecedent and bathymetric highs which formed irregularities on the sea floor and locally focused current energy. The shape and areal extent of these pinnacles, as well as the biofacies patterns, are directly related to the size, density and bathymetry of the initiating mounds (Kissling and Coughlin, 1979).

To date, eleven pinnacle reefs have been drilled in the North Dakota portion of the Elk Point Basin. These reefs formed where local carbonate production kept pace with the rapid rise in sea-level. The pinnacles vary in size both vertically and areally. Some reefs have approximately 230 feet (70 m) of vertical relief above the basin floor and expected symmetrical areal dimensions of approximately 160 acres.

The two pinnacle reef cores evaluated in this paper show a similar biologic and lithologic development sequence. Because of this similarity and the previous work of Perrin (1981; 1982) it is inferred that similar development occurred in all the pinnacle reefs within the original study area (Fig. 2).

The following sequential development scheme for the pinnacle reefs is a combination of the data from the Shell Oil--Greek #41-2 and Shell Oil--Osterberg #22X-1 reef cores. A detailed description is given in Figures 14 and 15.

Stage I - Stabilization

The first or basal stage represents an autogenic and intrinsic preparation of the substratum for reef growth (Fig. 16A). The rocks are comprised mainly of pelmatozoan and brachiopod debris with rare bryozoan, mollusk and trilobite fragments (Fig. 17A, 17B). This stage is representative of open marine sedimentation. These skeletal banks were colonized by pelmatozoans. Once colonization occurred, other benthic organisms and reef-building metazoans grew between the stabilizers (James and MacIntyre, 1985).

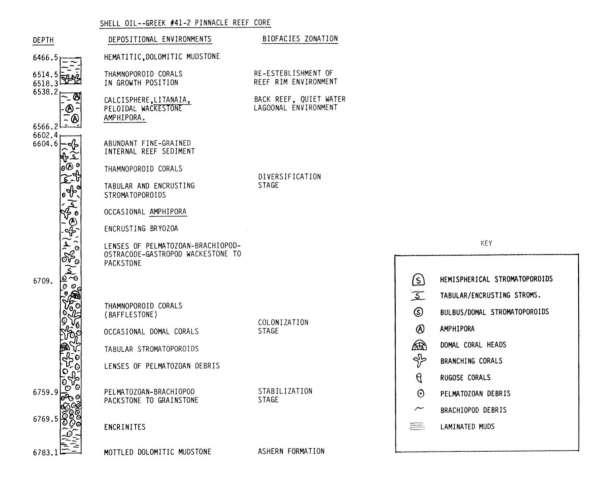

Figure 14. Geologic columnar section, Shell Oil--Greek #41-2 (NE NE 2-161-83, N.D.G.S. well #6535, Bottineau County)

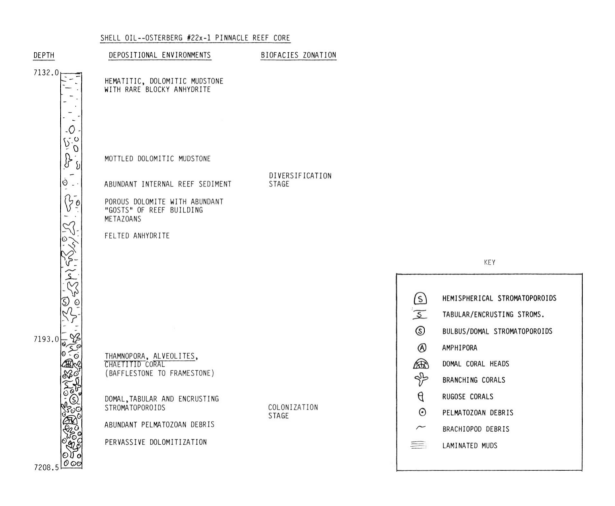

Figure 15. Geologic columnar section, Shell Oil--Osterberg #22x-1 (SE NW 1-161-85, N.D.G.S. well #6624, Renville County)

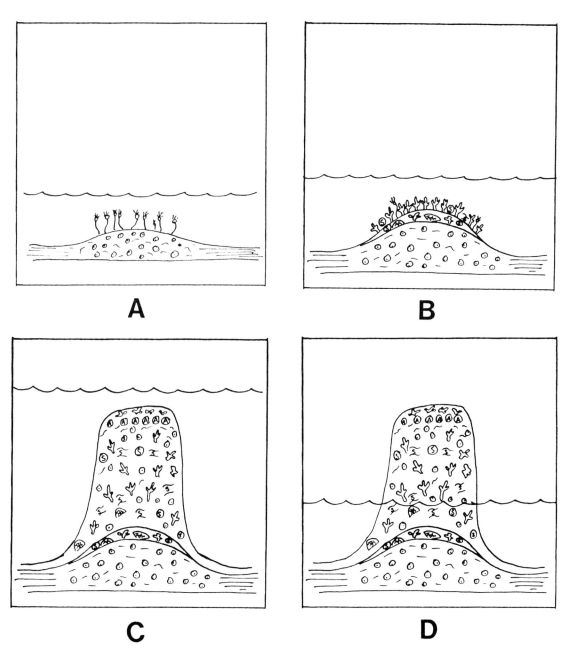

Figure 16. Successional buildup model, pinnacle reefs: A) stabilization; B) colonization; C) diversification; and D) subaereal exposure.

Figure 17. Shell Oil--Greek #41-2 well, core slab and thin section photographs:

- A. 6769 feet. Core slab photograph of brachiopod-pelmatozoan packstone with basinal muds below (stabilization stage).

- B. 6765 feet. Thin section photomicrograph of brachiopod-pelmatozoan packstone (stabilization stage.

- C. 6733 feet. Thin section photomicrograph of Thamnopora partially replaced by dolomite (colonization stage).

- D. 6736 feet. Core slab photograph of Thamnopora colonies suspended in a wackestone to packstone matrix (colonization stage).

- E. 6614 feet. Thin section photomicrograph of dolomitized wall and pillar structure of tabular stromatoporoid (diversification stage).

- F. 6742 feet. Thin section photomicrograph of blue-green algal filaments in matrix (diversification stage).

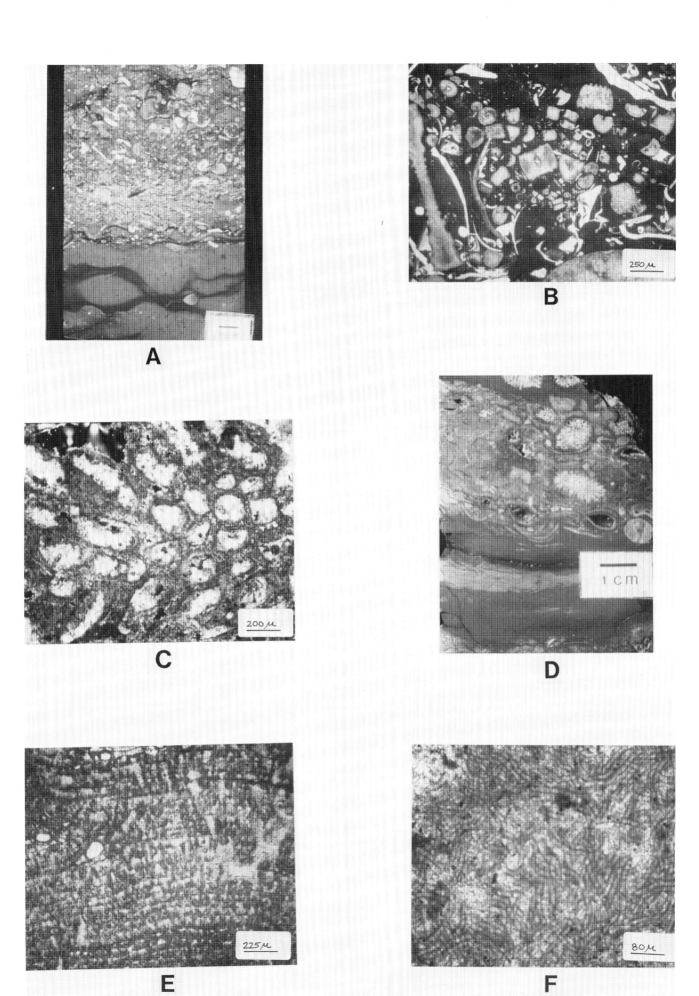

Stage II - Colonization

Initial colonization by reef-building organisms began with small domal and dendroidal, stick-like tabulate corals. The predominant forms are Thamnopora (Fig. 17C), Favosites and Hexagonaria. Less abundant are small tabular stromatoporoids. The stromatoporoids generally encrust other constituent grains. Rare rugose corals occur in the mudstone to wackestone matrix. Within the reef-building framework are pockets of pelmatozoan grainstones and packstones with some intraclasts.

Dense thickets of thamnoporid corals became established on the mounds and provided a baffle and trapping mechanism for fine-grained micritic sediment (Fig. 17D). Most of the corals observed were fragmented, but some in-situ colonies occur in a mudstone to wackestone matrix. These in-situ colonies represent coral growth which kept up with or exceeded the rate of local sedimentation.

The development of Stage II - Colonization is almost entirely intrinsic and autogenic. This is based on faunal evidence which shows that the vertical change in the lower portions of the reef are controlled by the organic community, independent of extrinsic environmental controls. This stage continued until reef growth reached wave-base (Fig. 16B).

Stage III - Diversification

The third stage of reef growth forms a large proportion of the total reef section. Although sea-level rise was most rapid at this time, vertical reef growth exceeded it. The upper portion of the reef extended above mean wave-base. Massive quantities of fine-grained internal reef sediment were trapped by the colonies of corals and stromatoporoids. This internal sediment comprises the bulk of the reef core.

As energy levels increased, more massive metazoan colonies developed. Alternating zones of tabulate corals and stromatoporoids were predominant. Stromatoporoids are laminar, domal and encrusting (Fig. 17E). Most specimens were detached and overturned. Fragmented remains of other organisms occur in the interstices within and between stromatoporoid colonies. Stromatoporoid-algal bindstones are commonly found encrusting each other and other fossil fragments.

Thamnopora is the most common coral. Other thamnoporoid-shaped tabulates in this biofacies are Coenites and Alveolites. The thamnoporid tabulates are in a mudstone to packstone matrix and are often associated with and encrusted by stromatoporoids and blue-green algal filaments (Fig. 17F).

Initial development of this stage was extrinsically and allogenically controlled, whereas continued growth was maintained by a combination of both intrinsic-autogenic and extrinsic-allogenic succession (Fig. 16C).

During diversification, a lateral biofacies zonation developed. Higher energy environments of an outer reef rim differentiated from a lower energy, backreef lagoon. This outer reef rim was probably much better developed and more continuous on the windward side of the pinnacle reef buildups, although this cannot be verified from the limited borehole information. The backreef deposits are dominated by fragments of the codiacean algae, Litanaia (Fig. 18A), abundant calcispheres (Fig. 18B), and peloids, within a packstone. Laminar

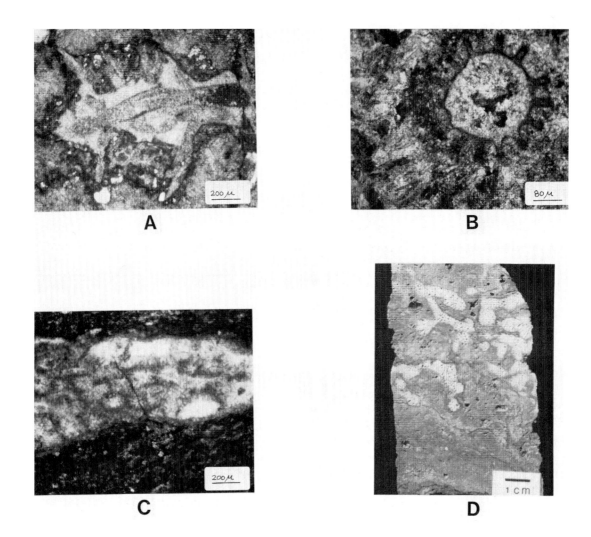

Figure 18. Shell Oil--Greek #41-2 well, core slab and thin section photographs:

    A. 6538.5 feet. Thin section photomicrograph of codiacean algae <u>Litanaia</u> (back reef, lagoonal environment).

    B. 6541 feet. Thin section photomicrograph of calcisphere showing radial spines (back reef, lagoonal environment).

    C. 6514.5 feet. Thin section photomicrograph of <u>Amphipora</u> colony in a mudstone matrix (back reef, lagoonal environment).

    D. 6515 feet. Core slab photograph of re-establishment of <u>Thamnopora</u> coral colonies capping back reef facies.

stromatoporoids also are common at the base of this lithofacies. The branching stromatoporoid Amphipora is found associated with laminar stromatoporoids (Fig. 18C). Amphipora, Litanaia and calcispheres are ubiquitous to quiet lagoonal settings floored by fine-grained sediments (Jamieson, 1971).

Directly above the lagoonal sediments, the reef rim organisms became briefly re-established in the previous backreef area (Fig. 18D). As sea-level began to fall, a high intertidal to supratidal environment, dominated by stromatolites, developed. This sea-level drop terminated pinnacle reef growth (Fig. 16D).

Patch Reef Biofacies

The patch reefs developed either at the shelf margin or on the shelf platform. They formed in a fashion similar to the pinnacle reefs utilizing antecedent highs to initiate reef growth. To date, the shape and overall areal extent of the Winnipegosis patch reefs are unknown.

Similar buildups of Devonian age have been studied in the Appalachian Basin by Eby, 1972; Smosna and Warshauer, 1979; Williams, 1980; and Precht, 1982. The latter patch reefs range in size from small mounds to true ecologic reefs. Maximum dimensions of these buildups are 40 feet (12 m) thick, by 200 feet (60 m) wide, by 2,000 feet (610 m) long. Assuming similar environmental conditions, growth rates and relative water depth, the size and extent of the Winnipegosis patch reefs can be inferred from one dimension, their thickness. Using the thickness information of the known patch reefs within the study area, the inferred maximum dimensions of the patch reefs are 55 feet (17 m) thick, by 275 feet (84 m) wide, by 2750 feet (838 m) long. These dimensions appear realistic when compared with Holocene analogs (Wallace and Schafersman, 1977; Precht, 1984).

To date, four Winnipegosis patch reefs have been drilled in North Dakota, two with cored reef intervals. The patch reef core described here, the McMoran Oil--Deraas #1, shows a distinct shallowing-upward ecologic and lithologic zonation. Although the lower 10 feet (3 m) of the reef was not cored, the lowermost stage can be inferred from the study of similar Devonian buildups. A detailed description of the patch reef core is given in Figure 19.

Stage I - Stabilization

This stage is inferred to have been comprised of pelmatozoan skeletal sands which formed shoals just below mean wave-base. These shoals formed local highs on the seafloor which subsequently could be preferentially colonized by reef-building metazoans. The first stage of patch reef growth, as with the pinnacle reefs, represents an autogenic and intrinsic preparation of the substratum for reef growth (Fig. 20A).

Stage II - Colonization

This is the initial stage of colonization by the larger reef-building metazoans. The major growth forms include large domal Chaetetes (Fig. 21A), Favosites and branching Thamnopora and Syringopora tabulate corals (Fig. 21B). In this biofacies there is an increase in abundance of stromatoporoids. The major growth forms of the

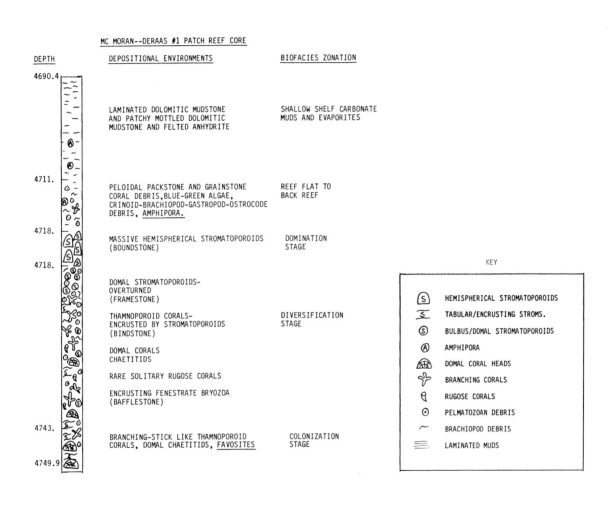

Figure 19. Geologic columnar section, McMoran Oil--Deraas #1 (SW SW 24-161-76, N.D.G.S. well #5280, Bottineau County).

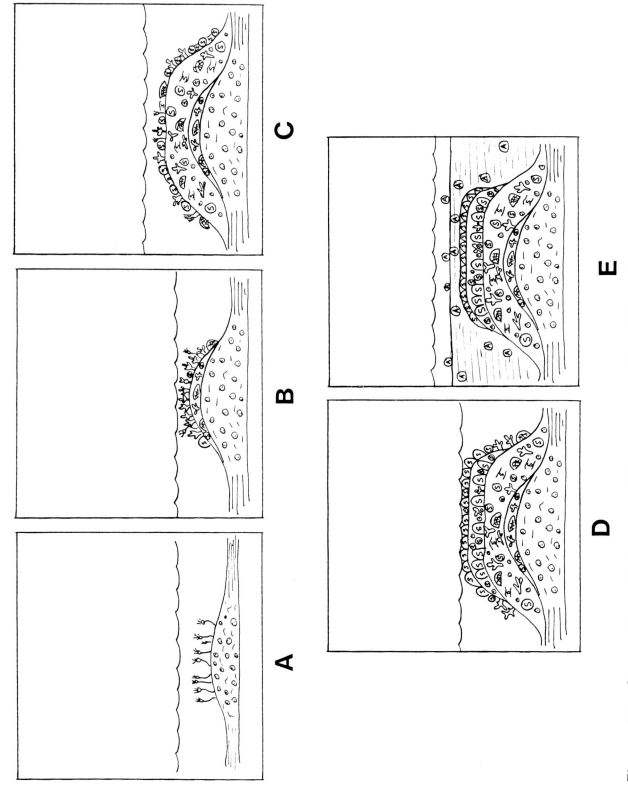

Figure 20. Successional buildup model, patch reefs: A) stabilization; B) colonization; C) diversification; D) domination; and E) capping by shelf carbonates and evaporites.

stromatoporoids include encrusting forms (Fig. 21C), small domal heads, and tabular colonies.

Reef growth during colonization developed quickly, resulting in upward growth to mean wave-base. The development of this colonization stage is considered to be community controlled and therefore intrinsic and autogenic (Fig. 20B).

Stage III - Diversification

The growth forms, sizes and diversity of the reef-building taxa reach a maximum during this stage. Growth shapes indicative of a higher energy (above mean wave-base) environment predominate. These include both domal and encrusting reef-building forms. Thamnoporid coral colonies are most robust in this biofacies.

Other tabulate corals present include Syringopora and fragmented Alveolites and Coenites. Chaetitids and domal stromatoporoids become more important in terms of reef framework, comprising approximately 50 percent of the constituent rock (Fig. 21D). Coarse grainstones and packstones comprised of fenestrate bryozoan, trilobite, ostracod, crinoid and coral debris fill nooks and crannies created by the larger metazoan colonies.

During this stage, a lateral biofacies zonation was probably present. This zonation would have been dependent upon numerous biological and physical factors. However, wave and current energy probably have had the greatest overall effect on this zonation.

The energy level of the patch reef during diversification was much greater than that of the pinnacle reefs. Because of this, diversification in the patch reefs is almost entirely environmentally

Figure 21. Core slab and thin section photographs, McMoran Oil--Deraas #1 well.

    A. 4750 feet. Core slab photograph of domal chaetitid colony with encrusting stromatoporoid at base (colonization stage).

    B. 4745 feet. Thin section photomicrograph of Bafflestone of Thamnopora, note gastropod in lower right corner (colonization stage).

    C. 4729 feet. Core slab photograph of Thamnopora colony encrusted both above and below by stromatoporoids (diversification stage).

    D. 4724 feet. Core slab photograph of overturned domal stromatoporoid (diversification stage).

    E. 4718 feet. Core slab photograph of massive hemispherical stromatoporoid (domination stage).

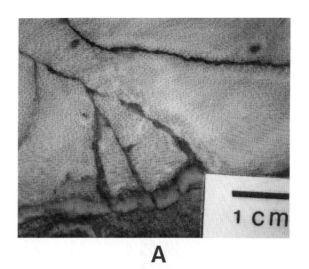

A

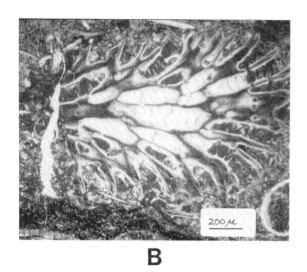

B

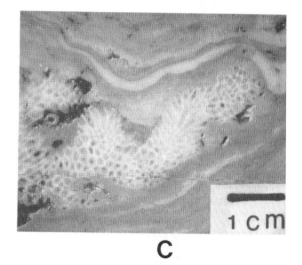

C

D

E

controlled. Therefore, the succession of this biofacies is considered extrinsic and allogenic (Fig. 20C).

Stage IV - Domination

There is an abrupt upward change from diversification to domination in the reef-building metazoans. Tabulate corals are absent and stromatoporoid diversity declined dramatically. Large massive, hemispherical stromatoporoids are the dominant organisms (Fig. 21E). In terms of size, they are the most important frame-building constituent of the patch reef mass. These growth forms are believed to have preferred turbulent high-energy conditions, commonly within the surf zone, near or at the reef crest.

The growth rate of these stromatoporoids appears to have been very rapid. These massive stromatoporoids probably had an affinity with massive modern-day scleractinian corals. Various authors have made comparisons of modern reefs with those of the Devonian (Klovan, 1974; Warshauer and Smosna, 1979; Williams, 1980; Precht, 1983). Possibly the most important application of such comparative studies is the ability to correlate biofacies relationships and zonation schemes of two-dimensional modern reefs with those of three-dimensions in the ancient (I. MacIntyre, pers. comm., 1985). Figure 22 shows a comparison of the successional stages in Winnipegosis patch reefs with the lateral ecologic zonation in a Holocene Bahamian patch reef (Precht, 1984).

As with the diversification stage of the patch reef, wave energies have a tremendous effect on the resulting biofacies development. This

VERY HIGH ENERGY (SURF-ZONE) MASSIVE STROMATOPOROID	PALMATA STAGE DEPTH, 0-3 METERS Acropora palmata	DOMINATION
MODERATE ENERGY (ABOVE MEAN WAVE-BASE) ABUND. STROMATOPOROIDS THAMNOPORID CORALS DOMAL CORALS	STAGHORN STAGE DEPTH, 4-15 METERS Acropora cervicornis PORITES STAGE DEPTH, 10-20 METERS Porites spp.	DIVERSIFICATION
LOW ENERGY (BELOW MEAN WAVE-BASE) OCC. STROMATOPOROID THAMNOPORID CORALS DOMAL CORAL HEADS	ANNULARIS STAGE DEPTH, 8-30 METERS Montastrea annularis Diploria spp.	COLONIZATION
LOW ENERGY (BELOW MEAN WAVE-BASE) PELMATOZOAN DEBRIS	PLEISTOCENE HARD-GROUNDS / HALIMEDA SAND SHOALS/ CORAL RUBBLE	STABILIZATION
WINNIPEGOSIS PATCH REEFS	HOLOCENE PATCH REEFS	

Figure 22. Generalized comparison of biofacies, Middle Devonian Winnipegosis patch reef and Holocene Bahamian patch reef. (Depth ranges are approximate and are not consistant for all Bahamian reefs).

environmental control on Stage IV - Domination is completely extrinsic and allogenic (Fig. 20D).

Above the massive stromatoporoid biofacies is a cap of reef-flat to backreef skeletal debris. This debris, in turn, is overlain by lagoonal muds. Inundation by shallow platform carbonate muds and evaporites caused the cessation of patch reef growth (Fig. 20E).

DIAGENESIS

Pinnacle Reef

Diagenesis of the pinnacle reefs includes partial to extensive dolomitization of the Stromatoporoid, Tabulate Coral Boundstone Lithofacies. The extent of dolomitization decreases downward in the pinnacle reefs. Where dolomitization is complete, the Porous Dolomite Lithofacies is the result. When the pinnacle reefs were exposed and vadose diagenetic conditions altered the original lithofacies, the Pisolite Dolomite Lithofacies developed. Where the Stromatoporoid, Tabulate Coral Boundstone Lithofacies can be recognized, cementation occurs in two stages; an initial, isopachous calcite probably of submarine origin followed by an equant calcite spar filling the remainder of the void. The original mud matrix has been altered to microspar and rarely to pseudospar. Mollusks have recrystallized to calcite through solution of the aragonitic shells and subsequent precipitation of equant calcite in the void.

Diagenesis of the Codiacean Algae, Calcisphere, Peloid Packstone Lithofacies in the pinnacle reefs is dominated by alteration of the lime mud to microspar and pseudospar and by dolomitization. Vugs and cavities are lined with fibrous isopachous calcite cement (Fig. 23B).

The center of the void may remain open or may be filled with equant calcite, dolomite, or rarely, by bladed anhydrite. Stylolites tend to follow the contact between matrix and cavity-fill or matrix and allochem, but may develop elsewhere.

Dominant diagenesis in the Crinkly Stromatolite Lithofacies of the pinnacle reefs and shelf margin is secondary dolomite replacement with rare replacement by felted anhydrite nodules.

Patch Reef

Diagenesis of the rocks from the patch reef environment is minor in comparison with that in the pinnacle reef environment. In the patch reef environment, diagenesis of the Stromatoporoid, Tabulate Coral Boundstone Lithofacies includes isopachous calcite and equant calcite spar cements and local dolomitization of allochems and the mud matrix. Compaction is indicated by rare broken and offset fossils (Fig. 10H) and by fractures, stylolites, and microstylolites which tend to form at the edges of massive allochems. Rarely, bladed or blocky anhydrite occurs in vugs and fractures and locally it replaces corals and stromatoporoids (Fig. 23A). Felted anhydrite nodules also have formed in some places.

DIAGENETIC LITHOFACIES

The Porous Dolomite Lithofacies and the Pisolite Dolomite Lithofacies are essentially "diagenetic" lithofacies which occur in the pinnacle reefs. Extensive dolomitization occurs in both lithofacies.

Figure 23. Core slab photographs and thin section microphotographs of diagenetic features from reef lithofacies.

    A. McMoran Oil--Fairbrother #1 5256.9 feet. Thin section photomicrograph (crossed polars) of anhydrite replacing both coral and its encrusting stromatoporoid.

    B. McMoran Oil - Deraas #1 4749.9 feet. Thin section photomicrograph of two stages of calcite cement, first isopachous and second equant.

    C. Phillips Oil - Brandvold #1 5059 feet. Core slab photograph of microstylites in Porous Dolomite Lithofacies.

    D. Marathon Oil - Adams #1 6670.1 feet. Thin section photomicrograph of zoned secondary dolomite rhombohedra.

    E. Amerada Oil - Trogstad #1 11712.8 feet. Thin section photomicrograph of replacement of coral by late secondary coarse-crystal dolomite.

    F. Shell Oil - Greek #41-2 6516.0 feet. Thin section photomicrograph of dolomitized coral, porosity pattern remains.

    G. Marathon Oil - Adams #1 6592/93 feet. Core slab photograph of vadose diagenesis in reef lithofacies.

    H. Marathon Oil - Adams #1 6537.8 feet. Thin section photomicrograph of vadose pisolite in reef lithofacies.

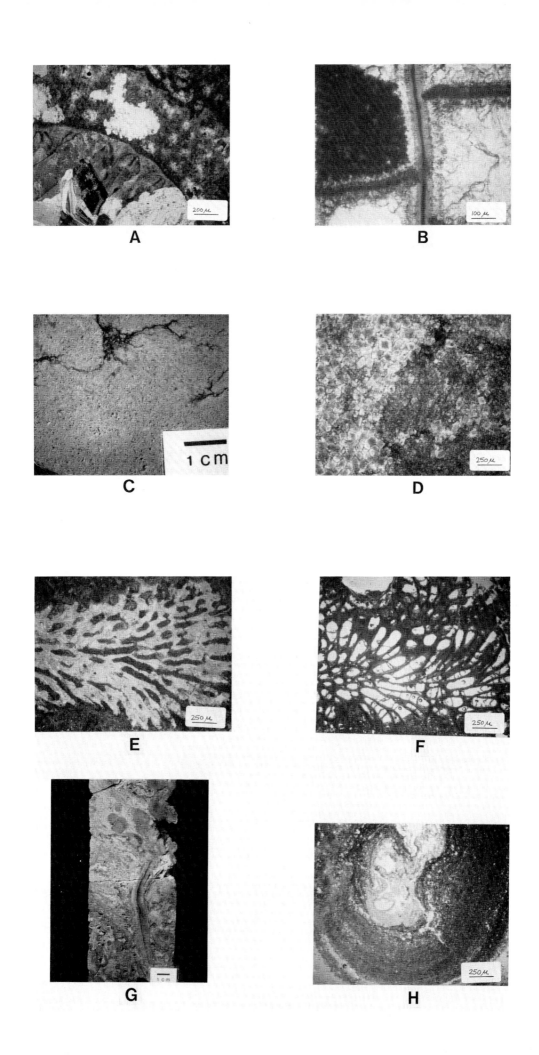

Identification of the reef lithofacies requires identification of allochems and textures, but as dolomitization proceeds, these criteria are progressively destroyed (Fig. 23C; 23D). The mud matrix and small allochems are the first to disappear. Some larger allochems are more susceptible than others and disappear next. Eventually, all traces of the original texture and allochems are gone. In this case, the rocks are assigned to one of two "diagenetic" lithofacies.

Porous Dolomite Lithofacies

Both previously described reef lithofacies, the Stromatoporoid, Tabulate Coral Boundstone Lithofacies and the Codiacean Algae, Calcisphere, Peloid Packstone Lithofacies, were nearly completely dolomitized in the upper part of the pinnacle reefs (Fig. 23E). Although dolomitization destroyed both matrix and allochems; the matrix, composed of small allochems and lime mud, seem to have been replaced first. The larger allochems were the last to be replaced. This dolomitization completely obscured many of the original constituents. Thus, this lithofacies is a bit of a "catch-all"; when it became impossible to assign a sample definitely to another lithofacies, it was assigned to the Porous Dolomite Lithofacies. Other diagenetic events in this lithofacies include stylolitization (horsetail and regular stylolites), growth of rare blocky or bladed anhydrite and lath-like crystals of anhydrite in the matrix, and precipitation of felted anhydrite nodules, disseminated pyrite, and zoned secondary dolomite rhombohedra (Fig. 23F).

Pisolite Dolomite Lithofacies

When sea level fell substantially, the tops of the pinnacle reefs were exposed and vadose diagenesis produced the Pisolite Dolomite Lithofacies (Fig. 23G). This lithofacies is composed of distinct, faintly laminated dolomite pisoliths occurring in a packstone or as a pisolitic crust. The pisoliths are composed of fine-grained dolomite, whereas the surrounding matrix varies from fine- to coarse-grained secondary dolomite (Fig. 23H). Both the Codiacean Algae, Calcisphere, Peloid Packstone Lithofacies and the Stromatoporoid, Tabulate Coral Boundstone Lithofacies have been identified as the primary depositional lithofacies for this rock. However, as the original fabric of the rock is rarely preserved, the Pisolite Dolomite Lithofacies is truly a "diagenetic" lithofacies. This lithofacies also has large felted to bladed anhydrite in the matrix between pisoliths.

CONCLUSIONS

(1) The Winnipegosis reefs of this study are excellent examples of Middle Devonian coral - stromatoporoid reefal buildups.

(2) Two types of reefs are identified: A) pinnacle reefs which developed as isolated mounds in the "deeper" basin region; and B) patch reefs which grew at the shelf margin or upon the shelf platform.

(3) The pinnacle reefs exhibit a tripartite successional biofacies development.

(4) The patch reefs exhibit a quadripartite, shallowing - upward succession.

(5) The pinnacle and patch reefs indicate that both extrinsic and intrinsic factors control reef facies development patterns.

(6) Porosity development and/or occlusion within the reefs is directly related to the amount, type and timing of the diagenetic fabrics observed. The best porosity is developed where there is evidence of extensive dolomitization.

(7) This study demonstrates that the development and diagenesis of these reefs is analogous with other models of reefs through time and therefore, can be used as an example for comparative studies of Devonian and modern reefal buildups.

ACKNOWLEDGEMENTS

The authors wish to thank Champlin Petroleum Company for permission to publish. Initial support for this study was given to N. A. Perrin from the Geology Department at the University of North Dakota, the North Dakota Geologic Survey, and from Gulf Oil Company. We would like to specifically thank Lee Gerhard, Colorado School of Mines and Sid Anderson of the North Dakota Geological Survey for access to the core materials. Mark Longman, Keith Shanley and Dave Eby edited the manuscript. Their comments, criticism and help were greatly appreciated. Joan Zabriskie, Jan Nall and Mary Bergmann assisted with the wordprocessing. Their generosity, graciousness, time and patience made the light at the end of the tunnel appear just a little bit closer.

REFERENCES

ALBERSTADT, L. P. and WALKER, K. R., 1973, Stages of ecological succession in Lower Paleozoic reefs of North America (discuss. paper): Geol. Soc. America Abs. with Programs, v.5, p. 530-531.

COPPER, P., 1974, Structure and development of Early Paleozoic reefs: Proc. Second Int. Coral Reef Symp., Brisbane., v.6, p. 365-386.

DAVIES, G. R., and LUDLAM, S. D., 1971, A basinal model for middle Devonian "laminates" Elk Point Basin of western Canada: Geol. Soc. America Abs. with Programs, v.3, p. 376.

DAVIES, G. R., and LUDLAM, S. D., 1973, Origin of laminated and graded sediments, Middle Devonian of western Canada: Geol. Soc. America Bull., v.84, p. 3527-3546.

DUNHAM, R. J., 1970, Stratigraphic reefs versus ecologic reefs: Am. Assoc. Petrol. Geol. Bull., v.54, p. 1931-1932.

EBY, D. E., 1972, Paleoecology and community structures of two bioherms from the Coeymans Formation (Lower Devonian) of central New York: Unpubl. M. S. Thesis, Brown University, 105 p.

EMBRY, A. F., and KLOVAN, J. E., 1972, Absolute water depth limits of Late Devonian paleoecologic zones: Geol. Rundschau, v.61, p. 672-686.

GERHARD, L. C., ANDERSON, S. B., LEFEVER, J. A. and CARLSON, C.G., 1982, Geologic development, origin and energy mineral resources of Williston Basin, North Dakota: Am. Assoc. Petrol. Geol. Bull., v.66, p. 989-1020.

GRAYSTON, L. D., SHERWIN, D. F. and ALLAN, J. F., 1964, Middle Devonian: in McCrossan, R. G., and Glaister, R. P. (eds.), Geological History of Western Canada, Alberta Soc. Petrol. Geol., p. 49-59.

HECKEL, P. H., 1972, Recognition of ancient shallow marine communities: in Rigby, J. K., and Hamblin, W. K. (eds.), Recognition of Ancient Sedimentary Environments, Soc. Econ. Paleont. and Mineral. Spec. Publ. 16, p. 226-286.

HECKEL, P. H., 1974, Carbonate buildups in the geologic record: a review: in Laporte, L. F. (ed.), Reefs in Time and Space, Soc. Econ. Paleont. Mineral. Spec. Publ. 18, p. 90-155.

HOFFMAN, A. and NARKIEWICZ, M., 1977, Developmental pattern of Lower to Middle Paleozoic banks and reefs: N. Jb. Palaont. Mh. v.5, p. 272-283.

JAMES, N. P., 1979, Reefs: in Walker, R. G. (ed.), Facies Models, Geoscience Canada, Reprint Ser. I, p. 121-132.

JAMES, N. P., 1983, Reefs: in Scholle, P. A., Bebout, D. G. and Moore, C. H. (eds.), Carbonate Depositional Environments, Am. Assoc. Petrol. Geol. Mem. 33, p. 346-440.

JAMES, N. P., and MACINTYRE, I. G., 1985, Zonation, depositional facies and diagenesis of modern and ancient reefs: unpublished notes for a short course, February 1985, Colorado School of Mines, 185 p.

JAMIESON, E. R., 1971, Paleoecology of Devonian reefs of western Canada: in Reef Organisms Through Time, North. Am. Paleon. Conv. Proc. part J., p. 1300-1340.

JOHNSON, J. H., 1961, Limestone-building algae and algal limestones: Colorado School of Mines, Dept. of Publications, 297 p.

KEPPER, J. C., 1974, Antipathetic relation between Cambrian trilobites and stromatolites: Am. Assoc. Petrol. Geol. Bull., v.58, p. 141-143.

KISSLING, D. L., and COUGHLIN, R. M., 1979, Succession of faunas and frameworks in Middle Devonian pinnacle reefs of south-central New York: Geol. Soc. America Abs. with Programs, v. 11, no. 1., p. 19.

KLOVAN, J. E., 1974, Development of western Canadian Devonian reefs and comparison with Holocene analogues: Am. Assoc. Petrol. Geol. Bull., v.58, p. 787-799.

LECOMPTE, M., 1970, Die Riffe im Devon der Ardennen und ihre Bildungsbedingung: Geol. et Palent., v.4, p. 25-71.

LONGMAN, M. W., 1981, A process approach to recognizing facies of reef complexes: in Toomey, D. F. (ed.), European Fossil Reef Models, Soc. Econ. Paleont. Mineral. Spec. Publ. 30, p. 9-40.

LOWENSTAM, H. A., 1950, Niagaran reefs in the Great Lakes area: Jour. Geol., v.58, p. 430-487.

LOWENSTAM, H. A., 1957, Niagaran reefs in the Great Lakes area: in Ladd, H. S. (ed.), Treatise on Marine Ecology and Paleoecology, Geol. Soc. America Mem. 67, part 2, p. 215-248.

MONTY, L. V., 1973, Precambrian background and Phanerozoic history of stromatolitic communities: Ann. Soc. Geol. Belg., v.96, p. 585-624.

NICOL, D., 1962, The biotic development of some Niagaran reefs - an example of an ecological succession or sere: Jour. Paleon., v.45, p. 172-176.

PERRIN, N. A., 1981, Reef facies of Winnipegosis Formation (Middle Devonian), Williston Basin, North Dakota (Abs.): Am. Assoc. Petrol. Geol. Bull., v.65, p. 969.

PERRIN, N. A., 1982, Environments of deposition and diagenesis of the Winnipegosis Formation (Middle Devonian), Williston Basin, North Dakoka: in Christopher, J. E. and Kaldi, J. (eds.), Fourth Int'l Williston Basin Symposium, Regina, p. 51-66.

PERRIN, N. A., 1985, Environments of deposition and diagenesis of the Winnipegosis formation (Middle Devonian) Williston Basin, North Dakota: in Peterson, J. A. (ed.), Am. Assoc. Petrol. Geol. Mem. (in press).

PRECHT, W. F., 1982, Paleoecology and structure of a Late Silurian - Early Devonian (?) patch reef, northwestern New Jersey (Abs.): Am. Assoc. Petrol. Geol. Bull., v.66, p. 1173.

PRECHT, W. F., 1983, Patch reef modeling - a comparison of Devonian and recent examples (Abs.): Am. Assoc. Petrol. Geol. Bull., v.67, p. 1459.

PRECHT, W. F., 1984, Zonation and development of the modern leeward reefs, San Salvador, Bahamas: Geol. Soc. America Abs. with Programs, v.17, p. 627.

SLOSS, L. L., 1963, Sequences in the cratonic interior of North America: Geol. Soc. America Bull., v.74, p. 93-114.

SMOSNA, R. A. and WARSHAUER, S. M., 1979, A very Early Devonian patch reef and its ecological setting: Jour. Paleon., v.53, p. 142-152.

SMOSNA, R. A. and WARSHAUER, S. M., 1983, Environment analysis of a Silurian patch reef, Lockport Dolomite of West Virginia: in Harris, P. M. (ed.), Carbonate Buildups: A Core Workshop, Soc. Econ. Paleon. Mineral. Core Workshop 4, p. 26-52.

TSIEN, H. H., 1974, Paleoecology of Middle Devonian and Frasnian in Belgium: Intern. Symp. Belg. Micropaleon. Limits, Namur, section 12, p. 1-53.

VIAU, C., 1983, Depositional sequences, facies and evolution of the Upper Devonian Swan Hills buildup, Central Alberta, Canada: in Harris, P. M., (ed.), Carbonate Buildups: A Core Workshop, Soc. Econ. Paleon. Mineral. Core Workshop 4, p. 112-143.

WALKER, K. R., and ALBERSTADT, L. P., 1975, Ecological succession as an aspect of structure in fossil communities: Paleobiol., v.1, p. 238-257.

WALLACE, R. J., and SCHAFERSMAN, S. D., 1977, Patch reef ecology and sedimentology of Glovers Reef Atoll, Belize: in Frost, S. H., Weiss, M. P., and Saunders, J. B., (eds.), Reefs and Related Carbonates - Ecology and Sedimentology, Am. Assoc. Petrol. Geol. Stud. in Geol. 4, p. 37-52.

WARSHAUER, S. M., and SMOSNA, R. A., 1979, Congruent patch reef biofacies: a comparison of the Mid-Appalachian Devonian with modern Florida analogs: Geol. Soc. America Abs. with Programs, v. 11, no. 1., p. 58.

WILLIAMS, L. A., 1980, Community succession in a Devonian patch reef (Onondaga Formation, New York) - physical and biotic controls: Jour. Sed. Petrol., v.50, p. 1169-1185.

WRAY, J. L., 1977, Calcareous Algae: Elsevier, Amsterdam, 185 p.

DEPOSITION, DIAGENESIS AND PALEOSTRUCTURAL CONTROL OF DUPEROW AND BIRDBEAR (NISKU) RESERVOIRS, WILLISTON BASIN

JAMES R. EHRETS AND DON L. KISSLING
Jackalope Geological, Ltd.
2019 19th Street
Boulder, CO 80302

ABSTRACT

The vertical facies sequence comprising the Upper Devonian Duperow and Birdbear (Nisku) formations in the Williston Basin reveals three major shoaling-upward depositional cycles, punctuated by more frequent salinity cycles expressed as remarkably widespread anhydrite and argillaceous marker beds. Identification of syndepositional structures, many coincident with structures, was gained through isopach mapping of intervals bounded by these marker beds. Excellent examples of reservoir development under paleostructural control are illustrated for Duperow Unit 4 in the Billing Nose area by the Tenneco Gawryluk #1-30 and Federal #2-30 cores and for the Birdbear (Nisku) in the Wolf Creek Nose area by the Murphy Sethre #1-B and Sletvold #1-B cores. In both examples, the paleostuctures represented relative paleotopographic highs and, as such, influenced the local development of favorable reservoir facies. Duperow Unit 4 skeletal banks, dominated by globular stromatoporoid floatstone, formed over the crest of the Billings Nose paleostructure. Birdbear (Nisku) skeletal banks, comprised of Amphipora wackestone, and packstone and platy stromatoporoid boundstone, formed along the flanks of the Wolf Creek Nose paleostructure, while the crest of the structure sustained peritidal deposition unsuited for bank development. Dolomitization of skeletal bank facies and portions of adjacent facies was the mechanism for reservoir development. Mg^{++}-enriched brines expelled from overlying evaporites during burial compaction provided the dolomitizing fluids in both examples, although the style and magnitude of dolomitization was regulated by the facies distribution and early burial history peculiar to each. Migration of dolomitizing fluids and hydrocarbons was facilitated by fracturing of intervening lithified strata. Sources for both fluids are believed to have been within readily defined stratigraphic intervals.

INTRODUCTION

The Upper Devonian Duperow and Birdbear (Nisku) formations exhibit prime examples of reservoir development through paleostructural control over the distribution and dolomitization of skeletal-rich facies within a shallow marine shelf setting. The producing area encompassed by the Billings Nose in Billings County, North Dakota and the Wolf Creek Nose in Roosevelt County, Montana are two paleostructural trends which significantly influenced the distribution of carbonate facies within the remote southeastern extremity of the Late Devonian cratonic shelf (Fig. 1).

Deposition of this Devonian carbonate sequence was controlled largely by three shelf-wide, gradually shoaling-upward depositional cycles in the Duperow and Birdbear (Nisku) formations, punctuated by more frequent shelf-wide fluctuations in salinity (Kissling and Ehrets; 1984, 1985). Episodic, shelf-wide cycles of progressive hypersalinity most likely reflect small eustatic sea-level fluctuations and restriction of the shelf by reef barriers fringing the Duvernay-Ireton and Winterburn basins in central Alberta. Culmination of these hypersalinity events is expressed by remarkably widespread, thin marker beds of anhydrite and argillaceous dolomite. These are used to subdivide the Duperow stratigraphic section (Fig. 2). Similar anhydrite beds also occur in the uppermost portion of the Birdbear (Nisku) Formation. Wilson (1967) and Dunn (1975) also recognized these cyclic marker beds, but postulated rather different depositional models.

Relative bathymetry across the Billings Nose and Wolf Creek Nose paleostructural trends and the superposition of hypersalinity events

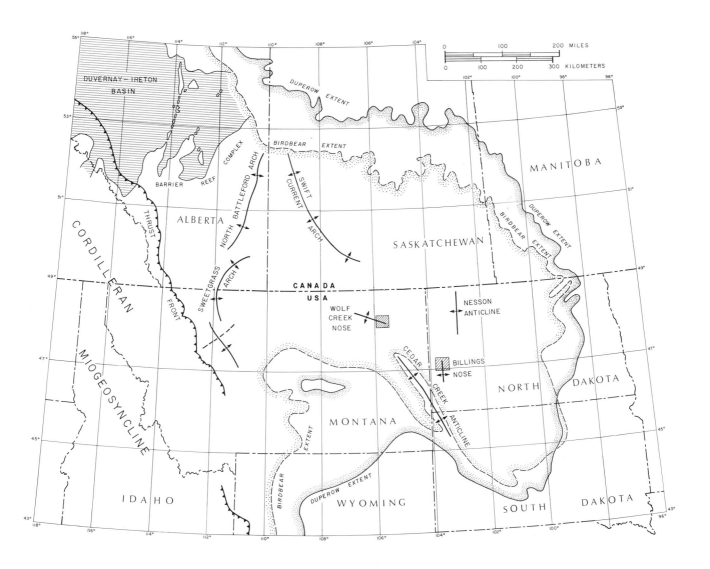

Figure 1. Present-day distribution of the Duperow and Birdbear (Nisku) formations and equivalent strata across the northern Great Plains, including trends of major Upper Devonian paleostructural elements. The northern and northeastern margins are erosional.

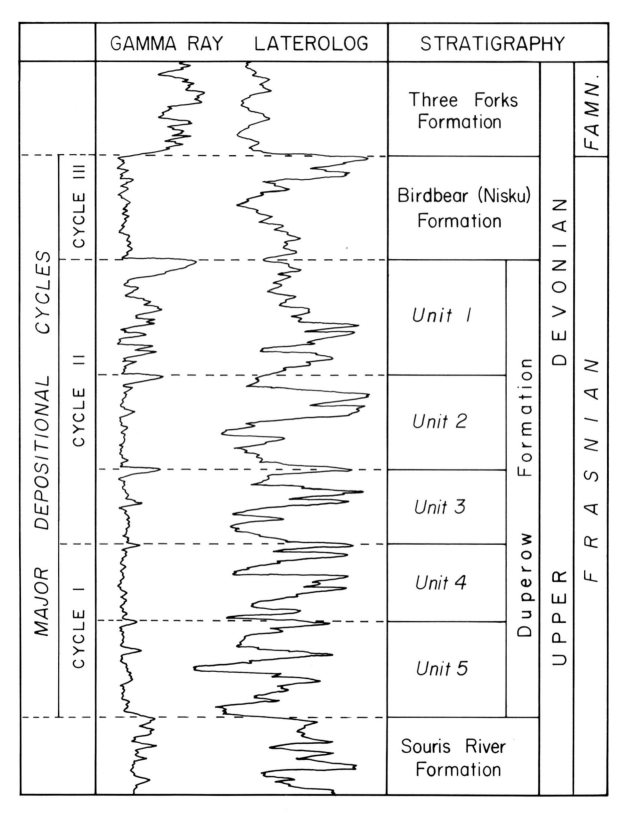

Figure 2. Upper Devonian stratigraphy with composite log characterestics and subdivisions of the Duperow and Birdbear (Nisku) formations.

within the broader shoaling-upward sedimentary cycles resulted in irregular distribution of skeletal facies and diagenetic fabrics on a local scale. This paper attempts to show the fundamental influence of the Billings Nose and Wolf Creek Nose paleostructures on the local distribution of depositional facies and dolomitization, and the manner in which these relate to reservoir development.

DUPEROW FORMATION

The Billings Nose hosts production from numerous Paleozoic formations, including more than twenty pools and fields in the Duperow. Production was first established near the structural axis at Four Eyes Field in 1978 (Fig. 3). Except for limited potential indicated for stratigraphically higher zones, Duperow production in most Billings Nose fields is from the "Four Eyes porosity zone," or the basal part of the interval designated Unit 4 (Fig. 2). Cores of this interval from two Tenneco wells in the Four Eyes area provide an opportunity to examine facies and reservoir characteristics. The Gawryluk #1-30 well was completed as a marginal producer and serves as the model core sequence for this discussion. Additional examples are drawn from the unproductive Federal #2-30 well which is located in a nearly identical structural position (Fig. 3).

Facies Sequence

Unit 4 in the Tenneco Gawryluk #1-30 cored interval records an asymmetrical depositional cycle that begins with normal (unrestricted)

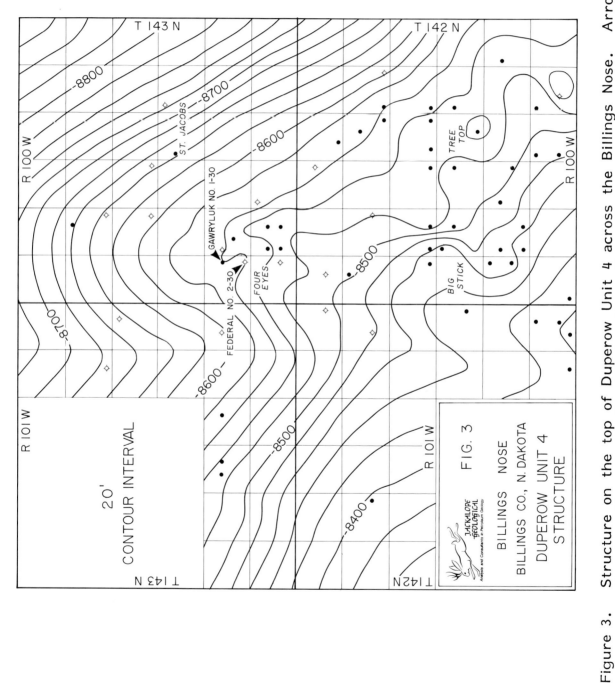

Figure 3. Structure on the top of Duperow Unit 4 across the Billings Nose. Arrows locate the two Tenneco wells in the Four Eyes Field area included for discussion. Producing well symbols include other Paleozoic formations.

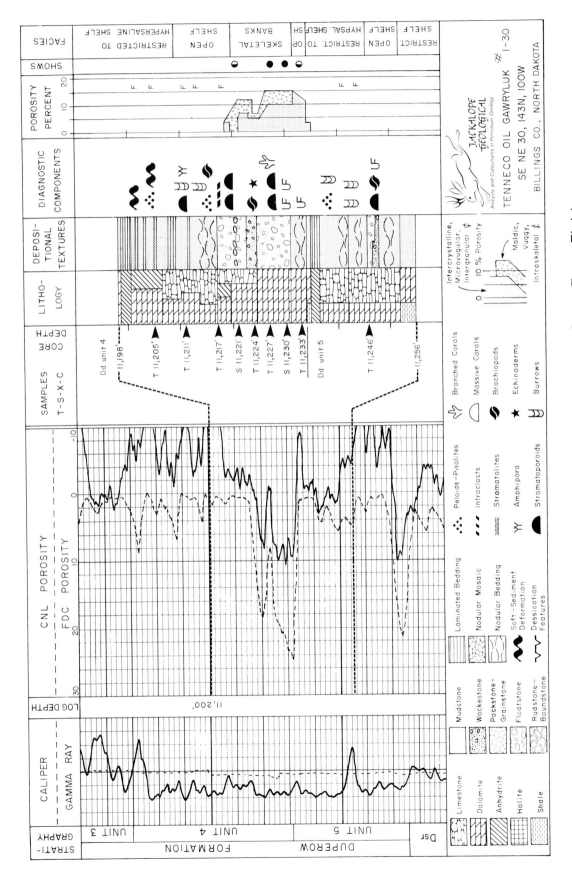

Figure 4. Graphic core column for Tenneco Gawryluk No. 1-30, Four Eyes Field area.

marine mudstone and skeletal bank facies and concludes with a relatively thick, interbedded sequence of anhydrite and anhydritic dolomitic mudstone representing an episode of shelf-wide hypersalinity (Fig. 4). The underlying Unit 5 displays normal marine facies (Figs. 5F, 6F) which includes a thin zone of skeletal bank facies. Unit 5 also culminates in a restricted shelf facies comprised of the anhydrite and argillaceous dolomite marker bed couplet, representing a shelf-wide hypersaline event (Figs. 5E, 6E). Overlying Unit 4 are open marine facies which dominate Unit 3 over large portions of the Williston Basin. Facies development similar to that of Unit 4 recurs within Unit 2, a stratigraphic interval more than 100 feet (30 m) above the Unit 4 skeletal banks.

The basal part of Unit 4 is characterized by a thin zone of nodular-bedded mudstone, marking the return to normal marine salinity (Fig. 6D). Overlying this thin basal mudstone facies are stomatoporoid-rich skeletal bank deposits which form the most conspicuous fabrics in Unit 4. Irregular, globular stromatoporoids up to 10 cm in diameter are particularly abundant (Fig. 5D). Together with less numerous laminar, tabular and domal stromatoporoids and rugose and tabulate corals, these form a floatstone fabric within a brachiopod-crinoid wackestone matrix (Figs. 5C, 5D, 6A, 6B and 6C). Bank facies are overlain by open shelf, nodular-bedded mudstone and wackestone which grade upward to burrowed and laminated mudstone and less abundant peloid-intraclast wackestone (Figs. 5B, 5G, 5H). The relatively rapid loss of normal marine fauna within this thin interval reflects the onset of hypersalinity which eventually resulted in the precipitation of bedded anhydrite (probably originally gypsum).

The entire Gawryluk core sequence, including the evaporite facies, was deposited under subtidal conditions (Kissling and Ehrets, 1984, 1985).

The coincidence of Unit 4 skeletal banks and thickness anomalies with the principal Billings Nose axis suggests that structure was a submarine topographic high during Duperow deposition, favored for colonization by stromatoporoids and associated fauna (Fig. 13A). Bank morphology was most likely broad biostromes having little relief. Thickness patterns, as shown by unit isopach maps, do not reflect organic buildups, but instead demonstrate rather subtle and uniform thinning of Unit 4 across the paleostructure (Fig. 7). Although skeletal bank deposits were decidedly less muddy than adjacent and underlying open shelf facies, carbonate mud nevertheless contributed significantly to their overall composition. Laminar and tabular stromatoporoids may have served to bind portions of this fossiliferous, muddy substrate into boundstone. However, boundstone fabrics are relatively unimportant in the Billings Nose area.

Reservoir Characteristics

Porosity in the Gawryluk well is almost exclusively limited to dolomitized, skeletal-rich facies within the basal 15 feet (4.5 m) of Unit 4, as well as in the thin zone of mudstone at the very base of the unit (Fig. 4).

The amount of porosity is related directly to the degree of dolomitization, with the highest porosity existing in the completely dolomitized facies in the basal 10 feet (3 m) of the unit.

Figure 5. Core photographs of Duperow facies from Tenneco Gawryluk #1-30 (A, D and F) and Tenneco Federal #2-30 (B, C and E). Composite facies sequences from top to bottom as illustrated in Figure 4. Slabbed core widths approxi- mately 10 cm. Thin section photomicrographs of Duperow facies from Tenneco Gawryluk #1-30 (G and H), in top to bottom facies sequence illustrated in Figure 4 (continued in Fig. 6). Photomicrograph widths represent approximately 6 mm.

(A) Laminated anhydrite and dolomite. Convolute bedding results from compaction and soft-sediment deformation.

(B) Dolomitic limestone; nodular-bedded mudstone.

(C) Very calcitic dolomite; stromatoporoid floatstone. A large tabular stromatoporoid and smaller globular stromatoporid fragments in a fine skeletal wackestone matrix.

(D) Calcitic dolomite; stromatoporoid floatstone. Dominantly globular and laminar stromatoporoid frag- ments in a wackestone matrix.

(E) Silty, argillaceous and anhydritic dolomite; deformed, laminated mudstone. Typical lithology and fabric of Duperow gamma-ray marker beds.

(F) Dolomitic limestone; Laminated and burrowed mud- stone.

(G) Dolomitic limestone; peloid-intraclast wackestone. Irregular, dark micrite peloids and intraclasts remain as calcite in otherwise dolomicrospar mudstone matrix. Core depth 11,211'.

(H) Dolomitic limestone; nodular-bedded skeletal wacke- stone. Dolomicrospar is scattered throughout the matrix and is concentrated under a brachiopod shelter. Core depth 11,217.

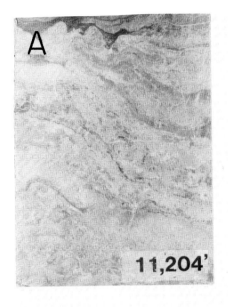
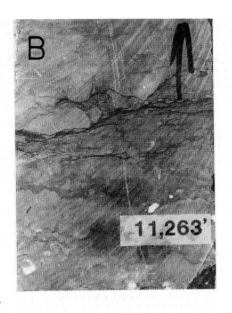
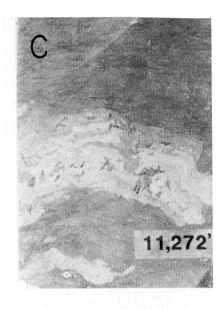
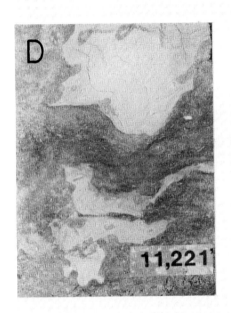
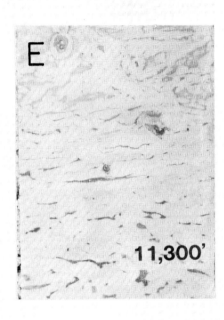
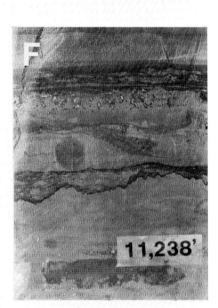
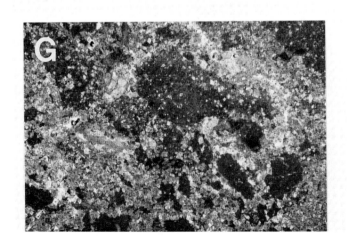
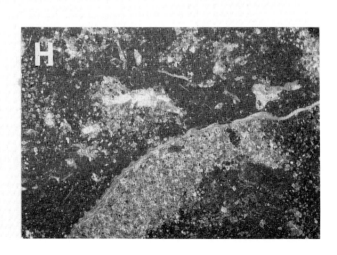

Figure 6. Thin section photomicrographs of Duperow facies (A-F), continuing sequence from Figure 5G. Samples from Tenneco Gawryluk unless otherwise indicated. Photomicrographs represent 6 mm. Scanning electron Photomicrographs of Duperow Unit 4 reservoir facies (G and H).

(A) Dolomitic limestone; stromatoporoid floatstone to boundstone. Apparently in situ laminar stromatoporoid retains calcite skeletal structure in largely dolomitized matrix. Core depth 11,266; Tenneco Federal #2-30.

(B) Dolomite; stromatoporoid floatstone. Stromatoporoid fragment as dolomicrospar in porous dolospar matrix. Some of the larger molds are filled by anhydrite cement (lower center). Core depth 11,227.

(C) Dolomite; stromatoporoid floatstone. Porous, dolomitized fabric preserves no original skeletal material, but molds record the prior existence of fine skeletal material in the matrix. Core depth 11,227.

(D) Slightly anhydritic dolomite; mudstone. Rare skeletal molds in microporous dolomicrospar matrix are partly filled by anhydrite cement. Dark matrix at top results from residual oil from the overlying reservoir zone. Core depth 11,233.

(E) Silty, argillaceous and anhydritic dolomite; crudely laminated mudstone. Quartz silt and larger nodules of anhydrite and sphalerite float in a matrix of dolomite mudstone. Unit 5 marker bed, Tenneco Federal #2-30; core depth 11,300.

(F) Dolomitic limestone; skeletal wackestone. Fine skeletal fragments, including a bridged specimen of the alga Parachaetetes from skeletal-rich facies in the upper portion of Unit 5. Core depth 11,246.

(G) Intercrystalline and microvugular porosity provides an open pore network connecting larger moldic pores in stromatoporoid floatstone. Core depth 11,227'. Photo width represents approximately 6mm. Photograph from Kissling and Ehrets, 1985.

(H) Coarse dolospar lining a moldic pore in wackestone. Oil residue coats most dolospar rhombs. Photograph width represents approximately 600 microns. Samples from Unit 4 in Roosevelt Field, Billings County. Photograph from Kissling and Ehrets, 1985.

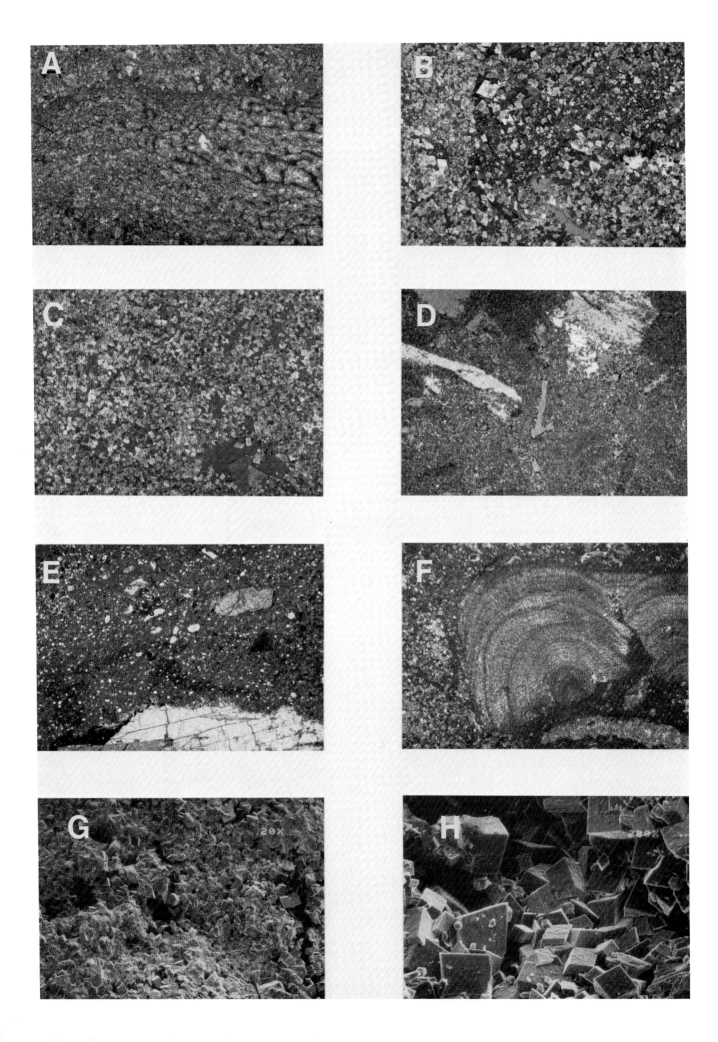

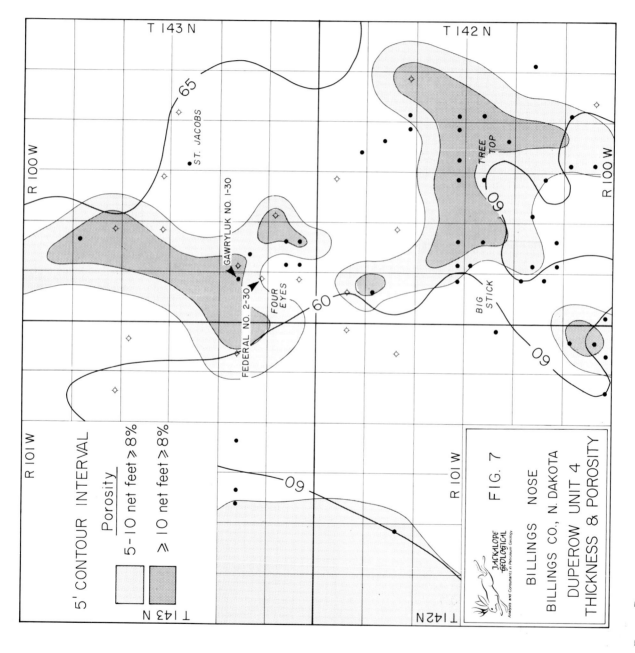

Figure 7. Duperow Unit 4 thickness and net porosity greater than 8%, Billings Nose area.

Intercrystalline and microvugular porosity are the most abundant pore types and likewise provide for reservoir permeability (Figs. 6B, 6C, 6D, 6G and 6H). Intraskeletal and moldic porosity are also present in amounts ranging up to 5% (Figs. 6C, 6G). The importance of these pores may be masked in thoroughly dolomitized fabrics because many larger microvugular pores may in fact be solution-enlarged molds or intraskeletal voids. Neomorphism of dolomicrospar to dolospar also tends to obscure former molds. Hence, observed moldic and intraskeletal porosity represents a minimum proportion of the total value. Elsewhere in the cored interval, less dolomitization or less fossiliferous textures severely inhibited secondary porosity development. In addition, rare peloidal textures originally possessing some intergranular porosity are now thoroughly cemented by anhydrite or fine calcite spar.

Reservoir-quality porosity (compensated neutron-density log cross-plot porosity of at least 8%) within Unit 4 shows an areally restricted distribution in the Billings Nose area, closely associated with the trend of the paleostructure (Fig. 7). Of particular interest is the development of thicker porosity zones (normally corresponding to more thoroughly dolomitized strata) along the eastern flank of the structural trend. Available log control indicates that, although at least some Unit 4 porosity exists for most wells in the area, zones of porosity and permeability suitable for fluid migration are invariably restricted and irregular in shape. Longman (1981) suggested that irregular porosity distribution in this area arising from patchy dolomitization, was responsible for stratigraphic traps in some Duperow fields. Whiskey Joe Field (south of the Big Stick/Tree Top field area) was cited as an example of an updip porosity pinchout. Although small structures undoubtedly play an important role in many Billings Nose fields, the non-uniform or decidedly patchy distribution of well developed Unit 4 porosity strongly indicates that many Duperow fields and pools may be stratigraphic in part, and that this model is basic to exploration for and development of Duperow fields.

BIRDBEAR (NISKU) FORMATION

Since the discovery of production in the Birdbear (Nisku) Formation during the early 1960's, more than a dozen small but prolific Birdbear (Nisku) fields have been established in the Wolf Creek Nose area (Fig. 8). Swenson (1967) cited multi-stage solution of the Middle Devonian Prairie salt and subsequent depositional compensation as the most probable mechanism in the creation of steep-sided structural traps in the area. Until recently, however, little information had been published regarding the facies sequence and reservoir characteristics of the formation (Ehrets and Kissling, 1983; Kissling and Ehrets, 1985). Evaluation of the numerous available Birdbear (Nisku) cores has demonstrated that paleostructural and facies controls were equally important to reservoir development. A core of nearly the entire formation from the Murphy Sethre #1-B is presented to illustrate the facies sequence and reservoir characteristics in the Tule Creek Field area. A cored interval from the nearby Murphy Sletvold #1-B provides additional examples for comparison.

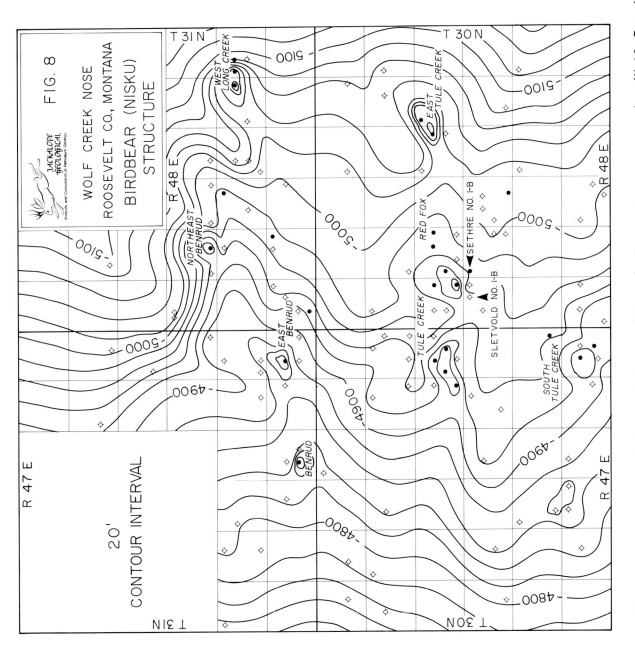

Figure 8. Structure on top of the Birdbear (Nisku) Formation across the Wolf Creek Nose. Arrows locate the two Murphy wells included for discussion. Wells producing only from the Birdbear (Nisku) Formation are indicated.

Facies Sequence

As indicated by both regional thickness trends (Fig. 12) and facies distribution in the Birdbear (Nisku) and uppermost Duperow, the Wolf Creek Nose had existed as a topographic high which was probably subaerially exposed along the paleostructural axis at the close of Duperow deposition. Initial Birdbear (Nisku) transgression of this feature was accompanied by a gradual increase in water depth. The resulting depositional cycle is decidedly more symmetrical than that of Duperow Unit 4 in the Billings Nose area. Shallow water shoals or skeletal banks, represented by Amphipora wackestone and packstone, were established along the flanks of the paleostructure early in the depositional sequence (Figs. 9, 10H, 10I, 11F). When maximum water depths and circulation were established, initial bank deposits were covered by nodular-bedded mudstone and brachiopod-crinoid wackestone of the open shelf facies (Figs. 10G, 11E). However, it is likely that much of the paleostructural crest remained within the intertidal environment. Following maximum transgression, widespread but discontinuous stromatoporoid biostromes, comprised of laminar and tabular stromatoporoid boundstone and overlain by Amphipora wackestone to packstone, flourished along the flanks of the paleostructure (Figs. 10E, 10F, 11D, 11F, 11H). Depending upon the relative position on the paleostructural flank, Amphipora bank and stromatoporoid biostrome facies account for 10 to 40 feet (3 to 12 m) of skeletal-rich strata within the middle of the formation.

Intertidal and lagoonal deposits of laminated and stromatolitic mudstone and peloidal packstone record a progressive loss in

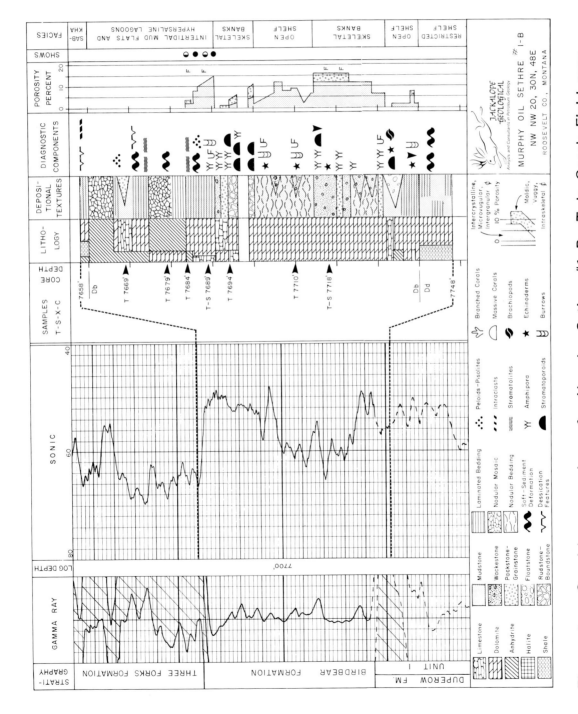

Figure 9. Graphic core column for Murphy Sethre #1-B, Tule Creek Field area.

Figure 10. Core surfaces of Birdbear (Nisku) facies from Murphy Sethre #1-B; shown in facies sequence, from top to bottom, as illustrated in Figure 9. Slabbed core widths approximately 10 cm.

   (A)   Nodular-mosaic anhydrite.

   (B)   Anhydritic limestone; interlaminated mudstone and peloid packstone. Anhydrite occurs as matrix-consuming laths as well as fracture-filling cement.

   (C)   Anhydritic dolomite. Mud-cracked, stromatolitic mudstone.

   (D)   Dolomitic anhydrite; laminated anhydrite and dolomicrite.

   (E)   Dolomitic limestone; Amphipora packstone, consisting of branching Amphipora fragments and fewer laminar stromatoporoids.

   (F)   Calcitic dolomite; stromatoporoid boundstone. Closely packed, in situ laminar stromatoporoids form boundstone in a matrix of Amphipora wackestone.

   (G)   Dolomite; nodular-bedded skeletal wackestone, grading to mudstone.

   (H)   Dolomite; Amphipora packstone and skeletal wackestone. Bedded Amphipora packstone and wackestone penetrated underlying wackestone by a high-amplitude stylolite.

   (I)   Anhydritic dolomite; mudstone. White anhydrite cement replaces dolomicrospar mudstone matrix containing few solitary rugose corals.

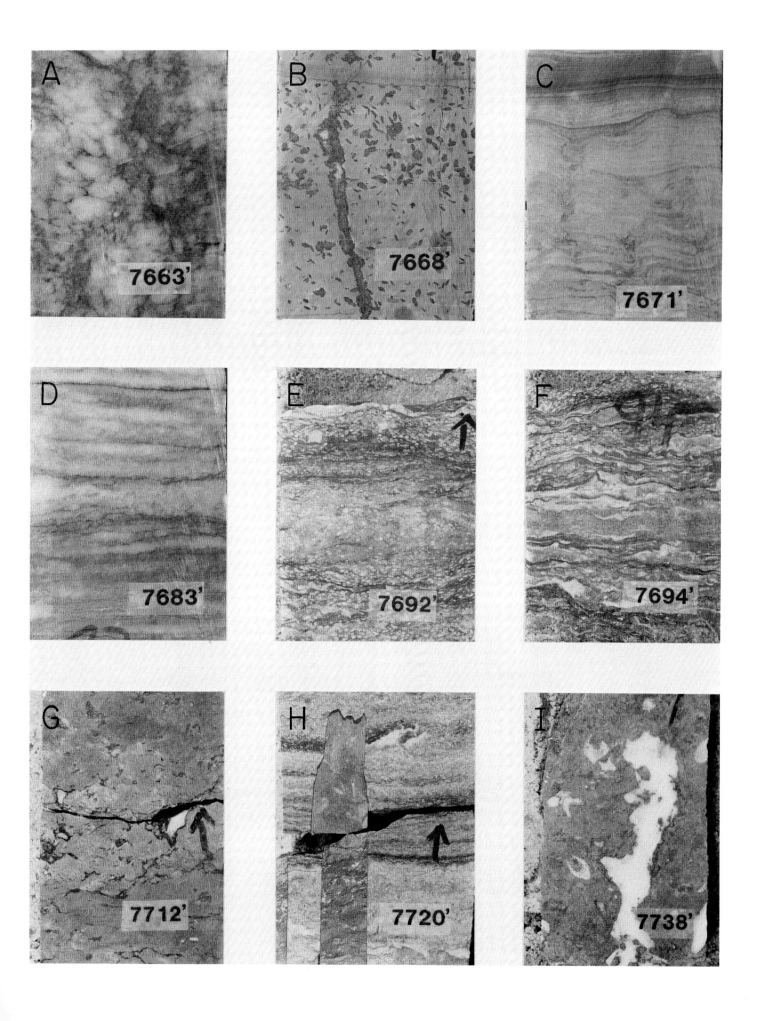

Figure 11. Photomicrographs of Birdbear (Nisku) facies, also representing the top to bottom facies sequence illustrated in Figure 9. Composite sequence from Murphy Sethre #1-B (C, E and F) and Murphy Sletvold #1-B (A, B and D). Scanning electron photomicrographs of Birdbear (Nisku) reservoir facies (G and H); samples from Murphy Bjorgen No. 1, Tule Creek Field.

(A) Argillaceous, anhydritic, silty dolomite; laminated and deformed mudstone. Three Forks Formation; core depth 7601'. Photo width represents 6 mm.

(B) Anhydritic dolomite; stromatolitic mudstone and algal-peloid packstone. Equant anhydrite fills pore spaces within stromatolite fabric (dark hued) and algal-textured packstone in lower portion of photo. Core depth 7621.5'; photo width represents 6 mm.

(C) Slightly calcitic dolomite; mudstone. Dolomicrospar and fine dolospar matrix bears appreciable intercrystalline porosity. Core depth 7689; photo width represents 3 mm.

(D) Calcite dolomite; Amphipora packstone. Amphipora and brachiopod fragments in incompletely dolomitized matrix. Core depth 7641'; photo width represents 6 mm.

(E) Dolomite; mudstone. Echinoderm fragments are preserved in dolomitized mudstone matrix. Core depth 7710'; photo width represents 3 mm.

(F) Dolomite; Amphipora wackestone. Typical preservation of Amphipora as relict dolomicrospar in an otherwise porous, dolospar matrix. Core depth 7718'; photo width represents 6 mm.

(G) Intercrystalline and microvugular porosity in skeletal wackestone matrix. Core depth 7688'. Photo width represents 2.5 mm. Photograph from Kissling and Ehrets, 1985.

(H) Dolospar-lined Amphipora molds in Amphipora packstone. Core depth 7724'. Photo width represents approximately 2 mm. Photograph from Kissling and Ehrets, 1985.

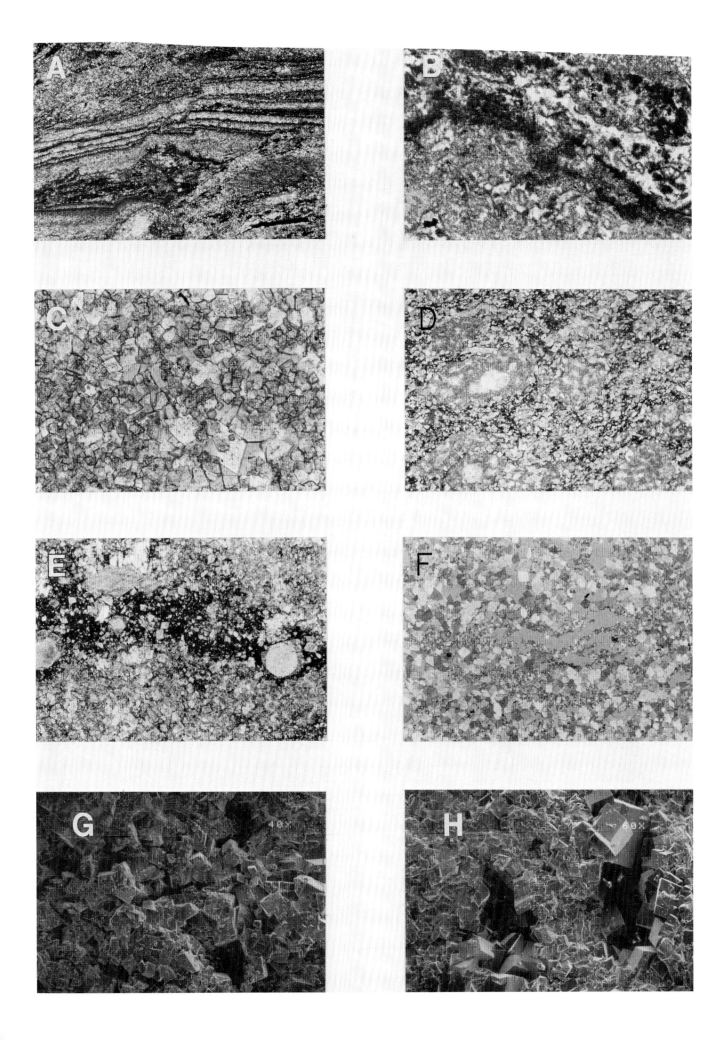

circulation and increase in salinity during continued regression and shoaling (Figs. 10B, 10C, 11B, 11C). Shelf-wide hypersalinity accompanying progresive shoaling is demonstrated by two to four laminated to nodular-mosaic anhydrite zones alternating with peritidal carbonates. Together, these comprise the upper third of the Birdbear (Nisku) across northern Montana and North Dakota (Figs. 10A, 10D). Sabkha facies may have been established before the close of Birdbear (Nisku) deposition over the Wolf Creek Nose. Reddish, anhydritic and argillaceous dolomite mudstone within the lower portion of the overlying Three Forks Formation certainly represents sabkha conditions (Fig. 11A).

Reservoir Characteristics

Porosity in the Sethre #1-B cored interval is developed in several dolomitized facies. Intercrystalline and microvugular porosity in medium to coarse dolospar fabrics represent the most important pore types (Figs. 9, 11C, 11F, 11G, 11H). Observed intraskeletal and moldic voids account for up to 5% porosity and are dominantly associated with partial to complete leaching of Amphipora fragments within skeletal bank facies (Figs. 11F, 11H). As with the Duperow Formation, pervasive dolomitization may mask the original distribution of pore types through neomorphism and the creation of solution-enlarged pores. Dolomitization and development of intercrystalline and microvugular porosity in less fossiliferous, open shelf facies, either stratigraphically or laterally adjacent to skeletal banks, is not uncommon along the flanks of the paleostructure. Porosity development in intertidal facies overlying the bank facies, aside from isolated preservation of intergranular pores in peloidal packstone, is far less common, although zones of reservoir-quality porosity (as exhibited by the Sethre #1-B core) may produce from structural closures (Fig. 8).

Areal distribution of Birdbear (Nisku) porosity in the Wolf Creek Nose area contrasts with Duperow Unit 4 porosity of the Billings Nose. Most wells in the Wolf Creek Nose area possess several feet of reservoir-quality porosity. However, in mapping only stratigraphically extensive zones of porosity (15 feet or more), the control of the Wolf Creek paleostructure over porosity development is well-illustrated (Fig. 12). The existence of thick porous intervals is clearly associated with the flanks of the structure where bank deposits are best developed. Along the crest of the trend, where intertidal conditions prevailed throughout much of Birdbear (Nisku) deposition and where bank facies are poorly developed, porosity is significantly less well-developed. Similarly, porosity is less well-developed farther off the principal structure, reflecting the diminishment of skeletal bank development and dolomitization on the outer flanks of the paleostructure. This distribution undoubtedly served as a template for the migration of hydrocarbons into the Wolf Creek structural trend by allowing adequate permeability in relatively broad areas of reservoir continuity.

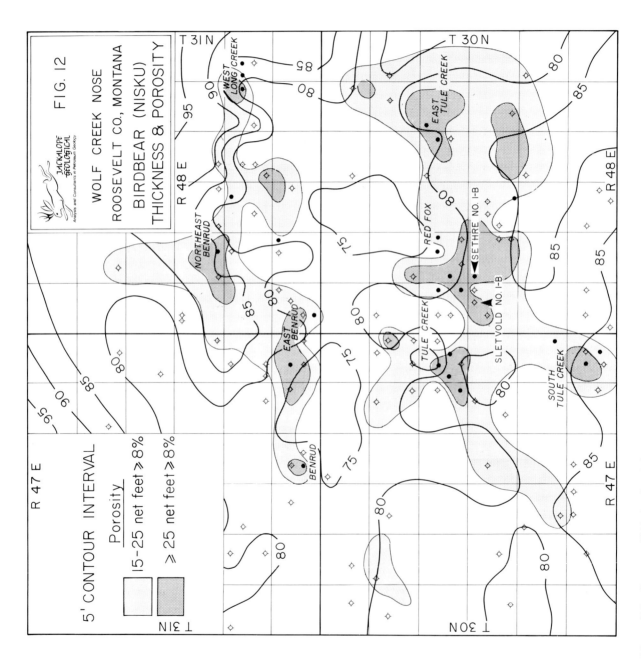

Figure 12. Birdbear (Nisku) thickness and net porosity greater than 8%, Wolf Creek Nose area.

DOLOMITIZATION AND POROSITY EVOLUTION

Skeletal bank facies in the Duperow and Birdbear (Nisku) formations are preferentially dolomitized despite their stratigraphic separation from overlying anhydrite beds by nonporous facies. These thick anhydrite zones probably served as the dominant source of magnesium-enriched brines in both formations. Additionally, there is a definite preference for porosity development within or adjacent to dolomitized Amphipora and stromatoporoid banks in both formations throughout the Williston Basin. In evaluating plausible mechanisms for dolomitization of these reservoirs, three conditions must be addressed: (1) the nature of the facies overlying and underlying the prospective reservoir facies; (2) the specific burial conditions established after deposition of the prospective reservoir facies; and (3) the source and migration pathways for dolomitizing fluids.

The stratigraphic position of Duperow Unit 4 skeletal bank facies between the Unit 4 and Unit 5 anhydrite and marker bed facies, places these banks in an essentially sealed system with respect to extrinsic dolomitizing fluids. Compaction of the bedded anhydrites (perhaps originally gypsum) and the accompanying expulsion of Ca^{++}-depleted and Mg^{++}-enriched brines provided an immediate source for dolomitizing fluids during shallow to deep burial stages (Fig. 13B). Although evidence points to the probable existence of an early lithified facies (dominantly dolomite mudstone) between the Unit 4 anhydrite and the underlying bank facies, fracturing of brittle, lithified mudstone upon burial may have allowed for enhanced flow of brine into the underlying bank facies.

Once brine migration was established, dolomitization of bank facies was aided by the permeability of fine skeletal debris and internally porous large skeletons which were partially to completely leached to form molds. Impermeable marker bed lithologies, within which fractures would readily heal, lie beneath the bank facies and served to direct brine migration laterally along the flanks of the Billings Nose paleostructure. The confinement of brines permitted the dolomitization of adjacent facies, including poorly fossiliferous mudstone at the base of Unit 4 (Figs. 6D, 13B). Deposition of Duperow Unit 3 open marine limestones further isolated the system from other brines. Fracturing of the Duperow during deep burial, as evident in most Duperow cores, would not have significantly influenced Unit 4 dolomitization because of the anhydrite and argillaceous marker bed seals bounding this interval.

More intensive late-stage dolomitization of skeletal bank facies, in particular neomorphism to dolospar, was necessarily the result of brines expelled from Unit 4 anhydrites. Structural enhancement of the Billings Nose feature during deeper burial may have had significant influence during this stage of dolomitization, perhaps accounting for better porosity development along the more steeply dipping eastern limb of the structure (Fig. 13C).

Within the Birdbear (Nisku) Formation, there is little difficulty in explaining reservoir development in the laminated mudstone zone immediately underlying the relatively thick interval of anhydrite in Murphy Sethre #1-B (Figs. 9, 11C). However, reservoir development in this facies is atypical for the Wolf Creek Nose area. Early

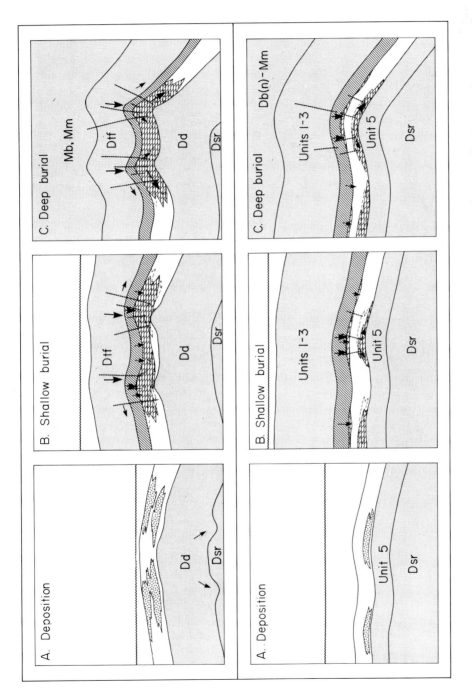

Figure 13. Depositional and diagenetic models for the Birdbear (Nisku) formation on the Wolf Creek Nose (upper panels) and for the Duperow Formation on the Billings Nose (lower panels). Arrows in uppermost left panel denote Prairie salt-solution sinks on the Wolf Creek Nose which developed and were compensated for during Duperow deposition. Arrows in all remaining panels indicate hypothesized flow directions of dolomitizing fluids into and through skeletal bank deposits during burial stages.

dolomitization or lithification of this facies typically resulted in an early loss of porosity and permeability. The low structural position of the Sethre #1-B well permits only the stratigraphically highest porous zone to be above the free water level (Fig. 8). More commonly, productive well are located higher within the structural closure, and well developed porosity within skeletal bank facies in the middle of the formation serves as a reservoir that may produce several hundred barrels of oil per day with reserves of several hundred thousand barrels per well. The open shelf, nodular-bedded mudstone and skeletal wackestone facies which lies between lower and upper skeletal bank facies thickens to the north, east and south of the paleostructure at the expense of the skeletal bank facies. On the outer flanks of the Wolf Creek Nose, skeletal bank facies are replaced entirely by non-porous, open shelf limestone. Dolomitization and accompanying porosity development in skeletal bank and intervening open marine facies are coincident with inner flanks of the paleostructure.

Deposition of the lower portion of the Three Forks Formation took place under sabkha conditions along the Wolf Creek Nose paleostructural high. Under these conditions, permeable or still-unlithified facies of the underlying Birdbear (Nisku) may have been dolomitized in part within a mixing zone of marine connate water and descending meteoric phreatic water. Upon progressively deeper burial, compaction of the thick overlying anhydrites and expulsion of Ca^{++}-depleted and Mg^{++}-enriched brines would have fostered significant dolomitization of the underlying skeletal-rich facies, as well as leaching of skeletal fragments to produce significant moldic porosity (Fig. 13B). Less skeletal-rich facies lower in the section and along the flanks of the paleostructure served as imperfect permeability barriers to migrating fluids, confining dolomitization to the immediate flank areas. The existence of early-lithified, nonporous limestone and dolomite mudstone between the anhydrites and the skeletal bank facies does not pose an insurmountable problem to downward migration of dolomitizing fluids. Brecciation and numerous fractures (now sealed by anhydrite) within these intervening, brittle strata indicate adequate passage for fluids during compaction and brine expulsion (Fig. 10B).

Compaction and expulsion of fluids from the Three Forks would have provided additional hydraulic head and brines if pathways through the upper Birdbear (Nisku) had been established through fracturing. Local structures resulting from multi-stage salt solution and depositional compensation in the Wolf Creek Nose area may have served as foci for fracturing during post-Birdbear (Nisku) structural development (Fig. 13C). However, is is likely that after compaction and lithification, the upper Birdbear (Nisku) anhydrites served as seals to downward-migrating fluids and later hydrocarbon emplacement. Later stage dolomitization during deeper burial is represented by neomorphism of dolomicrospar to dolospar and the continued leaching of unstable skeletal fragments. Previously dolomitized bank facies served as the conduits for sustained fluid migration within the system. This process improved porosity and permeability within skeletal bank facies, and also aided further dolomitization of intervening open shelf facies in the immediate area of the paleostructure.

CONCLUSIONS

These two examples, drawn from the Billings Nose and Wolf Creek Nose areas, demonstrate the influence of paleostructure over facies distribution and diagenesis and, in turn, over reservoir development in Upper Devonian carbonates of the Williston Basin. The coincidence of thickness anomalies with present-day structural trends and the concurrent distribution of skeletal bank facies with these trends reveals the paleotopographic expression of these structures during deposition. Relative bathymetry controlled the distribution of bank facies with respect to the paleostructural axes. Subtidal conditions along the crest of the Billings Nose paleostructure provided optimal development of Duperow Unit 4 skeletal banks dominated by irregular, globular stromatoporoids. On the other hand, shallow subtidal to supratidal conditions along the crest of the Wolf Creek Nose restricted development of skeletal bank facies (dominated by Amphipora and laminar and tabular stromatoporoids) to the flanks of the Wolf Creek paleostructure.

Dolomitization of skeletal bank facies and portions of adjacent facies was the mechanism for reservoir development in both formations. Brines expelled from overlying evaporite facies during burial provided the dolomitizing fluids in both systems; however, the distribution of bank and underlying facies, as well as differences in early burial histories, significantly controlled the style and magnitude of dolomitization within the two formations. Whereas paleostructure provided the local influence over facies distribution and reservoir development, the major shoaling-upward cycles punctuated by episodic salinity cycles created the particular depositional sequences that established the overall facies-reservoir relationships across the Upper Devonian shelf.

ACKNOWLEDGEMENTS

Depositional models and some aspects of diagenesis included in this paper were developed by the authors in the course of conducting a regional evaluation of Upper Devonian reservoirs while employed with and consulting for NL Erco Petroleum Services, Inc. of Houston, Texas. Much of the data on which this paper was based is embodied in the NL Erco report "Duperow and Birdbear (Nisku) Reservoirs of the Williston Basin". Scanning electron microscope photography was conducted by the authors and by Michael Chakarian of NL Erco. We gratefully acknowledge NL Erco for releasing this manuscript and the SEM photographs for publication. We also thank Thomas C. Michalski of the USGS Core Library, Denver, and his staff for providing the opportunity to re-examine and sample the cores discussed herein.

REFERENCES

Aultshuld, N. and Kerr, S.D., Jr., 1982, Mission Canyon and Duperow reservoirs of the Billings Nose, Billings County, North Dakota: in Christopher, J.E. and Kaldi, (eds.), Fourth Int'l Williston Basin Symposium, p. 103-112.

Dunn, C.E., 1975, The Upper Devonian Duperow Formation in southeastern Saskatchewan: Saskatchewan Dept. of Mineral Resources Rept. No. 179, 151 p.

Ehrets, J.R., and Kissling, D.L., 1983, Depositional and diagenetic models for Devonian Birdbear (Nisku) reservoirs, northeastern Montana (abs): Am. Assoc. Petrol. Geol. Bull., v. 67, p. 1336.

Kissling, D.L., and Ehrets, J.R., 1984, Depositional models for the Duperow and Birdbear (Nisku) formations: Implications for correlation and exploration: in Lorsong, J.A. and Wilson, M.A. (eds.), Saskatchewan Geol. Soc. Spec. Pub. No. 7, 6 p.

Kissling, D.L. and Ehrets, J.R., 1985, Upper Devonian Duperow and Birdbear (Nisku) reservoirs of the Williston Basin: NL Erco Commercial Report No. 3410008, 146 p. with appendices and maps.

Longman, M.W., 1981, Carbonate diagenesis as a control on stratigraphic traps (with examples from the Williston Basin): Am. Assoc. Petrol. Geol. Course Notes Series No. 21, 159 p.

Swenson, R. E., 1967, Trap mechanics in Nisku Formation of northeast Montana: Am. Assoc. Petrol. Geol. Bull., v. 41, p. 1948-1958.

Wilson, J.L., 1967, Carbonate-evaporite cycles in lower Duperow Formation of Williston Basin: Am. Assoc. Petrol. Geol. Bull., v. 51, p. 230-312.

RIVAL, NORTH AND SOUTH BLACK SLOUGH, FOOTHILLS AND LIGNITE OIL FIELDS: THEIR DEPOSITIONAL FACIES DIAGENESIS AND RESERVOIR CHARACTER, BURKE COUNTY, NORTH DAKOTA

ROBERT F. LINDSAY
GULF OIL EXPLORATION & PRODUCTION COMPANY
Gulf Exploration Technology Center
P. O. Box 36506
Houston, Texas 77236

ABSTRACT

Rival, North and South Black Slough, Foothills, and Lignite fields in northeastern portions of the Williston Basin had produced 24.2 million barrels of oil through mid 1982 from the Mississippian Lower Charles Formation. These fields were discovered in the late 1950's and unitized in the early to late 1960's. In 1984 all, except Rival, were deunitized. The Rival subinterval at the base of the Charles Formation serves as the reservoir in Rival, North and South Black Slough, and Foothills fields. It is 30 to 60 feet thick. This subinterval was named at Rival Field but is commonly called the "Nesson". The overlying Midale subinterval serves as the reservoir in Lignite Field and extends into southeastern Rival Field. It is 30 to 50 feet thick. Most oil production is from the Rival "Nesson" (21.3 MMSTBO, as of 1982). The study area is tilted basinward 2/3's of one degree, with three subtle anticlinal noses trending through the fields.

The Rival "Nesson" represents a rapidly prograding shoreline and coastal sabkha which ceased building basinward and began slowly retreating. Barrier island and intertidal buildups of sparsely skeletal to skeletal, oolitic, pisolitic, intraclastic packstones formed along the shoreline. Most particles were micritized and neomorphosed to microspar. Inner portions of the shoreline were tightly cemented by anhydrite, while outer portions were partially cemented. Later, anhydrite was leached in outer portions of the shoreline to enhance reservoir porosity. Inner portions of the shoreline remained tight and along with anhydrite beds provide the updip stratigraphic trap. Some pores were later partially to completely filled with dolomite and calcite cement, drastically reducing permeability.

The Midale records a transgression which flooded the study area. Restricted marine to tidal flat, sparsely anhydritic, spiculitic, pelletal wackestone/packstones were dolomitized and serve as the reservoir. Porous dolostone is aphanocrystalline to very finely crystalline. Leaching of sponge spicule monaxons and some anhydrite further enhanced porosity.

The Rival "Nesson" pore system is composed of: (1) moldic and solution enlarged pores (10 to 1000 microns in width); (2) interparticle pores (5 to 25 microns); (3) intraparticle pores (5 to 10 microns); and (4) intercrystal pores (approximately 1 to 3 microns in width).

The Midale pore system is composed of: (1) moldic pores (5 to 500 microns in width); and (2) intercrystalline pores which are, (a) polyhedral pores (3 to 10 microns), (b) tetrahedral pores (3 microns), and (c) interboundary-sheet pores (1 micron in width).

INTRODUCTION

Rival, North and South Black Slough, Foothills and Lignite oil fields are located north-northeast of the center of the Williston Basin in Burke County, North Dakota (Fig. 1). These fields are situated one to three townships south of Saskatchewan (Figs. 2 and 3) and all except Lignite, produce from the Rival "Nesson" subinterval at the base of the Mississippian Charles Formation (Figs. 4 and 5). Lignite Field produces from the Midale subinterval of the lower Charles, which lies directly on top of the Rival "Nesson". The Charles Formation forms the upper part of the Madison Group, with the Mission Canyon Formation and Lodgepole Limestone forming the middle and lower parts of the group. These fields were discovered in the late 1950's and were unitized in the early to late 1960's. Recently, by order at the North Dakota Industrial Commission, all, except Rival, were deunitized. Through mid-1982 these fields had a combined cumulative production of 24.2 MMSTBO, 24.2 MMSTBW, and 42.6 BCFG. Most production (21.3 MMSTBO) was from the Rival "Nesson".

In this paper the name Rival is used instead of the more commonly used term "Nesson" to refer to the basal limestone and anhydrite beds of the Charles Formation. This is done to avoid confusion with the original use of the term Nesson by Nordquist (1955) as a formation name for a 260 foot (80 m) thick Middle Jurassic (Bajocian) sequence of carbonates and evaporites that unconformably overlie the Triassic (?) Spearfish Formation.

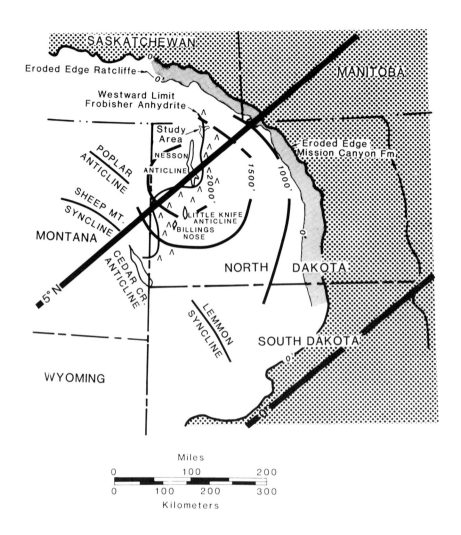

Figure 1. Index map of the Williston Basin displaying the study area, Ratcliffe and Mission Canyon subcrop edges (Proctor and Macauley, 1968), westward limit of the Frobisher Anhydrite (Peterson, 1985), isopach thickness of the Madison Group (Carlson and Anderson, 1965), Carboniferous paleolatitudes (Habicht, 1979), and major surface and subsurface structural features (Wittstrom and Hagemier, 1978 and 1979; Gerhard, 1982; Gerhard and others, 1982; and Lindsay and Roth, 1982).

During Rival deposition the study was part of a shoreline complex (Fig. 2). Total thickness of the Rival beds is 30 to 60 feet (10 to 20 m). Lateral facies changes from beds of porous, productive limestone into tight anhydrite-cemented limestone, and anhydrite beds take place along the eastern margin of these fields and provide an updip stratigraphic trap. Limestone beds deposited along the shoreline serve as the reservoir. Overlying tight limestone beds serve as the reservoir seal.

Lignite Field, located east and adjacent to Rival Field, is productive from the Midale subinterval of the lower Charles Formation (Fig. 3). The Midale is 30 to 50 feet thick and produces from beds of dolostone. Productive Midale beds extend southwestward into southeast parts of Rival Field. The underlying Rival beds at Lignite are of tight anhydrite-cemented limestone and anhydrite.

STRUCTURAL SETTING

This part of the Williston Basin dips to the southwest at a rate of 2/3° (Figs. 2, 3). The isopachous basin center during Madison Group deposition was located west of the southern half of Nesson Anticline in McKenzie County (Fig. 1).

Superimposed on regional dip are three subtle, southwest plunging anticlinal noses. These cut perpendicularly across the Rival paleo-shoreline (Anderson and others, 1960). The three noses

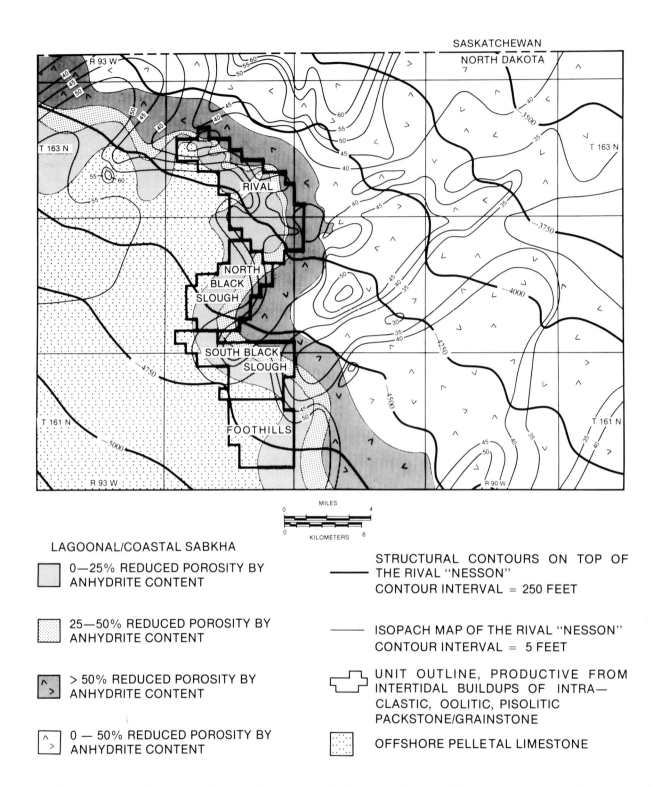

Figure 2. Map of Rival (Nesson) fields studied. Structural top of the Rival, isopach thickness of the Rival, original unitized outlines of each field, and the percentage of reduced porosity due to anhydrite cement and beds of anhydrite are shown. Modified after Anderson and others (1960).

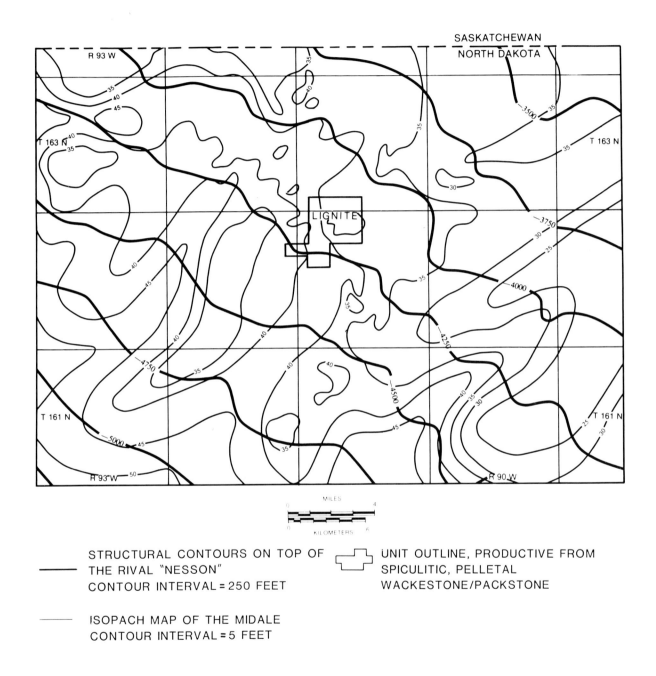

Figure 3. Map of Lignite Field which produces from the Midale. The field is productive from the Midale. Structural base of the Midale (top of Rival), isopach thickness of the Midale, and the original unitized outline of the field are shown. Modified after Anderson and others (1960).

help localize production in: (1) the north end of Rival, (2) Lignite, the south end of Rival and North Black Slough, and (3) Foothills fields. They are best expressed by isopach mapping the lower Charles (Ratcliffe) interval. They are further outlined by mapping the edge of the last (lowest) Charles "F" Salt, which wraps around each nose.

Anderson and others (1960) mapped surface lineaments throughout the study area and found average lineament trends of approximately N45°E and N50°W. These trends are similar to those Thomas (1974) found throughout the United States portion of the basin.

INDIVIDUAL FIELDS

Rival Field

The Rival discovery well was the Arrowhead No. 1 Probst (SE, SE SEC. 17, T163N, R92W) (Tyler and George, 1962) (Figs. 4, 5). The well was drilled to a total depth of 6425 feet and completed on October 24, 1957, with an initial potential of 56 BOPH (39.1° API). Average productive porosity is 9 percent. The reservoir had an original solution gas drive in a combination structural-stratigraphic trap. The field was unitized in June, 1961, with waterflood operations commencing in August of that year. By mid-1982 the field had produced 12.9 MMBO, 25.5 MMCFG, and 14.9 MMBW.

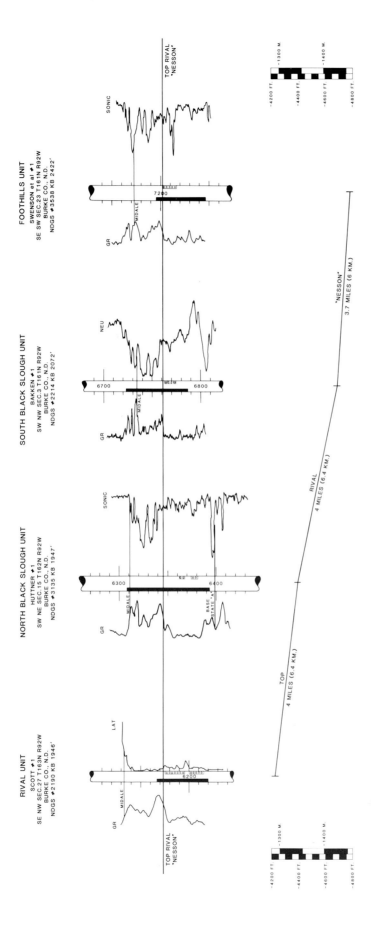

Figure 4. Stratigraphic cross section, hung on top of the Rival, through cored wells studied in Rival, North and South Black Slough, and Foothills fields. This cross section parallels the original shoreline. The structural position and distance between each well is displayed beneath. Subsea values are in feet and meters. The black bar in each wellbore represents cored intervals displayed in Figure 6. Boxes containing circles indicate perforated intervals.

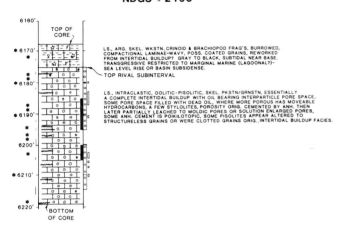

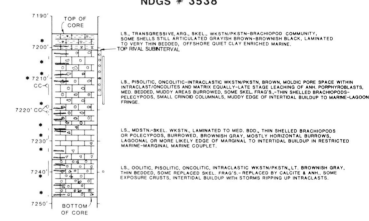

Figure 5. Core descriptions from Rival, North and South Black Slough, and Foothills fields.

North Black Slough Field

Black Slough field was discovered by the Anshutz Bakken No. 1 (SW, SW Sec. 3, T161N, R92W) (Tyler and George, 1962) (Figs. 4, 5). This well was drilled to a depth of 6,858 feet and completed on April 16, 1959 with an initial potential flowing of 315 BOPD (38.1° API). Later, Black Slough was divided into north and south portions, where a dog leg and a few dry holes divide the field (Fig. 2). Average porosity for North Black Slough is 9 percent and average permeability is 10 millidarcies. The trap for the reservoir is a combination stratigraphic and structural trap. North Black Slough is adjacent to and directly south of Rival. One dry hole lies between the fields. North Black Slough was unitized in January 1968 and waterflood operations started in April of that year. The field had produced 3.6 MMBO, 5.7 MMCFG, and 5.3 MMBW as of mid-1982.

South Black Slough Field

South Black Slough contains 7 percent average porosity and 2 millidarcies average permeability (Figs. 4, 5). The reservoir had an original gas-oil ratio of 1,270:1, and a solution gas and partial water drive. The trap for the reservoir is a combination stratigraphic and structural trap. The field is adjacent to and directly north of Foothills Field. The dividing line between the fields is a "tighter area". South Black Slough was unitized in May, 1968, with a waterflood commencing in March, 1970. The field had produced 2.6 MMBO, 441 MCFG, and 1.5 MMBW as of mid-1982.

Foothills Field

Foothills was discovered by the Pan American No. 1 Owings (SE, SE Sec. 13, T161N, R92W) (Tyler and George, 1962) (Figs. 4, 5). This well was drilled to a total depth of 7,250 feet and was completed on May 20, 1958, with an initial potential pumping of 41 BOPD (37.6° API) and 15 BWPD. Average porosity is 7 percent and average permeability is 7 millidarcies. The reservoir had an original gas-oil ratio of 1,000:1 and a solution gas drive. The trap for the reservoir is a combination stratigraphic and structural trap. Foothills field was unitized in March, 1969, and a waterflood commenced in March, 1970. The field had produced 2.2 MMBO, 2.4 MMCFG, and 1.9 MMBW as of mid-1982.

Lignite Field

Lignite was discovered by the Northwest Oil No. 1 Bunting (SE, SE Sec. 5, T162N, R91W) (Tyler and George, 1962) (Figs. 6, 7). The well was drilled to a total depth of 6,640 feet and was completed on July 30, 1957 with an initial potential flowing of 10 BOPH (37.2° API). Average porosity is 11 percent and average permeability is 1 millidarcy. The trap for the reservoir is a combination trap due to an updip loss of effective porosity along a structural nose. Lignite field was unitized in November, 1967, and a waterflood commenced in December, 1968. The field had produced 2.9 MMBO, 8.5 MMCFG, and 502 MBW as of mid-1982.

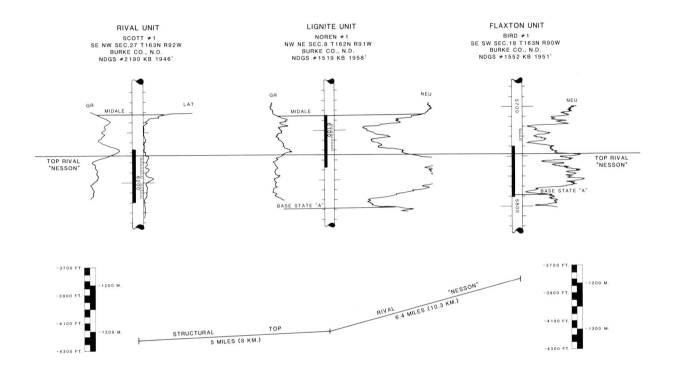

Figure 6. Stratigraphic cross section, hung on top of the Rival, through cored wells studied in Rival, Lignite, and Flaxton fields. This cross section is perpendicular to the original shoreline. The structural position and distance between each well is displayed beneath. Subsea values are in feet and meters. The black bars in each wellbore are cored intervals. Lignite's cored interval is displayed in Figure 7. Boxes containing circles indicate perforated intervals. Note how Rival beds are porous in Rival Field (left) and tight in Lignite and Flaxton fields.

DEPOSITIONAL SETTING

Regional Overview

The Rival beds at the base of the Charles Formation directly overlie the Mission Canyon Formation. The State "A" marker separates the two formations (Figs. 4, 6). This prominant marker bed extends throughout the North Dakota portion of the basin. The Rival is the top of the Frobisher-Alida, which are stratigraphically equivalent names for the Rival and Mission Canyon in southern Saskatchewan.

The Mission Canyon Formation represents a shallowing-upward carbonate to evaporite sequence. East of the study area it is composed of a series of small transgressive and regressive sequences, which represent shoreline advances and retreats (Fig. 8). These packages of genetically related sequences were recognized and described as "beds" by Harris and others (1966). Below the State "A" marker, they named these the Bluell, Sherwood, Mohall, Glenburn, Wayne, Landa and Tilston "beds". Southwest of the study area along Nesson Anticline, the Mission Canyon is composed of a single shallowing-upward sequence. This evaporitic shoreline did not prograde far enough into the basin center to cover Nesson Anticline with evaporites.

Once progradation of the Mission Canyon ceased, subaerial exposure allowed siliciclastic influx across the buried shelf (Peterson, 1985). These sediments were reworked by a subsequent marine transgression which flooded the exposed surface and deposited the State "A" marker.

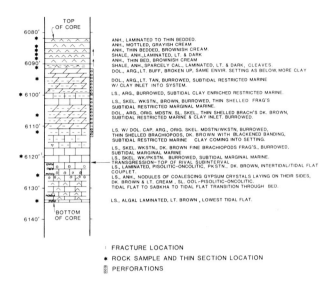

Figure 7. Core description from Lignite field.

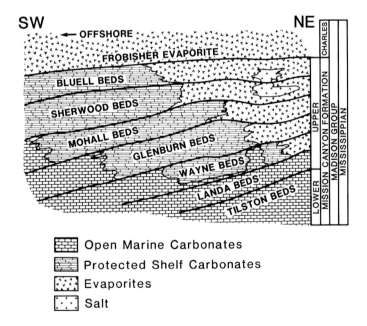

Figure 8. Cross section east of the study area through the Mission Canyon Formation and lowest Charles Formation, showing the position of individual prograding limestone/evaporite intervals. These are composed of offshore limestones, shoreline buildups, and coastal sabkha sequences. The overlying Frobisher anhydrite (Rival) prograded to the study area then began a slight retreat. Modified after Harris and others (1966).

Rapid progradation of the Rival shoreline followed. Barrier island to intertidal buildups, lagoons and tidal flats prograded westward from a coastal sabkha (Fig. 9). Anhydrite beds deposited in the lagoon to coastal sabkha are called the Frobisher Anhydrite (Fig. 8).

Shoreline progradation ceased during lower to middle Rival deposition when a subtle trangression within the Rival began to flood the shoreline. In the study area, this transgression slowed and allowed the shoreline facies to stack up with only slight retreat (Fig. 10). East of the shoreline, evaporites were deposited in lagoonal to coastal sabkha facies. Sulfate-rich brines moving seaward heavily cemented inner portions of shoreline carbonates with anhydrite to form the updip seal for the reservoirs. Outer portions of the shoreline were only partially cemented. Later leaching of anhydrite enhanced the reservoir porosity within carbonates deposited along the outer shoreline (Fig. 11).

Following deposition of the Rival interval a small transgression occurred and the Midale was deposited (Fig. 10). A thin basal offshore marine facies gave way to restricted marine and peritidal to tidal flat facies which were slightly anhydrite cemented and then dolomitized (Fig. 12). Mitchell and Petter (1958) state that reservoir potential at Lignite Field is directly related to the degree of dolomitization. Hartling (1983) suggests that as brines progressively move downdip away from their source, microcrystalline dolostone forms and grades downdip into porous finely crystalline, sucrosic dolostone. The tight updip dolostones may form a diagenetic seal.

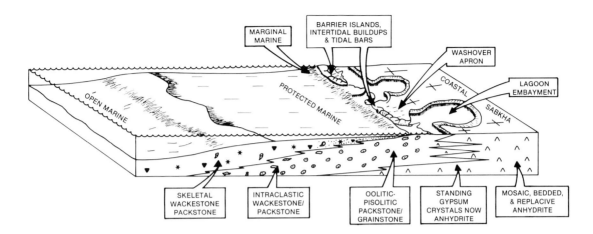

Figure 9. Depositional model for the Rival. The fields studied lie along the shoreline with lagoonal and coastal sabkha facies to the east. The offshore portion of the model is the depositional model for the Rival at North Tioga and Tioga fields, 20 miles to the southwest, in northern portions of Nesson Anticline. Total width of the model may be 40 to 50 miles (60 to 80 km).

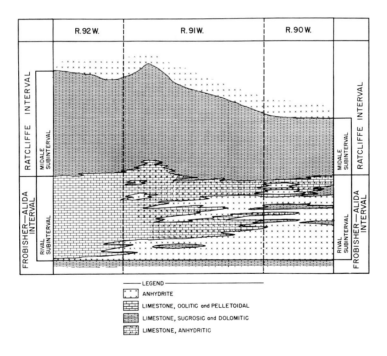

Figure 10. West to east schematic cross section of the Rival and Midale through Rival, Lignite and Woburn fields. the lower Rival prograded basinward, then began a gradual, slow retreat. The overlying Midale completely transgressed over the Rival. Modified from Anderson and others (1960).

Study Area

Rival

Rival beds in the fields studied, are laminated to very thin bedded and are composed of slightly skeletal to skeletal, intraclastic, oolitic to pisolitic packstone (Fig. 11). Some ooids and pisolites are partially to completely structureless and resemble peloids. Muddier burrowed intervals are locally interbedded with these deposits. Sparse nodules of anhydrite occur in some beds (Fig. 14E). Laminated interbeds sometimes contain crinkly algal laminations (Fig. 14D). A few thin beds contain anhydrite pseudomorphs after gypsum ax head-shaped crystals. Carbonate crusts are common in some localities (Fig. 14A). A few microstalactitic cements are present (Fig 13). One inclined bed appears to be a possible teepee structure (Fig. 14C).

The depositional setting for the Rival interval consisted of barrier island to intertidal buildups that graded into lagoons and tidal flats to the east, marginal marine areas to the west, and interbuildup areas north and south along strike. Rock building constituents are skeletal fragments of brachiopods, endothyrid foraminifers and calcispheres, with less common small crinoid columnals and gastropods, and sparse particles of algae (Fig. 13). Locally, storms washed skeletal particles onto buildups. Intraclasts were derived from offshore areas of the buildups (Fig. 14B). They are composed of skeletal fragments in micrite or pelletal muds. Ooids and pisolites formed subaqueously and were deposited on buildups by storms. Some ooids have intraclast nuclei, but most lack a nucleus. This is similar to Mescher and Pol's (1984) interpretation for Mission Canyon ooid and pisolite generation along the eastern basin margin, but differs from the interpretation of Gerhard and others (1978) who proposed a vadose origin for the Mission Canyon pisolites in Glenburn field.

In Foothills field, the producing core consists of intraclasts and skeletal particles washed in from marginal marine or muddy edges of a buildup. These rocks have very sparse microstalactitic cements (Fig. 13).

Some areas closer to or directly connected to the coastline contain nodules of anhydrite (Fig. 14B). Crinkly algal laminations are interpreted to have been deposited in a tidal flat setting on the coastal flat fringes. This was probably similar to present day Abu Dhabi portions of the Persian Gulf, where the outer edges of tidal flats are covered with blue-green algal mats (Kendall and Skipwith, 1969; Butler and others, 1982). A few beds contain ax head-shaped gypsum crystals, now anhydrite, which grew in muddy lagoonal sediments. These crystals are common in lagoonal beds behind buildups, and occupy similar positions in Little Knife Field (Lindsay and Roth, 1982). Burrows occur in some lagoonal muds and restricted to marginal marine muds. Some muddy lagoonal beds were replaced by anhydrite to form bedded anhydrite. Other anhydrite beds were originally selenite gypsum which grew as vertical crystals on lagoon floors and in ponds upon the sabkha (Fig. 14F). Similar subaqueous anhydrite after gypsum crystals has been found elsewhere in the Williston Basin, e.g., in Flaxton, Little Knife, and Whiskey Joe fields (Lindsay, 1985; Schreiber and others, 1982). Much of the anhydrite appears to be of a shallow subaqueous origin (Mescher and Pol, pers. comm., 1985).

Figure 11. Representative cores of the Rival from: (A) Rival Field (Scott #1), (B) North Black Slough Field (Huttner #1), (C) South Black Slough Field (Bakken #1), and (D) Foothills Field (Swenson #1). These are laminated to thin bedded deposits of porous, anhydritic, sparsely skeletal to skeletal, oolitic, pisolitic, intraclastic packstones from productive intervals. These cores are from portions of each field considered to contain typical reservoir characteristics. Photomicrographs and scanning-electron micrographs (SEM) from these cores are shown in Figures 13-17.

Figure 12. Representative core of the Midale from Lignite field (Noren #1). The Midale is burrowed, dolomitized, sparsely anhydritic, spiculitic, pelletal wackestone/packstone.

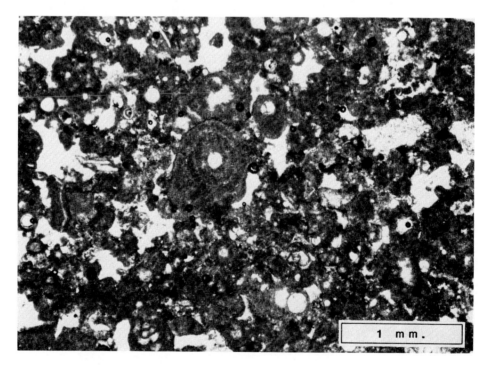

Figure 13. Anschutz Swenson #1, 7210 feet, Foothills Field, displaying micrite envelopes around probable calcispheres. Two grains have microstalactitic micritic cements. Calcispheres and some foram detritus are the only skeletal constituents. Plane light.

Subaerial exposure features in reservoir beds include carbonate crusts, sparse microstalactitic cements, and one teepee structure. In certain localities crusts are common but in most areas these features are not apparent.

Midale

A bed of skeletal wackestone occurs at the base of the Midale in the study area. This wackestone grades upward into thin bedded, slightly argillaceous spiculitic, pelletal dolostone, which can be partly calcareous (Fig. 12). A bed of limestone may be present, but is capped by more dolostone. These beds give way to fissile shales interbedded with anhydrite beds that cap the Midale (Figs. 15B, 15C).

Midale deposition began with a small trangression over Rival beds. The resulting depositional setting was an offshore marine skeletal wackestone facies. Deposition then shifted into a more restricted marine, low-energy, nearshore setting. Eventually, restricted marine to peritidal deposition dominated when the reservoir interval was deposited. Argillaceous material was probably supplied from the craton to the east. Restricted marine beds were burrowed while tidal flat bedding is laminated to very thin-bedded. Argillaceous beds form laminations. These are interbedded with lagoonal to tidal flat anhydrite beds. Some anhydrite appears to have totally replaced burrowed, lagoonal muds and laminated to thin-bedded tidal flat deposits.

Figure 14. Representative cores of the Rival deposition in Rival, North and South Black Slough, Foothills, Lignite and Flaxton fields. (A) Pan American Huttner #1, 6381 feet, North Black Slough Field, with several carbonate crusts developed and some microstalacttitie cements that are too small to see. (B) Anschutz Swenson #1, 7747 feet, Foothills Field, displaying large intraclasts composed of ripped up pisolites. (C) Pan American Huttner #1, 6364 feet, North Black Slough Field, where some carbonate crusts are developed and one possible teepee structure or a large intraclast is shown. (D) Mobile Noren #1, 6130-31 feet, Lignite Field, where two crinkly algal laminated intervals are separated by a storm deposit of laminated carbonate. (E) Mobile Noren #1, 6130 feet, Lignite Field, where laminated to thin bedded carbonate with flat-lying gypsum crystals, now anhydrite, grew and began to coalesce into larger nodules. (F) Central Le Duc Bird #1, 5747 feet, Flaxton Field, Frobisher Anhydrite, where three periods of gypsum crystal growth are separated by thin laminated intervals. Scales are 1 cm.

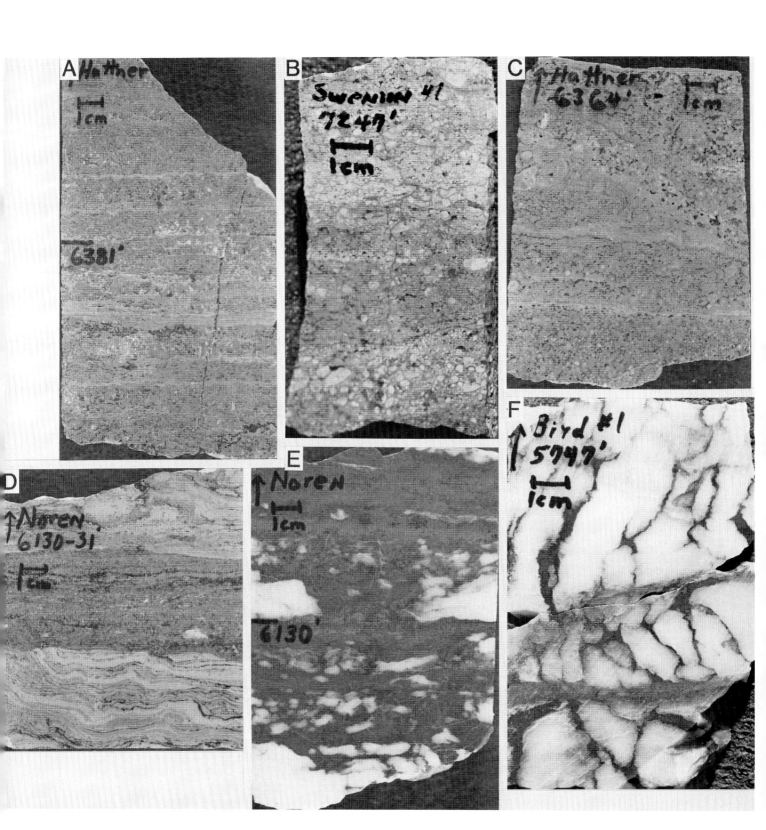

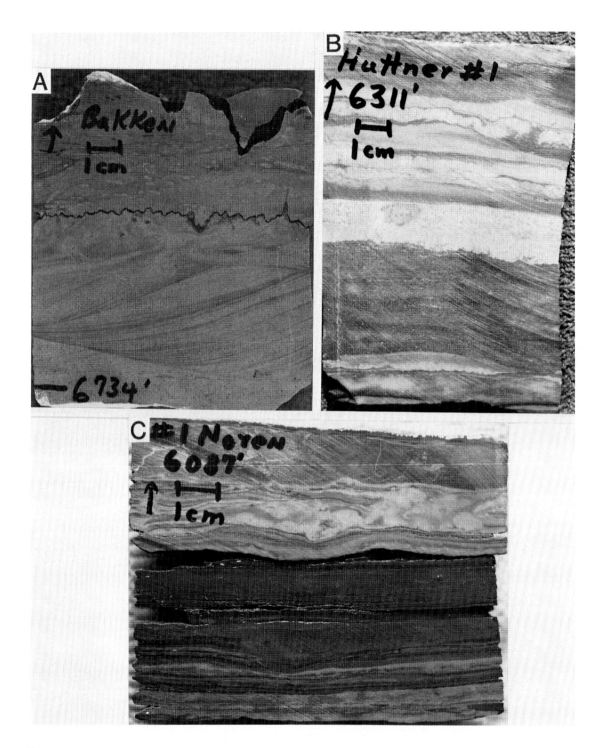

Figure 15. Representative cores of the upper Midale in North and South Black Slough and Lignite fields. (A) Anschutz Bakken #1, 6734 feet, South Black Slough Field, displaying a small trough cross-laminated ripple. (B) Pan American Huttner #1, 6311 feet, North Black Slough Field, illustrating original laminated to very thin bedded carbonate that is now completely replaced by anhydrite. (C) Mobile Noren #1, 6087 feet, Lignite Field, displaying interbedded black fissile shale capped by laminated to very thin bedded anhydrite. Scales are 1 cm.

DIAGENESIS

Rival

Several diagenetic changes occurred in porous and productive Rival beds (Table 1). Most of these were porosity destructive, but two created secondary porosity.

Early diagenetic events include: (1) micritization and aggrading neomorphism, (2) calcite cementation, and (3) anhydrite cementation. Later diagenetic events include: (1) partial leaching of anhydrite and solution enlargement of pores, (2) dolomite cementation, and (3) calcite cementation and calcitization of other cements. All diagenetic events are present in each field studied, but are not of equal importance.

Partial to complete micritization of some ooids, pisolites and intraclasts is common (Figs. 16-20). These underwent aggrading neomorphism to form microspar. Original pore space was sparsely cemented by calcite which appear as small acicular and blocky calcite crystals. Major early cementation was by anhydrite. Anhydrite porphyroblasts also formed in original pore space and replaced nearby matrix and particles. In places these porphyroblasts are poikilotopic (Fig. 16).

Later diagenetic modifications include leaching of anhydrite cement, with additional leaching, enlarging, and smoothing of original blocky-shaped pores (Table 1). Leaching of anhydrite porphyroblasts may give a solution-enlarged appearance to a pore when the pore is truly a moldic pore. Later cements partially filled these secondary moldic pores. Dolomite is a major late cement, particularly in North and South

RIVAL "NESSON" POROSITY DIAGENESIS

RIVAL	N. BLACK SLOUGH	S. BLACK SLOUGH	FOOTHILLS
ORIGINAL $\emptyset$	ORIGINAL $\emptyset$	ORIGINAL $\emptyset$	ORIGINAL $\emptyset$
		SI CALCITE CEMENT	
MICRITIZATION & NEOMORPHISM$^+$	MICRITIZATION & NEOMORPHISM$^+$	MICRITIZATION & NEOMORPHISM$^+$	MICRITIZATION & NEOMORPHISM$^+$
*^{1}ANHYDRITE CEMENT	*^{1}ANHYDRITE CEMENT	*^{1}ANHYDRITE CEMENT	*^{1}PARTIAL ANHYDRITE CEMENT
$^\emptyset$LEACHING & SOLUTION ENLARGEMENT	$^\emptyset$LEACHING & SOLUTION ENLARGEMENT	$^\emptyset$LEACHING	$^\emptyset$PROBABLE LEACHING
	CALCITE CEMENT		
*^{2}DOLOMITE CEMENT	*^{2}DOLOMITE CEMENT	*^{2}DOLOMITE CEMENT	TRACE DOLOMITE CEMENT
*^{2}CALCITE CEMENT	CALCITE CEMENT	CALCITE CEMENT	*^{2}CALCITE CEMENT

*^{1} MAJOR EARLY CEMENTATION
*^{2} MAJOR LATER CEMENTATION
$\emptyset$ POROSITY GENERATION
+ WITHIN OOIDS, PISOLITES & INTRACLASTS

Table 1. Rival "Nesson" porosity diagenesis in Rival, North and South Black Slough, and Foothills fields.

Black Slough fields. Late calcite cement is also common, particularly in Rival and Foothills fields (Figs. 16-20). Two types of dolomite are present. The first is dolomite that replaced anhydrite porphyroblasts. The second type of dolomite is saddle dolomite cement (Figs. 17 and 18). Calcite is another late cement. Equant calcite partially fills some moldic pores and replaces some unleached anhydrite cement and some dolomite cement (calcitization). Dolomite and calcite cements appear to have formed at about the same time. Most thin sections hint that calcite followed dolomite but a few locations show the opposite. No major diagenetic changes followed dolomite and calcite, probably because hydrocarbons were then trapped in the reservoirs.

Physical compaction of particles and muddy matrix and resulting stylolitization also effected the section. Specific timing of the various later cements to stylolitization was not attempted. Elliott (1982) showed that late dolomite cementation in the Mission Canyon Formation (Glenburn "beds") at Haas field post-dated stylolitization. He also concluded, based on petrography and oxygen isotopes, that calcite cementation followed dolomite cementation and that oil migration and entrapment in the field postdated the calcite. There is a striking similarity between the types of diagenetic porosity modifications found in this study and those Elliott (1982) reported, particularly because these two different stratigraphic intervals within the Madison Group are separated by more than 50 miles (80 km).

Figure 16. Photomicrograph of porous, productive Rival limestone from the Skelly Scott #1, 6180 feet, Rival Field. Sparsely skeletal, oolitic, pisolitic and intraclastic packstone is partially cemented by anhydrite (A), calcite (C), and some dolomite (D). Anhydrite cement (porphyroblasts) was partially leached producing moldic porosity (M). Some moldic pores (M) are solution enlarged. Some particles were micritized and all have undergone aggrading neomorphism. Up is to the left (arrow). Plain light.

Figure 17. Photomicrograph of porous, productive Rival limestone from the Pan American Huttner #1, 6377 feet, North Black Slough Field. Skeletal, intraclastic packstone serves as the reservoir. Anhydrite cement (A) (porphyroblasts) is nearly completely leached. Moldic pores (M) are solution enlarged and partially filled with dolomite (D) and some calcite (C) cement. Some anhydrite cement is partially replaced by calcite (A-C left of center). Photomicrograph is oriented up (arrow). Plain light.

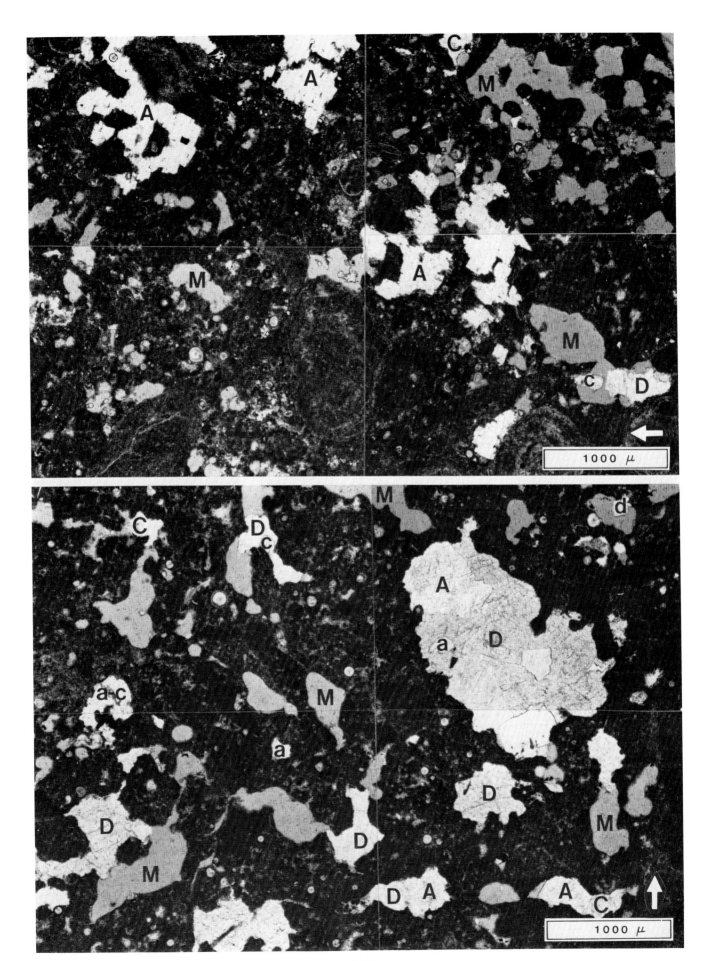

Figure 18. Photomicrograph of porous, productive Rival limestone from the Anschutz Bakken #1, 6764 feet, South Black Slough Field. Sparsely skeletal, oolitic, pisolitic, intraclastic packstone serves as the reservoir. Anhydrite cement (A) is partially leached away. Moldic and interparticle pores (M) are partially filled with calcite (C) and dolomite (D) cement. Compaction has crushed a micritized intraclast just beneath the center of the photomicrograph. Plain light.

Figure 19. Photomicrograph of porous, productive Rival limestone from the Anschutz Swenson #1, 7210 feet, Foothills Field. Skeletal, intraclastic packstone serves as the reservoir. Interparticle and intraparticle pores are common. Calcite cement (C) has partially cemented some pores and may have replaced some anhydrite. Photomicrograph is oriented up (arrow). Plain light.

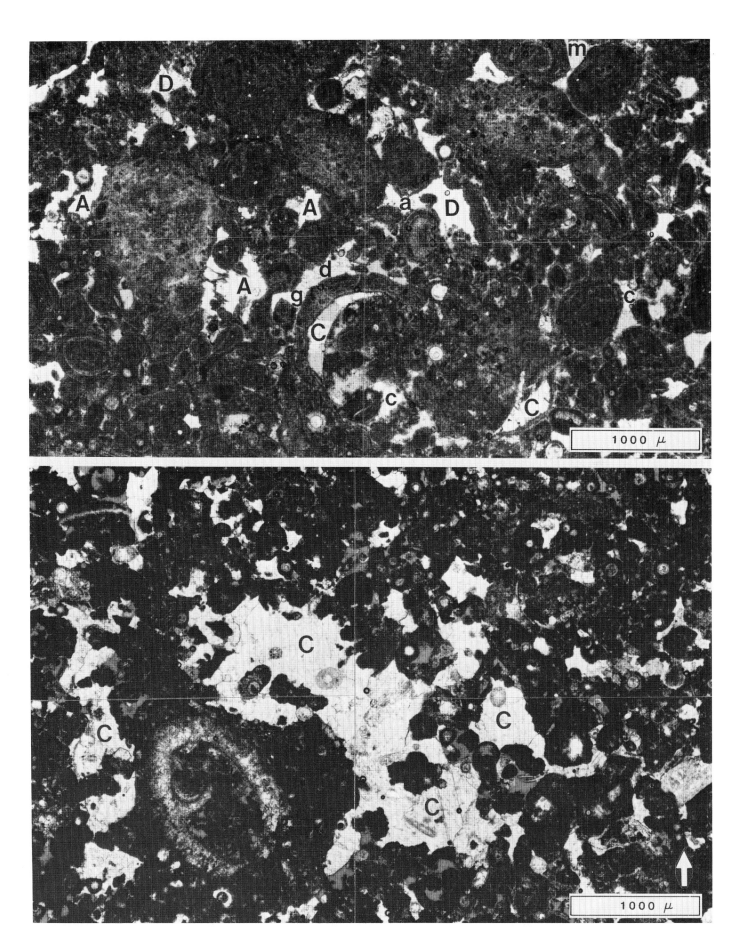

Figure 20. Scanning-electron micrographs (SEM) from porous and productive Rival intervals. A and B from Skelly Scott #1, 6180 feet, Rival Field. (A) Moldic pores partially filled with calcite cement. (B) Relief pore cast of pore and throat system. Small calcite crystals infilling pores have drastically reduced pore and pore throat configurations and significantly reduced permeability. C and D from Pan American Huttner #1, 6377 feet, North Black Slough Field. (C) Moldic pores slightly filled with calcite cement. (D) Relief pore cast where a pore is significantly reduced in size and contains drastically lowered permeability. E and F from Skelly Scott #1, 6180 feet, Rival Field. Both illustrate particles that have been micritized and neomorphosed to a microspar texture and contain intercrystal pores.

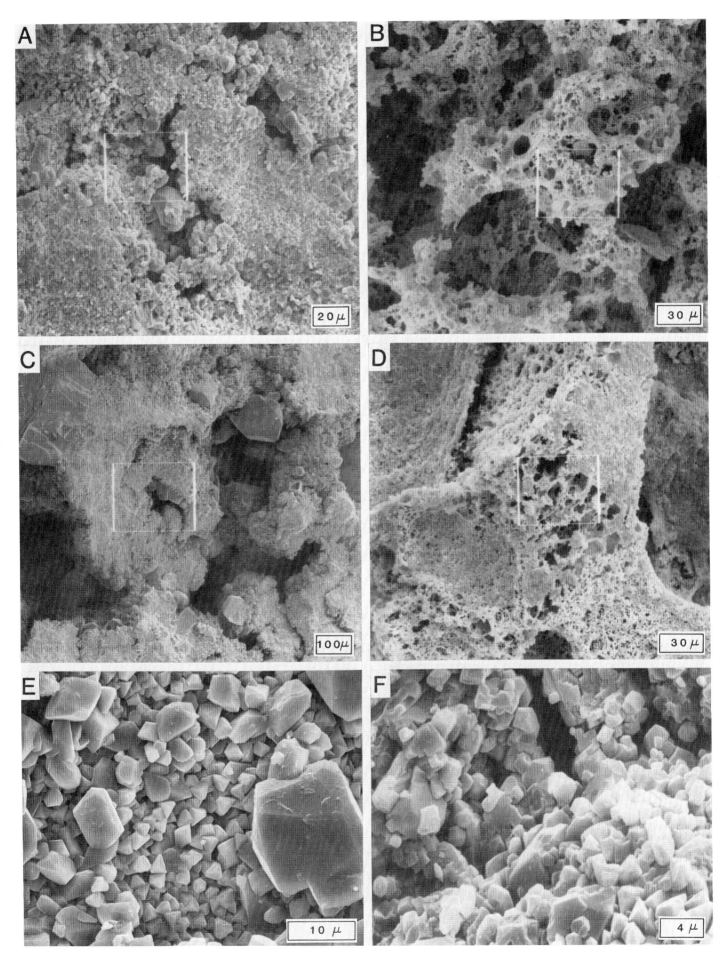

Midale

At least three types of porosity generation and three stages of porosity destruction have been documented in the Midale at Lignite Field (Table 2). These events are referred to as early or late, with no specific time implied. Early diagenetic events include anhydrite replacement, and dolomitization. Later diagenetic events are: (1) leaching of silica, (2) partial leaching of anhydrite cement, (3) saddle dolomite cementation, and (4) pyrite replacement. Timing of later events is difficult to interpret. Some may have occurred just before oil emplacement, which presumably occurred in the Cretaceous (Dow, 1974; Williams, 1974). Others may have been much earlier when subcrop edges of the Williston Basin were eroding and being covered in the Triassic.

Early small blocky anhydrite prophyroblasts are disseminated through the interval and replaced original pore space, pellets and matrix. Dolomitization of spiculitic, pelletal packstone created the reservoir (Fig. 21). Porous and permeable reservoir beds are those that were completely dolomitized (Mitchell and Petter, 1958). Outlines of pellets are preserved in a few places in spite of complete dolomitization. Dolomite crystals are euhedral and average 3 to 10 microns in width. Not all of the Midale was dolomitized. This may be due to a low volume of dolomitizing brines, or perhaps dolomitization was facies-specific, replacing pelleted lime mud facies preferentially.

Leaching of silica and anhydrite probably occurred at two different times. Sponge spicule monaxons were completely leached to produce rod-shaped pores. These spicules were probably leached while still in a hydrated opaline state. Leaching of anhydrite was incomplete with most crystals unaffected.

Late cements are dolomite and pyrite (Table 2). Sparse saddle dolomite filled some moldic pores. Pyrite has also formed in a few pores, and is either a cement or replaced an earlier cement.

PORE AND PORE THROAT SYSTEM

Rival

Leaching of anhydrite created most of the porosity in the Rival (Figs. 16-20). Moldic pores which were solution enlarged are the most common pore type and range in size from 10 to 1000 microns (Table 3). These pores are bimodal in their size distribution. Large moldic pores average 100 to 200 microns in width, while smaller moldic pores average 25 microns in width.

Some interparticle pores and sparse intraparticle pores are also present. These are significant only in Foothills Field where intraclasts are the most common rock-building constituent (Fig. 19). Interparticle pores are 5 to 20 microns in width. Intraparticle pores in centers of calcispheres and small ostracods are 5 to 10 microns in width and form the smallest petrographically visible pores.

Much smaller intercrystal pores occur within ooids, pisolites and possible intraclasts and can only be viewed by scanning-electron microscopy (SEM) (Fig. 20). These pores are less than 1 micron to 3 microns in width. They occur in micritized particles that have undergone aggrading neomorphism, forming porous aphanocrystalline to very finely crystalline microspar textures.

MIDALE POROSITY DIAGENESIS

LIGNITE

ORIGINAL ∅
*¹ANHYDRITE CEMENT
∅ DOLOMITIZATION
LEACHING OF SILICA
LEACHING OF ANHYDRITE
DOLOMITE CEMENT
PYRITE

*¹MAJOR EARLY CEMENTATION
∅ POROSITY GENERATION

Table 2.   Midale porosity diagenesis in Lignite Field.

RIVAL "NESSON" PORE SYSTEM

	RIVAL	N. BLACK SLOUGH	S. BLACK SLOUGH	FOOTHILLS
	MOLDIC & SOLUTION ENLARGED INTERCRYSTAL[+]	MOLDIC & SOLUTION ENLARGED INTERCRYSTAL[+]	MOLDIC & SPARSE INTERPARTICLE INTERCRYSTAL[+]	MOLDIC INTERPARTICLE INTERCRYSTAL[+] SPARSE INTRAPARTICLE
AVERAGE ∅	9%	9%	7%	7%
AVERAGE K	1md.	10md.	2md.	7md.

[+] WITHIN MICRITIZED & NEOMORPHOSED OOIDS, PISOLITES & INTRACLASTS

Table 3.   Rival "Nesson" pore system in Rival, North and South Black Slough, and Foothills fields. Each fields average porosity and permeability are listed below.

MIDALE PORE SYSTEM

	LIGNITE
	MOLDIC INTERCRYSTAL
AVERAGE ∅	11%
AVERAGE K	1md.

Table 4.   Midale pore system in Lignite Field. The fields average porosity and permeability is listed below.

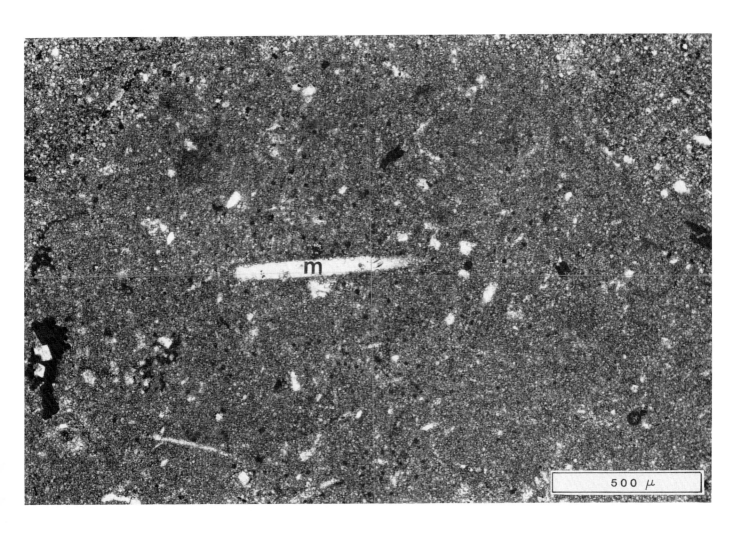

Figure 21. Photomicrograph of porous, productive Midale dolostone from the Mobile Noren #1, 6095 feet, Lignite Field. Very finely crystalline to slightly aphanocrystalline dolostone, originally sparsely anhydritic, spiculitic, pelletal wackestone/packstone is the reservoir. Porosity is moldic (M) from leaching sponge spicules, and intercrystalline, which cannot be seen at this magnification. Plain light.

Dolomite and calcite cements in some pores have reduced porosity and drastically reduced permeability (Figs. 16-20). Cements occur as large single crystals or as smaller crystals which dissected the original larger pore. In the latter case the resulting pore system resembles a spider web with the holes between the web being the cement crystals (Fig. 20). This heterogeneity within the pore system is one reason why waterflood operations in Rival porous intervals were not successful in banking much oil to producing wells. However, the pore system provided adequate communication for bottom-hole pressure maintenance.

Midale

The Midale pore system at Lignite field is dominantly intercrystalline (Table 4 and Figs. 21 and 22). Additional porosity was provided by leaching sponge spicule monaxons and anhydrite porphyroblasts. There are also a few original interparticle pores.

Dolomite crystals are aphanocrystalline to very finely crystalline (Fig. 22). Intercrystalline pores are small because individual dolomite crystals are only 3 to 10 microns in width. Moldic pores formed by leaching of sponge spicules average 25 to 60 microns in length and 5 to 10 microns in width. The longest spicule molds are up to 500 microns in length. Moldic pores created by leaching anhydrite porphyroblasts are 12 to 50 micron rectangular to equant pores.

Intercrystalline dolomite pores are composed of polyhedral, tetrahedral and interboundary-sheet pores (Wardlaw, 1976; Wardlaw and Taylor, 1976), and all are three-dimensionally interconnected (Fig. 22). Polyhedral pores are many-sided complex pores 3 to 10 microns in width. Tetrahedral pores are pyramid-shaped pores approximately 3

Figure 22. Scanning-electron micrographs (SEM) of porous, productive Midale dolostone from Mobile Noren #1, 6095 feet, Lignite Field. (A) Oblique-transverse view of a small moldic pore produced by leaching a sponge spicule. Note the surrounding dolomite intercrystalline pores. Note the very finely crystalline (20 to 4 microns) to slightly aphanocrystalline (4 microns) dolomite crystal sizes. Scale is 5 microns. (B) relief pore cast of various dolomite intercrystalline pore types, namely: polyhedral (P), tetrahedral (T), and interboundary-sheet (I) pores.

microns in width. Interboundary-sheet pores form sheets between adjacent crystals. These pores also act as pore throats and appear to be only 1 micron, or less, in width.

The dolomite matrix connects moldic and intercrystalline pores three dimensionally. Assuming the average pore throat size is 1 micron in width, pore to throat ratios for moldic pores are large and range from 500 to 5. Pore to throat ratios for intercrystalline pores are: 10 to 3 for polyhedral pores, approximately 3 for tetrahedral pores, and 1 for interboundary-sheet pores. The coordination number, which is the average number of throats connecting each pore, appears to be 3 to 5 in two-dimensional view.

GEOLOGIC IMPLICATIONS TO RESERVOIR PRODUCTION

Rival

In the fields studied, porous productive beds are distributed in a lenticular manner through each field, with interfingering less porous and tight beds. No continuously porous beds extend for any appreciable distance through individual fields. These heterogeneities formed by: (1) original deposition of laminated to very thin beds, which are not continuous; (2) incomplete anhydrite leaching; and (3) partial cementation of the pore system by dolomite and calcite. The result was a reservoir only partly able to react to secondary recovery operations. Reservoir studies also revealed that in all fields, except South Black Slough, the total amount of voidage created by producing wells was not replaced by water injection. Additional recovery of hydrocarbons did occur, but was not significant enough on field decline curves, except in Rival field, to show much incremental recovery.

To compare the effect of secondary (waterflood) recovery operations on a field by field basis, two sets of decline curves were prepared for each field. The first set was extrapolated to field abandonment using only production data prior to unitization and water injection. The second set included all production history after unitization. Comparison of the curves revealed incremental additional recovery of hydrocarbons in each field, but these increases were not, except in Rival field, significant. Thus, waterflooding did not generate much additional net revenue and the operations were considered failures. Pressure maintenance was achieved but significant banking of oil from injection to production wells was not effective.

CONCLUSIONS

Rival limestone and anhydrite (Frobisher Anhydrite) beds were deposited along a paleo-shoreline as barrier island to intertidal buildups which graded eastward into lagoonal, tidal flat and coastal sabkha settings. Within barrier island to intertidal buildups porous productive beds are composed of slightly skeletal to skeletal, intraclastic, oolitic to pisolitic packstone. These beds were cemented by some calcite and much anhydrite. Later the anhydrite was partially leached to produce reservoir porosity. Inner portions of the shoreline were cemented and not leached and combined with anhydrite beds to provide the updip seal.

The pore system is composed of moldic and solution enlarged pores. Ooids, pisolites and intraclasts were micritized and neomorphosed to microspar and contain intercrystal porosity.

The study area is regionally tilted at 2/3° toward the center of the basin. This provides the proper structural setting to form combination traps along the Rival paleo-shoreline. Small anticlinal noses, which trend perpendicular to the shoreline, provide additional assistance to structural portions of the trap.

The overlying Midale was deposited in a low-energy restricted marine to tidal flat setting, which graded upward into thin shales and evaporites deposited in lagoonal, tidal flat and coastal sabkha settings. Porous, productive beds are composed of dolomitized, slightly anhydritic, spiculitic, pelletal wackestone/ packstone. These beds were slightly cemented by anhydrite. Sponge spicule monaxons were completely leached to form rod-shaped moldic pores. Some anhydrite was leached to produce additional rectangular moldic pores. The originally pelleted mud was dolomitized to produce intercrystalline porosity which is composed of polyhedral, tetrahedral, and interboundary-sheet pores. The reservoir is a small anticlinal nose with stratigraphic trapping accomplished by an updip loss of effective porosity.

ACKNOWLEDGEMENTS

Thanks is given to Gulf Oil Corporation for granting permission to publish this paper. Mark W. Longman, Keith Shanley, Charles A. Ross and Ralph G. Stevenson provided excellent critical reviews of this manuscript. Their comments are great-ly appreciated. Yvan J. Beausoleil and Michael A. Cabirac were of great assistance in locating, cleaning, helping describe, sampling, and photographing cores from each field. Sidney A. Anderson, Randy B. Burke, and Rod A. Stoa of the North Dakota Geological Survey and Wes Norton and Clarence G. Carlson of the North Dakota Industrial Commission are thanked. They graciously provided space in the Wilson M. Laird Core and Sample Library to describe cores, obtain copies of logs and scout tickets in the survey offices and review unitization hearings and maps at the industrial commission. Most importantly, they freely gave of their time and technical understanding of Williston Basin geology. Michael Berry provided additional photography of core samples. Laurie English, David Morris, and John O'Brien drafted all illustrations.

REFERENCES

Anderson, S. B., Hansen, D. E., and Eastwood, W. P., 1960, Oil fields in the Burke County area, North Dakota: North Dakota Geol. Survey Rpt. Inv. No. 36, 71 p.

Butler, G. P., Harris, P. M., and Kendall, C. G. St. C., 1982, Recent evaporites from the Abu Dhabi coastal flats: in Handford, C. R., Loucks, R. G., and Davies, G. R. (eds.), Depositional and Diagenetic Spectra of Evaporites: Soc. Econ. Paleont. Mineral. Core Workshop No. 3, p. 33-64.

Carlson, C. G., and Anderson, S. B., 1965, Sedimentary and tectonic history of North Dakota part of Williston Basin: Am. Assoc. Petrol. Geol. Bull., v. 49, p. 1833-1846.

Dow, W. G., 1974, Application of oil-correlation and source-rock data to exploration in Williston Basin: Am. Assoc. Petrol. Geol. Bull., v. 58, p. 1253-1262.

Elliott, T. L., 1982, Carbonate facies, depositional cycles, and the development of secondary porosity during burial diagenesis: Mission Canyon Formation, Haas Field, North Dakota: in Christopher, J. E., and Kaldi, J. (eds.), Proc. Fourth Int'l Williston Basin Symposium, Regina, p. 131-151.

Gerhard, L. C., 1982, Geological evolution and energy resources of the Williston Basin: Univ. Missouri at Rolla Jour., v. 3, p. 83-120.

Gerhard, L. C., Anderson, S. B., and Berg, J., 1978, Mission Canyon porosity development, Glenburn Field, North Dakota Williston Basin: in Estelle, D., and Miller, R. (eds.), 1978 Williston Basin Symposium, Montana Geol. Soc. 24th Annual Conf., p. 177-188.

Gerhard, L. C., Anderson, S. B., Lefever, J. A., and Carlson, C. G., 1982, Geological development, origin and energy mineral resources of Williston Basin, North Dakota: Am. Assoc. Petrol. Geol. Bull, v. 66, n. 8, p. 989-1020.

Habicht, J. K. A., 1979, Paleoclimate, paleomagnetism and continental drift: Am. Assoc. Petrol. Geol. Studies in Geol., n. 9, 31 p.

Harris, S. H., Land, C. B. Jr., and McKeever, J. H., 1966, Relation of Mission Canyon stratigraphy to oil production in north-central North Dakota: Am. Assoc. Petrol. Geol. Bull., v. 50, n. 10, p. 2269-76.

Hartling, A., 1983, Early dolomitization: Its significance in creating subtle diagenetic hydrocarbon traps in Williston Basin (abs.): Am. Assoc. Petrol. Geol. Bull., v. 67, n. 8, p. 1341.

Kendall, C. G. St. C., and Skipwith, P. A. d'E., 1968, Recent algal mats of a Persian Gulf lagoon: Jour. Sed. Petrol., v. 38, p. 1040-1058.

Lindsay, R. F., 1985, Carbonate and evaporite facies, dolomitization and reservoir distribution of the Mission Canyon Formation, Little Knife Field, North Dakota: in Peterson, J. A. and others (eds.), Williston Basin a Cratonic Oil and Gas Province: Am. Assoc., Petrol. Geol. Memoir.

Lindsay, R. F., and Roth, M. S., 1982, Carbonate and evaporite facies, dolomitization and reservoir distribution of the Mission Canyon Formation, Little Knife Field, North Dakato: in Christopher, J. E., and Kaldi, J. (eds.), Proc. Fourth Int'l Williston Basin Symposium, Regina, p. 153-179.

Mescher, P. K., and Pol, J. C., 1984, Environmental significance of pisoliths, Mississippian Madison Formation of the Williston Basin, Bottineau, Renville, and McHenry Counties, North Dakota (abs.): Am. Assoc. Petrol. Geol. Bull., v. 68, n. 4, p. 507.

Mitchell, P. M., and Petter, C. K., 1958, The Lignite field, Burke County, North Dakota: Second Williston Basin Symposium, Saskatchewan Geol. Soc. and North Dakota Geol. Soc., p. 64-69.

Nordquist, J. W., 1955, Pre-Rierdon Jurassic stratigraphy in northern Montana and Williston Basin: Billings Geol. Soc. Guidebook, 6th Ann. Conf., p. 104-106.

Peterson, J. A., 1985, Subsurface strigraphy and depositional history of the Madison Limestone (Mississippian), Williston Basin, U.S., and adjacent area: in Peterson, J. A. and others (eds), Williston

Basin a Cratonic Oil and Gas Province: Am. Assoc. Petrol. Geol. Memoir.

Proctor, R. M., and Macauley, G., 1968, Mississippian of western Canada and Williston Basin: Am. Assoc. Petrol. Geol. Bull., v. 52, p. 1956-1968.

Schreiber, B. C., Roth, M. S., and Helman, M. L., 1982, Recognition of primary facies characteristics of evaporites and the differentiation of these forms from diagenetic overprints: in Handford, C. R., Loucks, R. G., and Davies, G. R. (eds), Depositional and Diagenetic Spectra of Evaporites: Soc. Econ. Paleont. Mineral. Core Workshop No. 3, p. 1-32.

Thomas, G. E., 1974, Lineament-block tectonics: Williston-Blood Creek Basin: Am. Assoc. Petrol. Geol. Bull., v. 58, p. 1305-1322.

Tyler, C. D., and George, R. S. (eds), 1962, Oil and gas fields of North Dakota-a symposium: North Dakota Geol. Soc. 227 p.

Wardlaw, N. C., 1976, Pore geometry of carbonate rocks as revealed by pore casts and capillary pressure: Am. Assoc. Petrol. Geol. Bull., v. 60, n. 2, p. 245-257.

Wardlaw, N. C., and Taylor, R. P., 1976, Mercury capillary pressure curves and the interpretation of pore structure and capillary behavior in reservoir rocks: Canadian Petr. Geol. Bull, v. 24, n. 2, p. 225-262.

Williams, J. A., 1974, Characterization of oil types in Williston Basin: Am. Assoc. Petrol. Geol. Bull., v. 58, p. 1243-1252.

Wittstrom, M. D. Jr., and Hagemeier, M. E., 1978, A review of Little Knife field development, North Dakota: Williston Basin Symposium, Montana Geol. Soc., p. 361-368.

Wittstrom, M. D. Jr., and Hagemeier, M. E., 1979, A review of Little Knife field development: Oil and Gas Jour., v. 77, n. 6, p. 86-92.

DEPOSITION AND DIAGENESIS OF THE MISSISSIPPIAN CHARLES "C" (RATCLIFFE) RESERVOIR IN LUSTRE FIELD, VALLEY COUNTY, MONTANA

Mark W. Longman
Consulting Geologist
701 Harlan, #E69
Lakewood, Colorado 80214

Kenneth H. Schmidtman
Santa Fe Energy Company
2600 Security Life Bldg.
Denver, Colorado 80202

ABSTRACT

Lustre Field, the westernmost commercial Mississippian field in the Williston Basin, produces mainly from dolomites in the Charles Formation. Stratigraphic terminology is still debated, but the major reservoir occurs in the Charles "C" or Ratcliffe Zone just above the Richey Shale.

Depositional facies indicate that the Charles "C" interval represents a generally shallowing-upward depositional sequence. Fossiliferous open shelf facies predominate near the base, and grade upward through restricted shelf mudstones into peloidal, oolitic, and oncolitic grainstones near the middle of the zone. The upper part of the Charles "C" consists of anhydritic, laminated, dolomitic mudstones deposited in a sabkha to hypersaline lagoon environment.

The major shallowing upward cycle can be subdivided into five smaller cycles separated by minor periods of transgression. Each of these subcycles is incomplete, i.e., only a few of the facies occur in each, but isopach maps and facies distribution within these subcycles suggest that a northwest-southeast trending structural nose extended through the Lustre Field area and exerted an influence on deposition.

The depositional facies can be traced far beyond the field's productive limits, but the degree of dolomitization in the lower part of the Charles "C" correlates directly with production. The best wells in the field contain up to 35 feet (10 m) of dolomite with more than 14% intercrystalline porosity and 0.5 md permeability. In the best reservoir zones, porosity reaches 30% and permeability (in unfractured rocks) is typically 1 to 10 md. Sparse vertical fractures enhance production.

The porous dolomites formed when brines generated in the sabkha to hypersaline lagoonal facies of the upper Charles "C" seeped down into the burrowed mudstones and wackestones of the lower Charles "C". Subtle paleotopography, permeability of the lower Charles "C" rocks, and original composition of the Charles sediments interacted to determine porosity development.

INTRODUCTION

Lustre Field, discovered in 1982 by Exxon Corporation, lies on the west flank of the Williston Basin (Fig. 1). The field represents the westernmost commercial Mississippian production in the basin and produced 1.3 million barrels of oil from 55 wells in its first two years on production. Decline curves for the wells are highly variable, but it is estimated that ultimate recovery in the field will exceed four million barrels of oil.

The discovery well, the #1 Tieszen-Toews (Section 3, T30N-R44E), flowed 254 BOPD with no water from the Mississippian Charles "C" or Ratcliffe Zone at a depth of about 5,750 feet (1.75 km). Stratigraphic terminology is still controversial, but two of the most widely used sets of terms are shown in Figure 2. After 18 months on production, the discovery well was still producing 2,300 barrels of oil per month with no water. Decline curve analysis suggests that the well will eventually yield more than 300,000 barrels of oil. Development drilling has resulted in some production from the Charles "D" and Mission Canyon formations of the Madison Group, but the most important producing zone remains the Charles "C". It is this zone that is the focus of this study.

Lustre Field was reportedly found by drilling a seismic high beneath expiring leases. Extensive development drilling has confirmed the presence of a structural nose in the field area, with subtle structural closures.

Study of cores and samples from the Charles "C" in 24 wells (7 producing, 17 dry holes) in and around Lustre Field indicates that the

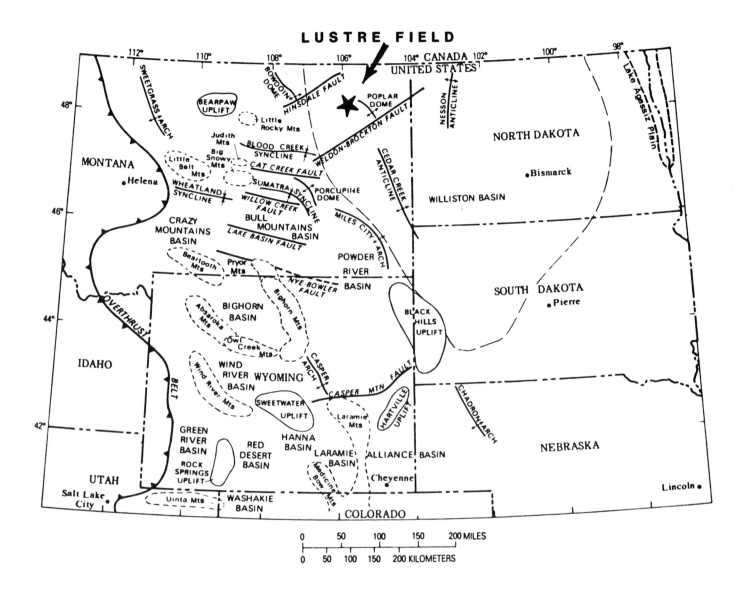

Figure 1. Present day structural and physiographic features of the Northern Great Plains. The location of Lustre Field is indicated west of Poplar Dome. From Peterson, 1984, p. A5.

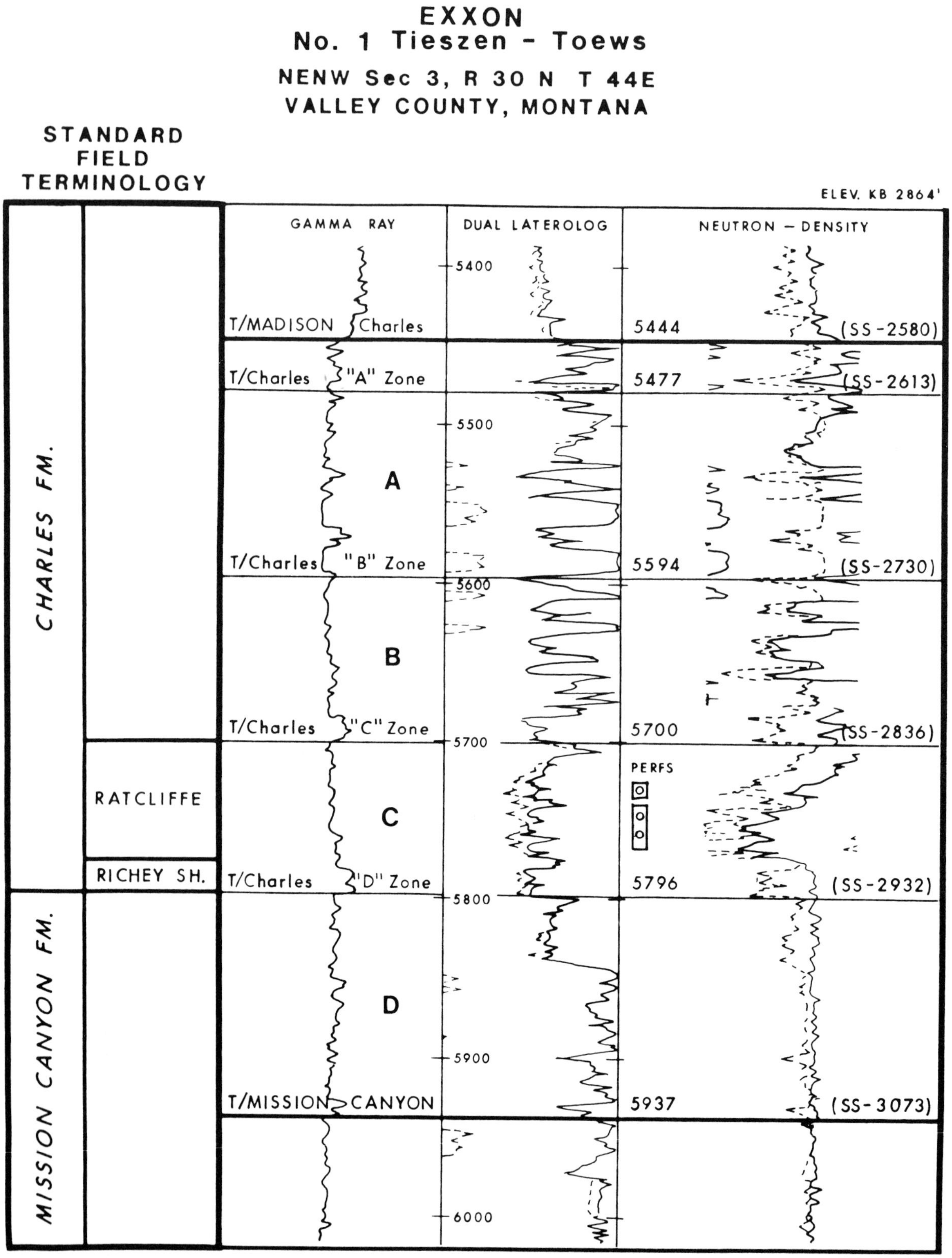

Figure 2. Logs from the Lustre Field Discovery well showing the stratigraphic terminology for the Mississippian section and the traditional terminology followed by many in the area. The producing zone (shown by the perforations) is within the Charles "C" or Ratcliffe Zone.

producing zone is part of a major shallowing-upward depositional sequence. Individual facies from a variety of depositional environments are widespread and can be correlated throughout the field area. Even the reservoir facies, a burrowed and dolomitized mudstone to wackestone, can be traced far beyond the field's limits. Thus, depositional facies appear to exert only a secondary influence on production.

The primary control on production is degree of dolomitization of the burrowed mudstone to wackestone facies. The best reservoir occur in finely crystalline (30 to 50 micron) dolomites with up to 30% porosity and several millidarcies permeability. Vertical fractures enhance production locally.

PREVIOUS WORK AND REGIONAL SETTING

Lustre field lies about 35 miles west of Poplar Dome, one of several positive paleogeographic features on the western flank of the Williston Basin (Fig. 1). Other important paleogeographic features were the Blood Creek Syncline to the south and the Bowdin Dome to the west. However, more subtle paleogeographic structures such as the Wolf Creek Nose controlled deposition in the immediate Lustre Field area. Position of the Wolf Creek Nose is revealed in a structure map of the Charles "C" (Fig. 3).

The Charles "C" in Lustre Field differs somewhat from many of the Mississippian fields in the basin in that the reservoir consists of rather low energy sediments that have been dolomitized rather than pisolitic/oolitic grainstones with preserved primary porosity. How-

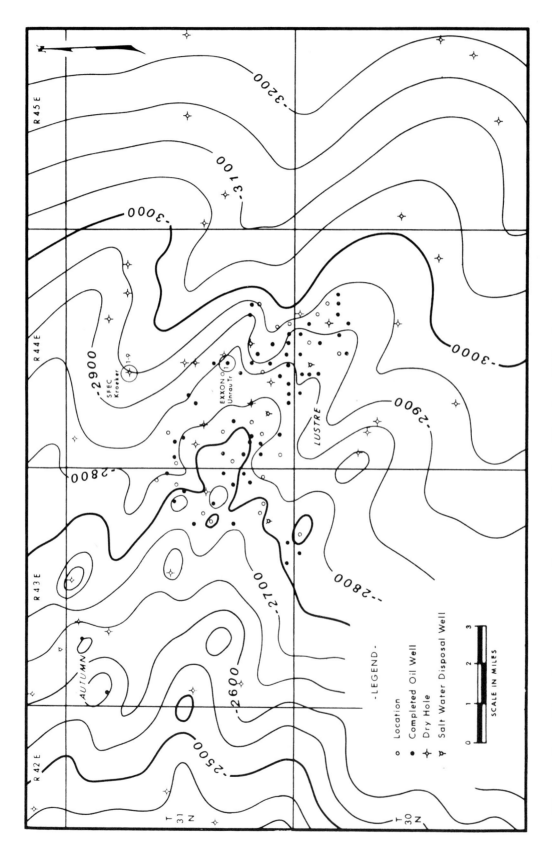

Figure 3. Structure map on the top of the Charles "C" showing position Lustre Field on the Wolf Creek Nose. Locations of the Exxon #1 Unrau Trust and Santa Fe #1-9 Kroeker wells which provided the cores shown in the workshop are also shown.

ever, it is similar to some of the Midale fields in Saskatchewan (Edie, 1958; Kaldi, 1982), the Oungre (Ratcliffe) beds of southern Saskatchewan (Hartling and others, 1982), the Mission Canyon in Little Knife Field (Lindsay and Roth, 1982; Lindsay, 1985) and the Mission Canyon in the Billings Nose area (Altschuld and Kerr, 1982).

DEPOSITIONAL CYCLES IN THE RATCLIFFE

Five depositional cycles can be recognized in cores (and logs) of the Charles "C" from the Lustre Field area (Fig. 4). Each cycle consists of part of an upward-shoaling sequence. Deeper-water open shelf facies dominate in the lower cycles whereas sabkha to hypersaline lagoonal deposits dominate in the upper cycles. Characteristics of each cycle are described below.

Cycle 1

Lithology

This cycle consists of the Richey Shale and the lower 15 to 20 feet (4.5 to 6m) of the Charles "C" carbonate. The top of the cycle is marked by a black, shaly wackestone that yields a moderate gamma ray kick (about 50 API Units or half that of the Richey Shale) on logs. Total thickness of the cycle ranges from a thin of 26 feet (8 m) in the Autumn Field area to the northwest of Lustre Field (Fig. 3) to 46 feet (14 m) south of the field. An elongate isopach thin trends from Autumn Field southeastward through Lustre Field and indicates the position and size of the subtle Wolf Creek Nose during deposition.

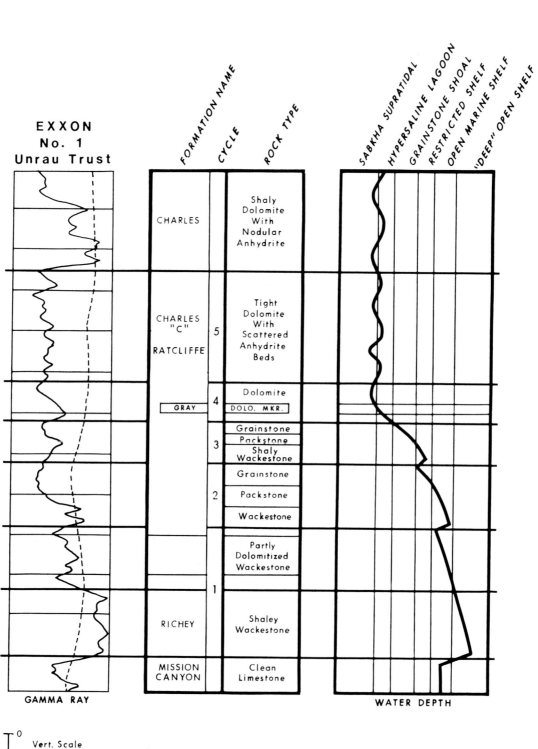

Figure 4. Gamma ray and caliper logs for the Exxon #1 Unrau Trust well showing the distribution of the depositional cycles. Rock type, relative water depth, and inferred depositional environments are also indicated.

No significant facies changes occur in cycle 1 across this nose, however.

Rocks types in Cycle 1 (Figs. 5A, 5B), in ascending order, are: (1) the Richey "Shale," (2) fossiliferous wackestones and packstones containing common wispy stylolites (Figs. 5C, 5D), (3) partially dolomitized wackestones to packstones (Fig. 6A), and (4) burrowed dolomitic mudstone to wackestone commonly containing more than 14% porosity (Figs. 6B, 6C, 6D). This latter facies is one of the most important producing horizons in the field, but it can be traced far beyond the producing area.

Color mottling in this reservoir facies reveals a burrowed fabric (Fig. 6B). Other fossils are less common, although fragments of ostracods, echinoderms, and brachiopods are present in thin sections from the tighter dolomites. Few fossils are visible in the most porous dolomites, but this is almost certainly a function of the dolomitization rather than an absence during deposition.

The burrows which characterize this unit range in diameter from a millimeter or so to about a centimeter (Fig. 6C). Most are horizontal or oblique. Ichnogenera include _Thalassinoides_ and _Chondrites_, but a variety of other nondescript burrows are also present.

Thin sections show that dolomite rhombs ranging in size from one to fifty microns dominate (Fig. 6D). Scattered fossil fragments, local coarse anhydrite crystals, sparse vertical fractures, and burrow mottling can also be detected. The dominant form of porosity is intercrystalline, but small vugs, presumably formed by dissolution of grains, can also be found. Permeability varies markedly in the

Figure 5. Rocks of Cycle 1 including the Richey Shale.

    A. Core slab of the Richey Shale showing the laminations and storm beds composed of ostracods and echinoderm fragments. From the Santa Fe #1-9 Kroeker; Depth = 5,961 ft.

    B. Thin section photomicrograph of an ostracod-rich storm bed in the Richey Shale from the core slab shown in "A".

    C. Core slab from just above the contact between the Richey Shale and the overlying open shelf carbonates. Some shale is interbedded within the limestone near the contact. Santa Fe #1-9 Kroeker; Depth = 5,958 ft.

    D. Relatively clean carbonate near the middle of Cycle 1. The coral visible near the bottom of the slab is *Syringopora*. A chert nodule (dark area) is present near the middle of the slab. Santa Fe #1-9 Kroeker; Depth = 5,944 ft.

A

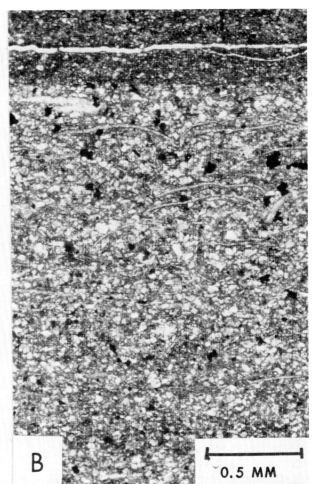

B

C

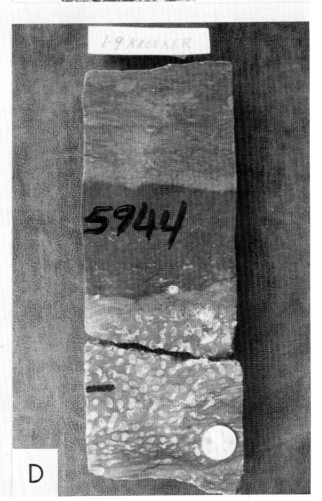

D

Figure 6.   Samples of the porous dolomite at the top of the Cycle 1.

    A.   Thin section photomicrograph of a partly dolomitized open shelf limestone from the upper part of Cycle 1. All of the micrite matrix has been replaced by finely crystalline dolomite, but crinoid and brachiopod fragments are still calcite (although the brachiopod just above the "A" has been silicified). Thin section from the Exxon #1 Unrau Trust core; Depth = 5,796 ft.

    B.   Core slab of the burrowed dolomitic mudstone to wackestone at the top of Cycle 1. This is one of the best reservoir zones in the field. Exxon #1 Unrau Trust; Depth = 5,782 ft.

    C.   Thin section photomicrograph of burrowed porous dolomite from a dry hole about 20 miles south of Lustre Field. An open fracture partly filled with anhydrite is also visible. Santa Fe #1-9 Tribal Federal; Depth = 5,455 ft.

    D.   Thin section photomicrograph showing the intercrystalline porosity that forms the Lustre Field reservoir. Some plugging by anhydrite (the whitest crystals) is also visible. Porosity in this sample is about 22% with a permeability of about 5 md. Exxon #1 Unrau Trust, Depth = 5,783 ft.

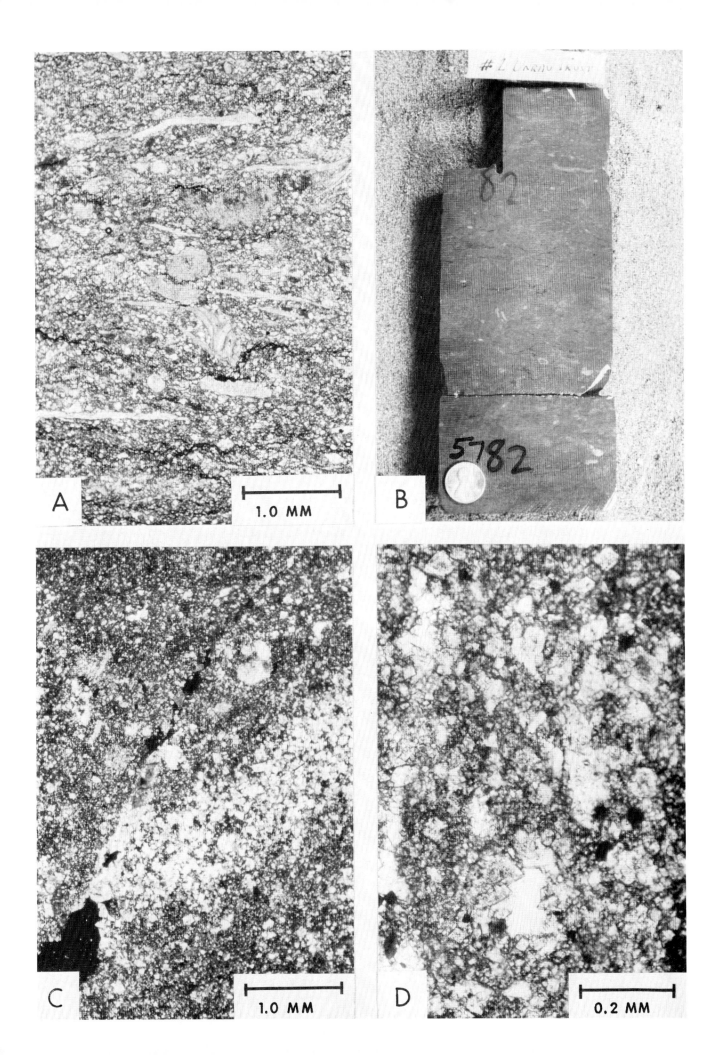

dolomites of this burrowed facies, and thin sections reveal that where permeability is low, small crystal size and incomplete dolomitization are responsible. Where dolomitization has gone to completion, rhombs 30 to 40 microns wide with well-developed intercrystalline pores are present. Typical permeabilities are 0.5 to 7 md.

An isoporosity map of the porous dolomite in Cycle 1 (Fig. 7) indicates that porosity is best developed in the heart of the field. This same area is also extensively dolomitized in the overlying cycles and is the best producing area in the field. The geometry of this porous dolomite suggests that downward-migrating Mg-rich brines generated higher in the Ratcliffe caused dolomite to form. An original abundance of Mg-calcite micrite may have enhanced dolomitization, however, by providing an internal source of magnesium.

Immediately below the burrowed mudstone facies are partly dolomitized wackestones and packstones containing common crinoid fragments, brachiopods, and ostracods. This lithology ranges in thickness from 1 to 8 feet (0.3 to 2.4 m). Burrows are present, although they are not as in obvious as the overlying facies, and wispy stylolites are common. Porosity values up to 12% occur in some of the more dolomitic beds, but permeability seldom exceeds 0.01 md.

Within this unit is a persistent bed of slightly more porous dolomite. Considering the low permeability of the rocks, the lateral extent of this dolomite is puzzling. Deposition under hypersaline conditions is unlikely considering the normal marine fauna and the position of the dolomite in the stratigraphic sequence. Mixing zone dolomitization of such a thin, continuous bed would be virtually impossible. Thus, the most likely explanation for this dolomite is a

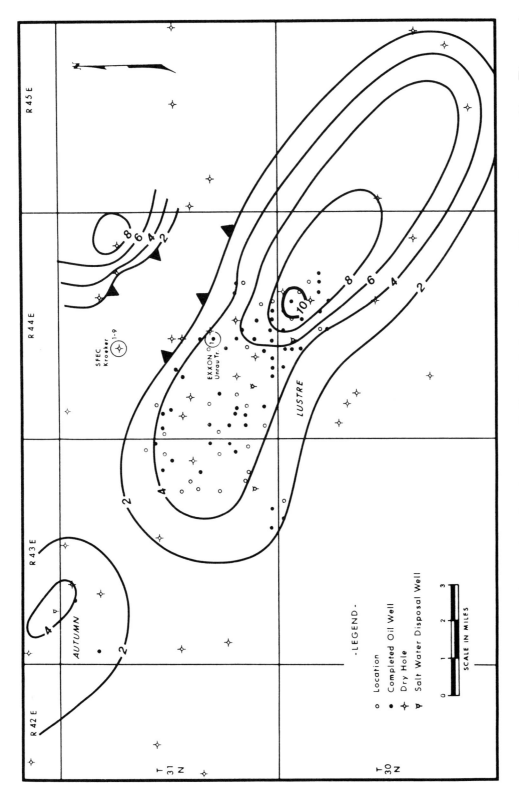

Figure 7. Isoporosity (porosity-foot) map of Cycle 1 based on a 14% porosity cutoff. The pod of porous dolomite in the best part of the field apparently formed by seepage reflux of magnesium-rich brines from the overlying sabkha facies.

difference in composition of the original carbonate mud. A bed of Mg-calcite carbonate mud interbedded within wackestones dominated by aragonite (or possibly by low-magnesium calcite) and subjected to neomorphism during early burial could explain the origin of this dolomite through a solution—cannibalization process.

The normal marine wackestones and packstones that comprise the lower part of the Charles "C" contain abundant crinoid and brachiopod debris. Pellets are locally common. Wispy stylolites are almost ubiquitous within this unit, and vertical fractures occur locally. Few fossils are _in situ_ and the lack of orientation of elongate grains suggests that burrowing has been extensive. However, the color mottling seen in the dolomitic mudstone lithofacies is not preserved within these limestones.

The Richey "Shale" is acutally a laminated to thin-bedded silty, slightly argillaceous carbonate mudstone to wackestone. Thin beds of packstone containing abundant ostracod valves and echinoderm fragments occur within the mudstones. The abundance of transported echinoderm fragments in certain beds indicates proximity to a normal marine carbonate shelf, but the absence of _in situ_ fossils and the paucity of burrows show that few or none of the Mississippian marine fauna could survive in the actual site of deposition. The beds of crinoid and ostracod debris are interpreted as storm deposits. Turbidity currents could produce the same textures, but the homogeneity and continuity of the Richey throughout the study area indicates that no significant depositional slope was present.

Interpretation of Depositional Environments

Cycle 1 reflects deposition during gradually shallowing conditions. Initial deposition began under anoxic bottom conditions on a relatively deep marine shelf. This is indicated by the paucity of burrows in the Richey and the storm beds dominated by a completely allochthonous fauna of crinoids and ostracods. When deposition of the fossiliferous burrowed carbonates of the lower Charles "C" began, conditions on the sea floor must have become more "normal". Good circulation allowed a diverse marine fauna to develop. However, even during deposition of these clean carbonates, water depth apparently exceeded normal wave base because most of the carbonates are mud-rich and poorly sorted.

The burrowed porous dolomite at the top of Cycle 1 was apparently deposited in an open to slightly restricted shelf setting. Remains of brachiopods and echinoderms and the abundant burrows indicate proximity to a normal marine environment, but fossils appear less common than in the underlying beds and this may indicate some slightly anomalous condition such as minor hypersalinity. Minor hypersalinity could also have favored precipitation of Mg-calcite mud over aragonite muds. In any case, the important fact is that this burrowed dolomite reservoir facies was not deposited in a unique environment; the same facies occur in many of the dry holes surrounding the field.

Figure 8. Rocks of Cycle 2.

    A. Black shaly bed at the base of Cycle 2. Extensive microstylolitization has concentrated quartz silt, clay, and organic material in this shale. Some crinoid fragments are also present. Part of the porous dolomite at the top of Cycle 1 is also visible. Exxon #1 Unrau Trust; Depth = 5,778 ft.

    B. Thin section photomicrograph of the shale bed shown in "A". Abundant quartz silt, some dolomite, fossil fragments, and clay are visible.

    C. Peloidal grainstone from the upper part of Cycle 2. Much local replacement of this limestone by anhydrite (the darker patches) has occurred. Santa Fe #1-9 Kroeker; Depth = 5,926 ft.

    D. Thin section photomicrograph of the same peloidal grainstone in the Exxon #1 Unrau Trust core. Large anhydrite crystals (white in this photo) have replaced the limestone. Depth = 5,767 ft.

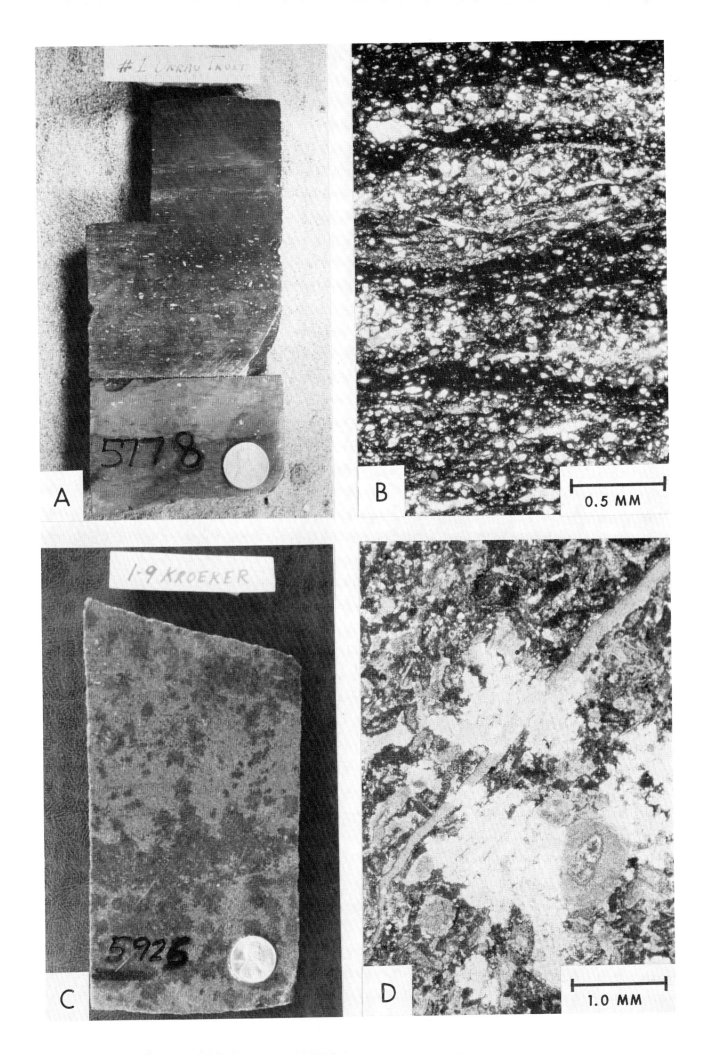

Cycle 2

Lithology

Cycle 2 shows significant facies variations across the study area. Its base is marked by a black "shale" (actually an argillaceous wackestone) (Figs. 8A, 8B) and it grades upward through wackestone and mudstones into peloidal grainstones in the wells on the north side of Lustre Field (Figs. 8C, 8D). Normal marine fossils are common in the lower part of the unit, particularly in wells south of the field, but only ostracods and peloids are common in the rocks at the top of the cycle. Burrows are common in many of the mudstones and wackestones. Chondrites is the most easily recognized ichnogenus.

Cycle 2 ranges in thickness from 6 feet (2 m) in the wells far south of the field such as the Santa Fe #1-9 Tribal Federal to 18 feet (5.5 m) in the Unrau Trust well. Apparently the development of a large peloidal shoal created the thick in the area north of Lustre Field.

The peloidal grainstones at the top of Cycle 2 are very fine-grained (Fig. 8D). The dominant grains are peloids, ostracods and foraminifers. Most of the grainstones were cemented by equant calcite prior to compaction. Minor intergranular and intragranular porosity is preserved locally, but most of the grainstones have less than 8% porosity and contain no significant permeability. The early cementation by calcite apparently inhibited dolomitization, but vertical fractures cut through the grainstones and allowed brine migration as shown by anhydrite precipitation in and along the fractures.

Dolomitization in this cycle is highly variable. Within the field area, most or all of the mudstones have been dolomitized. Thus these rocks are important reservoirs in some of the wells. Porosity ranges up to 25% where dolomitization is complete and permeability in the best dolomites is typically from 0.5 to 5 md. Dolomitization is less common in the wackestones containing a more normal marine fauna south of the field.

Interpretation of Depositional Environments

The shaly bed at the base of Cycle 2 was apparently deposited in relatively deep water under anoxic conditions. Thus, a minor marine transgression ended Cycle 1 and initiated Cycle 2. Water depth shallowed quickly during deposition of the cycle, however, and the lower mudstones and wackestones were apparently deposited in an open to restricted shelf setting. More restricted conditions prevailed on the structural nose extending through Lustre Field, whereas conditions farther south were apparently more normal as indicated by the more abundant marine fossils. Anoxic conditions apparently developed sporadically during deposition of the mudstones and wackestones because certain layers within the lower part of the cycle are laminated and unburrowed.

By the end of Cycle 2 deposition, a peloid shoal had developed along the north edge of Lustre Field. Open water was apparently to the south and a protected area behind the shoal provided the site for deposition of much fine peloidal debris. The end of Cycle 2 occurred when another minor marine transgression drowned the shoal allowing deposition of mudstones and wackestones over the peloidal grainstones.

Cycle 3

Lithology

The boundary between Cycle 2 and 3 is a slightly shaly wackestone to mudstone that yields a subtle gamma ray kick on logs. The contact is sharp in the Kroeker well (Fig. 9A) but is much more gradational away from the Cycle 2 peloidal shoal. In wells in the offshoal area south of the field, burrowed mudstones at the base of Cycle 3 sit directly on burrowed mudstones of Cycle 2 and the boundary is difficult to recognize.

Cycle 3 grades upward from burrowed mudstones and wackestones at the base into grainstones containing peloids, ooids, and oncolites (Figs. 7C, 7D). Although the mudstones and wackestones are laterally quite consistant, the grainstones at the top of the cycle are variable. The degree of dolomitization also varies considerably. Throughout most of the field area, the cycle is extensively dolomitized; only some of the grainstones remain undolomitized. In more distant wells, dolomitization was incomplete with most mudstone remaining as limestone.

The thickness of Cycle 3 ranges from only 4 feet (1.2 m) in wells directly on top of the Cycle 2 peloid shoal about 20 feet (6 m) in the Autumn Field area. Apparently progradation of the grainstones at the top of Cycle 3 created the anomalously thick section in the areas west and immediately south of Lustre Field.

Interpretation of Depositional Environments

Except for the fact that the grainstones are much more widespread and diverse than the grainstones in Cycle 2, Cycle 3 is quite similar to Cycle 2. Deposition of both began on a relatively low energy restricted shelf on which water depth gradually shallowed until grainstones were deposited. Because of the greater extent of the grainstones at the top of Cycle 3, a variety of facies including channels, ooid shoals, oncolite banks, and so on can be recognized.

Cycle 4

Lithology

The base of Cycle 4 is marked by the lowest laminated dolomitic mudstone in the Charles "C". In the field, this mudstone is typically about 2 feet (0.6 m) thick and is overlain by another grainstone unit similar to those found near the top of Cycle 3 (Fig. 10A). The key bed in Cycle 4 is a mottled to massive very finely crystalline gray dolomite about 2 feet (0.6 m) thick. This distinctive bed is informally called the Gray Dolomite Marker. It can be easily recognized in cores and consists of detrital quartz and dolomite silt in a very finely crystalline dolomite matrix (Fig. 10B). This marker is overlain by more laminated dolomitic mudstone (Figs, 10C, 10D). Locally, particularly west and southwest of the field, a peloidal grainstone occurs near the top of Cycle 4.

Cycle 4 ranges from 8 to 16 feet (2.4 to 5 m) thick and thickens gradually southward across the field. It can be recognized on gamma ray logs by the first significant kick above the clean grainstone shoal

Figure 9. Rocks of Cycle 3.

    A. Contact between the peloidal grainstones of Cycle 2 and the overlying slightly shaly wackestones at the base of Cycle 3. Santa Fe #1-9 Kroeker; Depth = 5,924 ft.

    B. Thin section photomicrograph of a fossiliferous wackestone from the lower part of Cycle 3. Dolomitization of the micrite has created a dolomite with 25% porosity and 1 md. permeability. Sample from the Exxon #1 J. Reddig; Depth = 5,721 ft.

    C. Partially dolomitized oncolites from the grainstone near the top of Cycle 3. Exxon #1 Unrau Trust; Depth = 5,754 ft.

    D. Ooids from a grainstone at the top of Cycle 3 in the Santa Fe #1-9 Kroeker; Depth = 5,914 ft. A nodule of silica has replaced the ooids near the center of the photo.

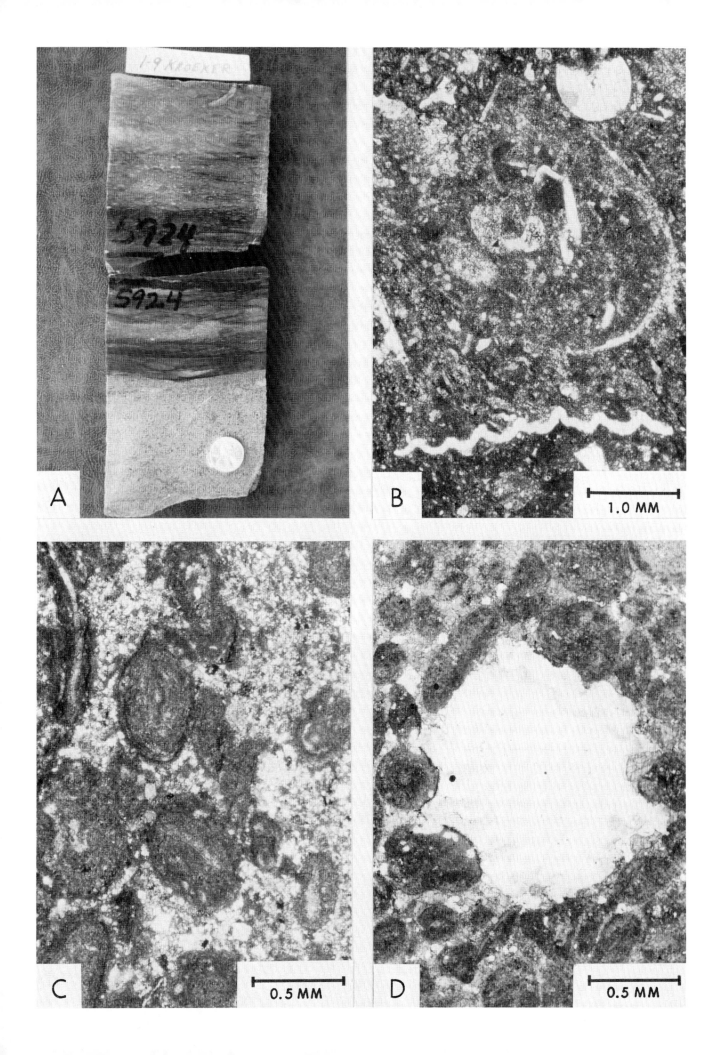

Figure 10. Rocks of Cycle 4.

    A. Laminated dolomitic mudstone at the bse of Cycle 4 and the overlying peloidal/oncolitic grainstone. These rocks lack significant porosity and permeability. Exxon #1 Olfert Cattle; Depth = 5,688 ft.

    B. Thin section photomicrograph of the dolomite in the Gray Dolomite Marker and the edge of an anhydrite nodule. Many of the larger dolomite rhombs have cloudy centers suggesting that these are detrital grains with overgrowths. The matrix is very finely crystalline dolomite that may be a primary precipitate. Santa Fe #1-32 Nashua Tribal core; Depth = 5,441 ft.

    C. Contact between the Gray Dolomite Marker and the overlying laminated dolomitic mudstone. The mudstone was probably deposited in a hypersaline lagoon. Exxon #1 Unrau Trust; Depth = 5,752 ft.

    D. Same contact as seen in the Santa Fe #1-9 Kroeker well located three miles north of the Unrau Trust well. Depth = 5,909 ft.

A

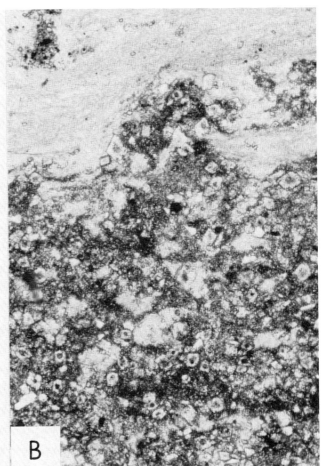

B

C

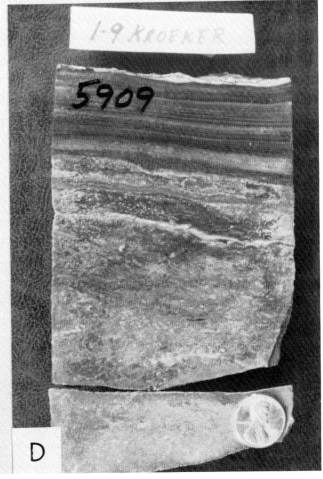

D

of Cycle 3. The clean grainstone above the first laminated dolomite generally yields a clean trough in the gamma ray, but the overlying laminated dolomites and the Gray Dolomite Marker also yield a significant gamma ray kick. The stratigraphically highest significant gamma ray kick within the upper Ratcliffe generally corresponds to the uppermost porous dolomite which is also the Gray Dolomite Marker.

Although most of Cycle 4 consists of dolomite, the rocks offer little reservoir potential. Permeability in the porous gray dolomite marker seldom exceeds 0.1 md, even within the field area. Fractures are rare and these grainstones are always tight. Thus, this cycle contributes little or nothing to production, even though the Gray Dolomite Marker is often perforated along with the lower porous dolomites.

Interpretation of Depositional Environments

The laminated dolomitic mudstones at the base of Cycle 4 represent deposition in a restricted anoxic lagoon or sabkha pond. Burrows are rare or absent, but the continuity of laminae indicates subaqueous deposition. There were probably periods of complete desiccation and subaerial exposure, but mudcracks are not preserved and it is difficult to pinpoint exposure surfaces.

The lagoonal deposits gave way to a bed of peloidal to oncolitic grainstone in the field area. This suggests that a minor transgression allowed shoals to redevelop on the Lustre Field paleohigh.

The Gray Dolomite Marker is a particularly interesting unit because of its lateral continuity throughout the study area and its unusual (for the Charles "C") gray color. Thin sections indicate that it generally contains no fossils, but calicspheres do occur in a few samples. Sedimentary structures include mottling, sparse laminations, and some anhydrite nodules. No evidence of subaerial exposure can be found within the unit. Thus, this bed is interpreted as a hypersaline lagoonal deposit. The lateral extent of this dolomite, even in wells which contain little dolomite suggests the aphanocrystalline matrix is quite likely a primary precipitate.

The laminated mudstones above the dolomite were deposited in both sabkha and hypersaline lagoonal environments. Algal filaments indicative of stromatolites occur in some layers. Anhydrite nodules occur locally and exposure surfaces are present. Mudcracks are surprisingly rare, but may have been more abundant during deposition. The grainstone above the Gray Dolomite Marker west and north of the field probably reflects a minor marine transgression which allowed a shoal to develop on the distal flanks of the Lustre Field paleohigh.

Cycle 5

Lithology

The upper 26 to 36 feet (8 to 11 m) of the Charles "C" consists of tight, partly to completely dolomitized mudstone with some associated anhydrite. Laminated dolomites and massive beds with patterned bedding dominate this interval (Figs. 11A, 11B). Nowhere does this unit contain sufficiently porous dolomite to be productive, so only a brief description is given here.

Figure 11. Rocks of top of Cycle 4 and Cycle 5.

    A.   Laminated dolomitic mudstone from the upper part of Cycle 4 showing continuity of laminae. A minor erosional surface may be present near the top of the slab. Thin sections suggest that some of the laminae are stromatolitic. Santa Fe #1-9 Kroeker; Depth = 5,908 ft.

    B.   Laminated dolomitic mudstone interbedded with dolomitic mudstone containing patterned bedding. This is part of the sabkha to hypersaline lagoon deposits in Cycle 5. Exxon #1 Unrau Trust; Depth = 5,740 ft.

    C.   Dolomitic mudstone with bedded anhydrite from the upper part of Cycle 5. Exxon #1 Reddies; Depth = 5,176 ft.

    D.   Nodular anhydrite from a laterally persistant anhydrite bed at the top of the Charles "C" (Ratcliffe) Zone. Exxon #1 Reddies; Depth - 5,170 ft.

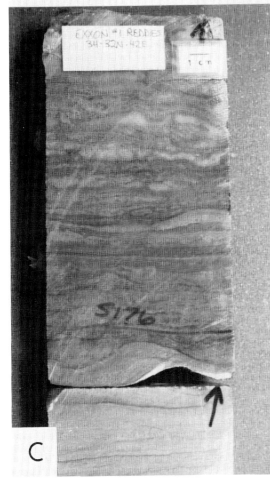
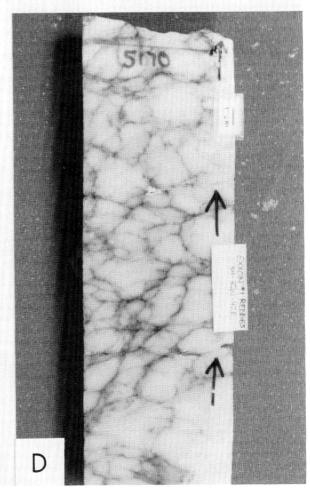

Thickness of the cycle varies in a complex fashion across the Lustre Field area. It is thinnest, 26 feet (8 m), on the west side of Lustre Field and in Autumn Field and thickens to 36 feet (11 m) to the northeast of the field. This suggests that the Wolf Creek Nose exerted a minor influence on deposition.

Fossils are rare to completely absent in Cycle 5 rocks. Only ostracods, stromatolites, and calcispheres have been noted. Anhydrite nodules, mudcracks, deflation surfaces, and patterned bedding (as described by Dixon, 1976) are the dominant and most interesting depositional features.

At the top of Cycle 5 are two thin (1 to 2 foot thick) beds of bedded to nodular anhydrite which can be correlated throughout the Lustre Field area (Figs. 11C, 11D). These anhydrites apparently formed during a prolonged period of sabkha (subaerial) deposition at the end of Ratcliffe time. Dolomite is closely associated with these anhydrites and suggests that brines generated during anhydrite precipitation may be at least partly responsible for the dolomite in Cycle 5.

In spite of the abundance of dolomite in this Cycle, there is virtually no porosity. The dolomites are very finely crystalline and tight with much anhydrite plugging. There is no significant permeability. Furthermore, the dolomites are generally light-colored and do not appear to have any source potential. However, if unfractured, as they appear to be in the cores, they may play an important role in sealing the reservoir.

Interpretation of Deposition Environments

The common anhydrite nodules, mudcracks, deflation surfaces, possible stromatolites, and patterned bedding indicate that deposition of the Cycle 5 rocks occurred on a sabkha which was periodically flooded by several feet of water to become a hypersaline lagoon. Laminated mudstones represent both sabkha and restricted subaqueous deposition, whereas the more massive dolomites with patterned bedding were probably deposited in hypersaline lagoons.

The most important aspect of the Cycle 5 sabkha deposits was probably their role as a source for magnesium-rich brines which seeped downward to form the dolomites in the lower Ratcliffe. There is a good correlation between degree of dolomitization of Cycle 5 rocks and dolomitization in the underlying rocks.

RATCLIFFE DEPOSITIONAL MODEL

The extensive work on Mississippian reservoirs has led to the proposal of a variety of depositional models in the literature. These range from Hansen's (1966) algal reef model for the Ratcliffe reservoirs in northeastern Montana (including Flat Lake Field) to a tidal flat-supratidal pond model for the Nesson in the Billings Nose (Altschuld and Kerr, 1982).

Lindsay's study of Little Knife Field in North Dakota (Lindsay and Roth, 1982; Lindsay, 1985) led him to develop a depositional model based partly on modern facies in the Persian Gulf (Fig. 12). Textures and fossils in the Ratcliffe rocks indicate that deposition occurred on a broad carbonate ramp similar to that proposed by Lindsay for Little

Knife Field. Subtle differences in paleotopography caused facies changes within the cycles, and at one time or another the Lustre Field area was a deep, anoxic shelf, an open marine shelf, a restricted shelf or lagoon, a shoal, a hypersaline lagoon, and a sabkha. It is unlikely that all these environments co-existed within the study area at any one time, but a depositional model can be created as if this were the case (Fig. 12).

Critical to understanding this depositional model is understanding the scale, i.e., the lateral distances, between the various facies. Some idea of the scale can be obtained from examination of modern carbonate ramps and sabkhas. The best known example of a modern carbonate ramp/sabkha system occurs in the Persian Gulf. Work by Purser and Evans (1973) and Butler and others (1982) indicates that many facies on the Persian Gulf ramp/sabkha are similar to those recognized in the Ratcliffe. Thus, their maps of facies distribution (Figs. 13, 14) give some indication of how the Ratcliffe facies might have been distributed. Not only do these figures show the complexity of facies patterns, but they also show that individual shoals, lagoons, and so on may be miles to ten or so miles long. Scale of the various facies on the Ratcliffe carbonate ramp may have been somewhat greater simply because of the extremely flat surface on which the ramp developed.

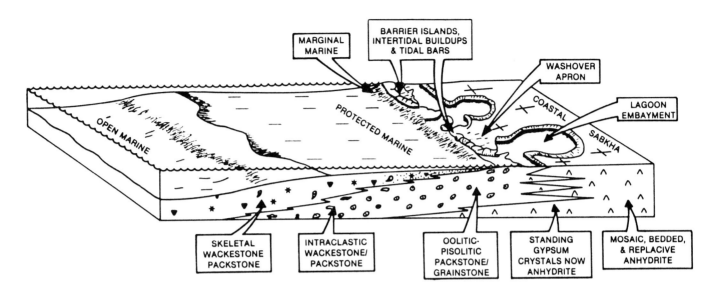

Figure 12. Depositional model proposed for the Mission Canyon Formation in Little Knife Field. This model, based partly on modern depositional facies relationships in the Persian Gulf, is probably a good analog for Ratcliffe deposition in the Lustre Field area. From Lindsay, 1985.

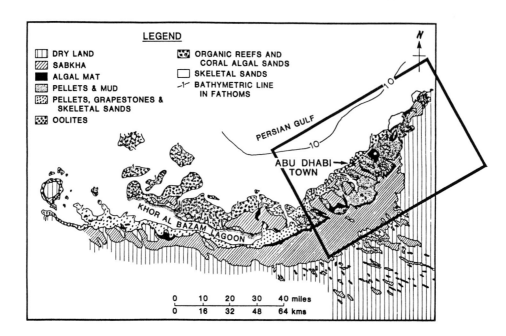

Figure 13. Recent facies distribution along the Trucial Coast near Abu Dhabi, Persian Gulf. This map gives at least a hint of how the Ratcliffe facies may have been distributed during deposition in the Lustre Field area. Of equal importance is the scale of the various facies. The area outlined is shown in more detail in Figure 14. From Butler and others, 1982, p. 45.

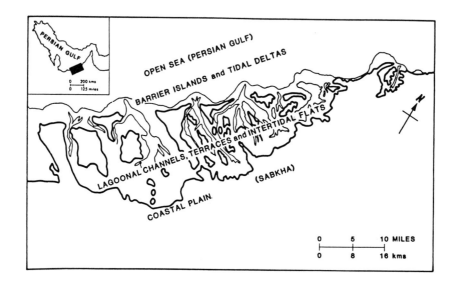

Figure 14. Closer view of depositional facies and geometry of the barrier island buildups, tidal channels, and intertidal facies along the modern Trucial Coast. Many of the shoals are several miles long and separated by irregular channels from a half to several miles across. The shoals formed during deposition of Cycle 3 in the Ratcliffe may well have had a similar complex geometry. From Butler and others, 1982, p. 45.

DIAGENESIS

Six major diagenetic events can be recognized in the Charles reservoir rocks. There are: (1) calcite cementation of the grainstones, (2) precipitation of (and replacement by) anhydrite, (3) silicification and formation of chert nodules, (4) dolomitization, (5) stylolitization, and (6) fracturing. Of these, only calcite cementation and dolomitization have played a significant role in shaping the reservoir.

Study of the producing rocks shows that dolomitization is the primary control on porosity and production (Fig. 15). Primary intergranular and intragranular porosity also exist in limestones in some wells, but this porosity is seldom associated with good permeability and contributes little to the production.

Calcite Cementation

As with all carbonate sediments, the Charles rocks contained excellent porosity at the time of deposition. Many of the grainstones now contain about 25 to 40% sparry calcite, but less than 6% porosity.

Loose grain packing in most grainstones and lack of broken grains or other compaction features suggest that the calcite formed soon after deposition of the grainstones. The textures of the calcite cement suggest that it precipitated in a fresh water phreatic environment (cf. Longman, 1980). Such an interpretation is consistent with the depositional model of grainstone shoals; subtle fluctuations in sea level could have exposed the shoals and allowed development of a

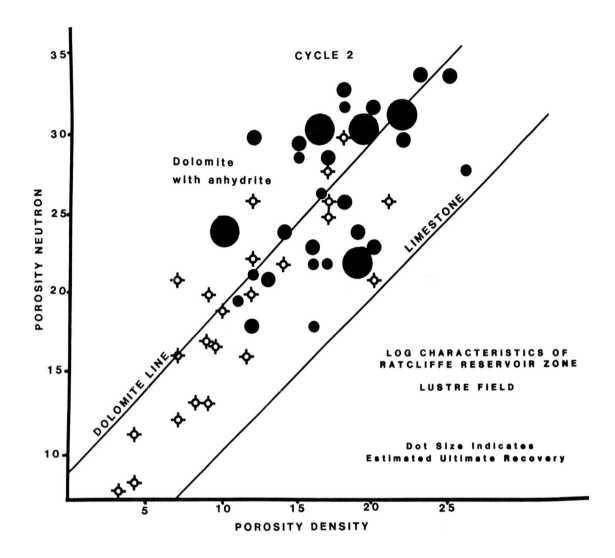

Figure 15. Cross-plot of neutron porosity and density porosity for the most porous bed within Cycle 2, a major reservoir zone within the field. Wells in and surrounding Lustre Field were used (thus, the numerous dry holes). Size of the dot indicates estimated ultimate recovery of each well. Most good wells fall near the dolomite line, although a few seem to be dominantly calcite. Examination of core from the well which yielded the large dot closest to the Limestone Line (the #1 Olfert Cattle) indicates that the reservoir there is also completely dolomite (the reason the point falls so close to the limestone line is unknown). Note that most wells with good porosity produce, though in varying amounts, whereas most dry holes are tight. The highly porous dry holes fall within a mile of the best producing wells.

fresh water lens much like that found today under many exposed shoals in the Bahamas (e.g., Halley and Harris, 1979).

Weakening the interpretation of precipitation in a fresh water environment is the close association of the calcite and anhydrite. Such anhydrite could not survive in a fresh water environment, and its presence indicates that the rocks were never exposed to significant amounts of fresh water following anhydrite precipitation. Thus, the presence of fresh water calcite and hypersaline brine anhydrite together as early cements presents a paradox that can only be solved if: (1) the calcite was not precipitated in fresh water, or (2) all anhydrite precipitated after all the calcite. Considering the fact that the rocks overlying the grainstones include many sabkha deposits filled with anhydrite, the latter interpretation seems more likely.

The ubiquity of equant calcite cement in the grainstones is important not only because it destroyed the reservoir potential of this facies, but also because it gives clues to the topographic relief, diagenetic conditions, and timing of dolomitization. If significant relief had existed during deposition (and subsequent subaerial exposure), localization of the equant calcite cement probably would have occurred. However, some of the structurally lowest wells near the field (e.g., the #1-9 Kroeker) contain equant calcite in amounts comparable to the highest wells in the field. Thus, paleotopographic relief across the Cycle 3 shoals during deposition was apparently low.

Most of the grainstones remain limestone and it appears that the calcite cement prevented hypersaline brines from dolomitizing the peloids. This indicates that calcite cementation preceded dolomitization.

Dolomitization

Dolomites in the Charles "C" formed in at least four different ways: (1) as a primary precipitate (in the Gray Dolomite Marker); (2) by solution-cannibalization of Mg-calcite micrite, particularly in the rocks of Cycle 1; (3) by replacement of micrite and grains in downward-seeping Mg-rich brines; and (4) as baroque or saddle dolomite inside some grains and in vugs. Of these, only replacement of micrite and grains has created significant porosity and permeability, but recognizing the other types is important when interpreting well logs. The primary dolomite in particular often appears porous, but seldom contains sufficient permeability to be a good reservoir.

Primary Dolomite

The earliest form of dolomite in the Ratcliffe was that which precipitated directly from a hypersaline water column. Although primary dolomite may be common, it has been recognized only in the Gray Dolomite Marker in Cycle 4. Grain size is typically 1 to 4 microns and permeability in this bed is generally low. Because of this low permeability, the primary dolomite has little reservoir potential and probably yields little of the oil produced in the field, even though this bed is generally perforated along with the other porous dolomites.

Similar primary dolomite has been recognized in Holocene rocks of Baffin Bay, Texas by Behrens and Land (1972). Evidence for their interpretation included fine grain size, depositional setting and geometry of the dolomite and in many ways their dolomite is identical to that found within the Gray Dolomite Marker. Certainly the key clues to a primary origin for the Gray Dolomite Marker are the geometry, position in the stratigraphic sequence, and textures of this bed. Some primary dolomite may also be present within Cycle 5, but this is difficult to prove because of the associated secondary dolomitization.

Detrital dolomite, recognizable because of its relatively large size, cloudy centers, and clear overgrowths, is common within the primary dolomite matrix. This detrital dolomite probably originated on some distant sabkha and was carried into the hypersaline lagoon by winds. Quartz silt of about the same gain size was also carried in by winds.

Solution-Cannibalization Dolomite

Probably the second form of dolomite to develop within the Ratcliff derived its magnesium from Mg-calcite micrite muds through solution-cannibalization of diagenetic unmixing. This process has been frequently recognized, but is particularly well-documented by Kendall (1977). He recognized that dolomitized burrows in the Ordovician of Saskatchewan formed by dolomitization of micrite in the burrows and that the magnesium came from Mg-calcite within the rocks.

Dolomite formed by cannibalization in the Ratcliffe is most apparent in the wackestones of Cycle 1, but probably occurs in other rocks as well. This type of dolomite can be recognized by its regional geometry, textures, and contacts with the adjacent limestones. Magnesium was derived from both the micrite and grains such as the echinoderm fragments. The resulting dolomite was generally very fine grained and contained much associated remnant calcite. Porosity and permeability in these dolomites are low.

Seepage-Reflux Dolomite

The most important form of dolomite for creating the reservoir in Lustre Field formed in downward-seeping magnesium-rich brines. This model for dolomitization was originally described by Adams and Rhodes (1960) who gave it the name "seepage-reflux". Seapage-reflux dolomites formed in the lower Ratcliffe when brines generated in the sabkha to hypersaline lagoon facies of Cycles 4 and 5 seeped downward into the underlying limestones. Both gravity and hydrostatic head probably drove the brines.

Closely associated with seepage reflux is dolomitization upon the sabkha by brine generation during storm recharge (Patterson and Kinsman, 1982). This process was also important in the Ratcliffe and resulted in much dolomitization of the rocks of Cycles 4 and 5. However, while brine generation upon the Ratcliffe sabkhas created much dolomite, it created little porous dolomite of reservoir quality. Only where brines were able to seep downward into the underlying limestones of the lower Charles "C" Zone did significant porosity develop.

By the time these brines migrated into the lower part of the Charles "C" the peloidal grainstones at the top of Cycles 2 and 3 had already been cemented by calcite, but the mudstones and wackestones retained enough porosity to act as local pathways for brine migration. Dolomite rhombs from 10 to 50 microns in size formed in these brines. Micrites were selectively dolomitized, probably because their small grain size allowed much surface area for the brines to attack.

The strongest evidence for seepage-reflux dolomitization is the geometry of the porous dolomites. Figure 7 shows the thickness of porous dolomite in Cycle 1. The lenticular shape of the dolomite, which cuts across the depositional facies in the Cycle, precludes a depositional (primary) origin, and the precompaction timing of dolomitization (burrows are not compressed in the dolomites) confirms an early origin. A mixing zone origin is improbable in light of the abundant anhydrite present within the grainstones.

CONCLUSIONS

The discovery of Lustre Field proves that major oil fields can occur in Valley County, not only in the Charles, but also in the Mission Canyon. Thus, both thermally mature source rocks, possibly in the Lodgepole, and migration paths for getting the oil into the Mission Canyon and Charles reservoirs must exist.

Study of cores form the Charles "C" Zone in the Lustre Field area reveals that depositional facies with the potential to become good reservoirs extend throughout the southern part of Valley County and probably into adjacent parts of Daniels County. Dolomitization is required for reservoir development, but the extensive sabkha and hypersaline lagoons which generated the brines which caused the dolomitization also extend throughout the area. Thus, all that is required to form good Charles reservoirs is a means of getting enough brine from the sabkha deposits down into the mudstones and wackestones on the lower Charles "C".

Paleotopography, faulting penecontemporaneous with deposition, and fracturing all provide possible mechanisms for getting dolomitizing fluids into the lower Charles. It is also possible that localized brine migration could occur without any of these causes. This random factor both helps and hinders exploration in Valley County. It helps because serendipity may come into play when drilling toward other reservoirs, but it hinders because it makes it difficult to predict where dolomitization may occur.

ACKNOWLEDGEMENTS

This paper was made possible through the support and cooperation of Santa Fe Energy Company and Quintana Petroleum Corporation. Our thanks to these companies for granting permission to publish the results of our study of Lustre Field. Thanks also to Greg Stanbro, David Eby, Bill Precht, and Keith Shanley for their comments on an earlier version of this manuscript.

The drafting department of Santa Fe Energy provided much help with the figures. The staff of the U.S.G.S. Core Library and Exxon Corporation provided us with the core from the #1 Unrau Trust well. Our thanks to these people for their help.

REFERENCES

Adams, J.E., and Rhodes, M.L., 1960, Dolomitization by seepage refluxion: Am. Assoc. Petrol. Geol. Bull., v. 44, p. 1912-1920.

Altschuld, N., and Kerr, S.D., Jr., 1982, Mission Canyon and Duperow reservoirs of the Billings Nose, Billings County, North Dakota: in J.E. Christopher and J. Kaldi (eds.), Proc. Fourth Int'l Williston Basin Symposium, Regina, p. 103-112.

Behrens, E.W., and Land, L.S., 1972, Subtidal Holocene dolomite, Baffin Bay, Texas: Jour. Sed. Petrol., v. 42, p. 155-161.

Butler, G.P., Harris, P.M., and Kendall, C.G. ST. C., 1982, Recent evaporites from the Abu Dhabi coastal flats: in C. R. Handford, R.G. Loucks, and G.R. Davies (eds.), Depositional and Diagenetic Spectra of Evaporites-A Core Workshop: Soc. Econ. Paleont. Mineral. Core Workshop No. 3, p. 33-64.

Dixon, J., 1976, Patterned carbonate - a diagenetic feature: Bull. Canadian Petr. Geol., v. 24, p. 450-456.

Edie, R.W., 1958, Mississippian sedimentation and oil fields in southeastern Saskatchewan: Am. Assoc. Petrol. Geol. Bull., v. 42, p. 94-126.

Halley, R.B., and Harris, P.M., 1979, Fresh-water cementation of a 1,000-year-old oolite: Jour. Sed. Petrol., v. 49, p. 969-989.

Hansen, A.R., 1966, Reef trends of Mississippian Ratcliffe zone, northeast Montana and northwest North Dakota: Am. Assoc. Petrol. Geol. Bull., v. 50, p. 2260-2268.

Hartling, A., Brewster, A., and Posehn, G., 1982, The geology and hydrocarbon trapping mechanisms of the Mississippian Oungre Zone (Ratcliffe beds) of the Williston Basin: in J.E. Christopher and J. Kaldi (eds.), Proc. Fourth Int'l Williston Basin Symposium, Regina, p. 217-223.

Kaldi, J., 1982, Reservoir properties, Depositional environments and diagenesis of Mississippian Midale beds, Midale Field, southwestern Saskatchewan: in J.E. Christopher and J. Kaldi (eds.), Proc. Fourth Int'l Williston Basin Symposium, Regina, p. 211-216.

Kendall, A.C., 1977, Origin of dolomite mottling in Ordovician limestones from Saskatchewan and Manitoba: Bull. Can. Petr. Geol., v. 25, p. 480-504.

Lindsay, R.F., 1985, Carbonate and evaporite facies, dolomitization and reservoir distribution of the Mission Canyon Formation, Little Knife Field, North Dakota: Am. Assoc. Petrol. Geol. Memoir. In Press.

Lindsay, R.F., and Roth, M., 1982, Carbonate and evaporite facies, dolomitization and reservoir distribution of the Mission Canyon Formation, Little Knife Field, North Dakota: in J.E. Christopher and J. Kaldi (eds.), Proc. Fourth Int'l Williston Basin Symposium, Regina, p. 153-180.

Longman, M.W., 1980, Carbonate diagenetic textures from nearsurface diagenetic environments: Am. Assoc. Petrol. Geol. Bull., v. 64, p. 461-487.

Patterson, R.J., and Kinsman, D.J.J., 1982, Formation of diagenetic dolomite in coastal sabkha along Arabian (Persian) Gulf: Am. Assoc. Petrol. Geol. Bull., v. 66, p. 28-43.

Peterson, J.A., 1984, Stratigraphy and sedimentary facies of the Madison Limestone and associated rocks in parts of Montana, Nebraska, North Dakota, South Dakota, and Wyoming: U.S. Geol. Surv. Prof. Paper 1273-A, 34 p.

Purser, B.H., and Evans, G., 1973, Regional sedimentation along the Trucial Coast, SE Persian Gulf: in B.H. Purser (ed.), The Persian Gulf, New York: Springer-Verlag, p. 211-231.

DEVONIAN HARE INDIAN - RAMPARTS (KEE SCARP) EVOLUTION, MACKENZIE MOUNTAINS AND SUBSURFACE NORMAN WELLS, N.W.T.: BASIN-FILL AND PLATFORM-REEF DEVELOPMENT

Iain Muir
University of Ottawa
Ontario, K1N 6N5

Pak Wong
Esso Resources Canada
237 4th Ave. SW
Calgary, Alberta T2POH6

Jack Wendte
Esso Resources Canada
237 4th Ave. SW
Calgary, Alberta T2POH6

ABSTRACT

A joint surface-subsurface study of the Devonian Hare Indian-Ramparts (Kee Scarp) Formation examined the nature of basin-fill and platform-reef development in the Mackenzie Mountains and Norman Wells area, Northwest Territories.

The Givetian-Frasnian (?) Hare Indian and Ramparts (Kee Scarp) strata consist of repeated shoaling-upward sequences. These sequences or cycles of sedimentation were initiated in response to accelerated rates of relative sea-level rise. Two major first-order cycles (each greater than two hundred metres thick) were discerned. The lower cycle consists of prograding shale banks of the Hare Indian Formation and the immediately overlying "shale ramp" sequence of the Ramparts Formation. The upper cycle commenced with deposition of the widely correlatable, dark argillaceous, carbonaceous limestones of the "Carcajou Marker". Overlying platform and reefal limestones of the Ramparts (Kee Scarp) Formation, off-reef and fondothem shales of the Canol Formation and interbedded clinothem shales and sandstones of the Imperial Formation make up the remainder of the upper cycle.

The first-order cycles consist of a number of smaller second-order cycles. These cycles are best defined in shallow-water platform and reef complexes where they can be traced across the entire complex, but also are recognized in the Hare Indian shale clinothem. In platforms and reefs, these second-order cycles contain a wade range of facies, varying from shallow reef and platform to deeper foreslope limestones that can be traced, in certain locations, even further out into adjacent basinal limestones. In reef interiors, these second—order cycles are made up of even smaller third-order cycles consisting of subtidal and tidal-flat limestones.

This cyclic arrangement of strata is well-defined in both the Mackenzie Mountains and the subsurface Norman Wells Field, enabling correlation of time-equivalent growth stages between both complexes over a distance in excess of one hundred kilometres. This and the small-scale periodicity of the second and third-order cycles argue for eustatic sea-level control. The following eight major growth stages are recognized in both complexes: (1) drowning of the Hume limestone, (2) progradation of Hare Indian shale banks, (3) drowning of Hare Indian shale banks, and deposition of the "Carcajou Marker", (4) carbonate platform inception, initial growth and subsequent localized upbuilding, (5) reef inception, progradation and subsequent aggradation, (6) reef backstepping and subsequent aggradation, (7) drowning of the Norman Wells reef complex and backstepping of the Mackenzie Mountain reef complex and (8) drowning of the Mackenzie Mountain reef complex.

Evidence to support the presence of the previously interpreted pre-Canol unconformity above the Ramparts and underlying Formation is lacking. Instead, the Canol shale is observed to intertongue with, onlap onto and drape over (at Norman Wells) the Ramparts and Kee Scarp reef complexes.

INTRODUCTION

This paper describes, compares and contrasts the Devonian Hare Indian and Ramparts (Kee Scarp) evolution from the subsurface in the Norman Wells field and outcrops in the Mackenzie Mountains, approximately 100 kilometres west of Norman Wells. The locations of the Norman Wells field and the Mackenzie Mountain outcrops are shown in Figure 1.

This work is based on outcrop studies by Iain Muir and subsurface core and wireline log investigations by Pak Wong and Jack Wendte. Muir's study is part of a Ph.D. investigation at the University of Ottawa, while Wong's and Wendte's study is aimed at providing a sophisticated reservoir model.

STRATIGRAPHY

Our interpretations of the stratigraphic relationships and relative ages of surface and subsurface units are shown in Figure 2. These interpretations are based on the stratal relationships of the various units, particularily whether units intertongue, and therefore are correlative, or onlap and drape over other units, and thus postdate them.

Surface and subsurface terminology is similar, but differ in one important way. The Hare Indian Formation in the Norman Wells subsurface is time-equivalent to both the Hare Indian Formation plus the lower portion of the Ramparts Formation, and interbedded limestone-shale sequence (Muir's shale ramp) in the Mackenzie Mountain

LOCATION MAP OF STUDY AREAS

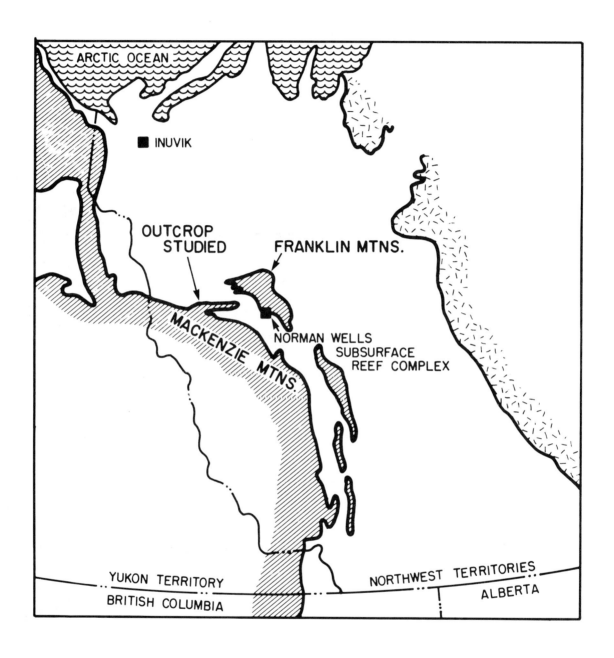

Figure 1. Location map of study areas.

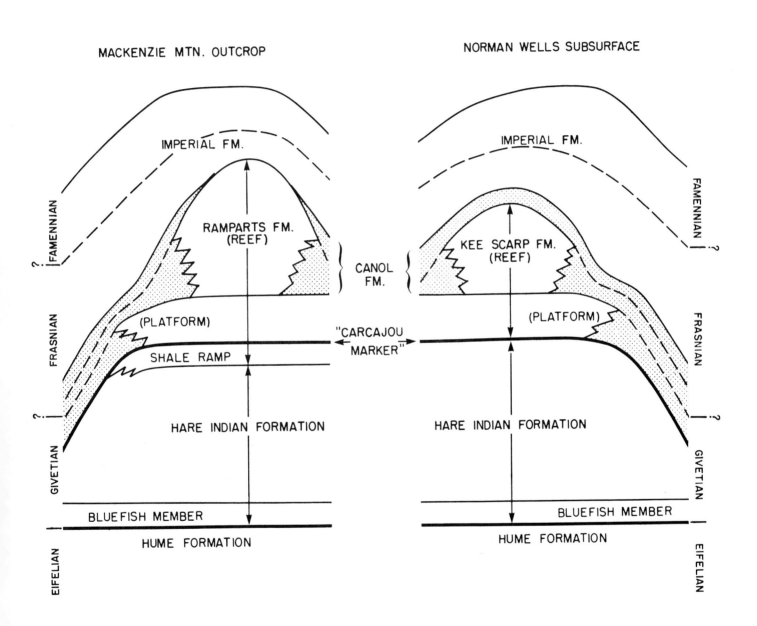

Figure 2. Middle and Upper Devonian Stratigraphy of the Norman Wells Subsurface and MacKenzie Mountain Outcrop.

outcrops. To simplify discussion, surface terminology is used in this paper, unless otherwise noted.

The stratigraphic relationships purposed here disagree with those of other workers. In particular, Warren and Stelck (1962), Bassett and Stout (1968), Gilbert (1973) and Braun (1966, 1977 and 1978) have interpreted a pre-Canol unconformity. This interpretation places the Canol shale as entirely post-Ramparts reef. Braun's (1977) interpretation of a pre-Canol unconformity is based on field relationships and paleontology. He cites evidence of Canol shale sitting directly on middle and basal (Bluefish Member) Hare Indian shales at Mountain River and directly on Hume carbonates in the vicinity of the Snake River. These occurrences, he notes, correspond to missing faunal zones. Braun interprets the apparently missing section as a result of pre-Canol erosional truncation of the Ramparts and the Hare Indian Formations.

We offer a different interpretation, supported by stratal relationships in both the surface and subsurface. We observe the Canol shale to both intertongue with and onlap onto the Ramparts limestone in both the surface and subsurface. At Norman Wells, the Canol also drapes over the subsurface reef complex.

These stratal relationships indicate that the Canol is both time-equivalent to and postdates the Ramparts (Kee Scarp) Formation. We interpret the thinning of the Hare Indian Formation at Mountain River and in the vicinity of the Snake River not to be due to erosional truncation but merely caused by depositional thinning within a shale clinoform. The missing faunal zones reported by Braun (1977) simply reflect a "condensed interval" caused by very slow sedimentation in deep-water, clinoform and fondoform environments. Evidence supporting this interpretation is discussed later in this paper.

STRATIGRAPHIC EVOLUTION

Givetian-Frasnian strata in the Mackenzie Mountains and the Norman Wells area overlie the Hume Formation of Eifelian age. The uppermost part of the Hume Formation consists of interbedded shales and nodular lime wackestones containing inarticulate and articulate brachiopods, solitary rugose corals, tentaculitids and stylolinids. These rocks are interpreted as open-marine, downslope limestones and shales deposited in a few tens of metres of water.

The overlying Givetian-Frasnian strata include limestones, shale and sandstones of the Hare Indian, Ramparts, Canol and Imperial Formations in excess of 400 meters thick. Examination of these deposits show that they consist of two major or first-order cycles of sedimentation that are on the same level as "depositional sequences" defined from seismic stratigraphic studies (Vail and others, 1977).

The use of these and smaller order cycles permits the definition and delineation of true time stratigraphic stages. Thus, cycle boundaries should preserve a normal lateral sequence of facies. Furthermore, knowing that limestones present along the undaform portions of large banks or upper parts of reef complexes were deposited in shallow water, we can approximate the paleobathymetries, along cycle boundaries, of deeper water facies. Because some of the shallow-water limestones were deposited somewhat below sea level, these measurements may represent only minimum estimates of the water depths of the deeper water facies. Wendte and Stoakes (1982) describe this technique in more detail.

Lower First-Order Cycle

The lower first-order cycle includes the surface Hare Indian shales and the lower "shale ramp" portion of the Ramparts Formation (or the entire subsurface Hare Indian Formation).

Hare Indian Formation - This Formation includes a lower Bluefish Member and a much thicker Upper Member.

The lower Bluefish Member is 12 to 17 metres thick and consists of black, laminated, organic-rich shale (Fig. 3). These shales contain a planktonic-nektonic fauna of planktonic bivalves, stylolinids, and plant fragments. We interperet these shales as having been deposited in deep, euxinic conditions on the basin floor (or fondoform) in water as much as 150 meters deep.

The Upper Member of the Hare Indian Formation attains a maximum thickness of 177 meters in the study area and consists of interbedded green shales and thin-bedded limestones. These shales and limestones occur in a cyclic manner. These smaller or second-order cycles are tens of meters thick and have a shaley lower portion and a more predominant limestone upper portion (Fig. 4). The limestone interbeds are normally one to two centimeters thick and consist of silt to very fine sand size, peloidal (altered skeletal) packstones with a high percentage of clay minerals and only rare identifiable skeletal constituents (brachiopods, crinoid columnals). These beds have sharp basal contacts, common normal grading, some slump features and syndepositional folds, and rare sole marks. We interpret these limestones as turbidite deposits on a shale clinoform.

"Shale Ramp" of the Ramparts Formation - Shale of the underlying Hare Indian Formation grade up into the "shale ramp" of the Ramparts Formation in the Mackenzie Mountains (Fig. 5). The lower boundary of the Ramparts Formation is located where limestone beds become predominant over shales in the vertical succession. The "shale ramp" varies in thickness from less than 10 meters in off-reef sections to up to 26 meters under the reefs. The lower transitional limestone beds of the "shale ramp" consist of bioturbated, nodular mudstones and wackestones. Limestones in the upper portion of the "shale ramp" contain stromatoporoids, corals, brachiopods and crinoids. These limestones in the upper part of the "shale ramp" are interpreted to represent relatively shallow-water, in situ carbonate growth along the top of the Hare Indian shale clinoform. Time-correlative deposits to the "shale ramp" in the Norman Wells subsurface are greenish coloured, argillaceous, crinoidal lime wackestones and packstones of the upper Hare Indian Formation.

Examination of the vertical succession of lithologies in both the Lower Bluefish and Upper Member of the Hare Indian Formation and the lower "shale ramp" of the Ramparts Formation indicate all are facies components of large shale banks. The relative proportion of these facies depends upon the position within the shale banks. In situ carbonate deposits of the "shale ramp" are restricted in occurrence to upper parts of the shale banks. Conversely, the dark coloured shale of the Bluefish Member comprise all or nearly all of the Hare Indian

Figure 3. Basinal laminites (bituminous shales) of the Bluefish Member, Hare Indian Formation. Gradationally overlain by calcareous shales and argillaceous limestones of the Upper Member. Possibile coarsening-upward cycles indi- cated by arrows. Bluefish Member is 16 metres thick. Location at Section 15, Gayna River.

Figure 4. Coarsening- and thickening-upward sequence in clino- them facies, Hare Indian Formation. Alternation of dark green-grey calcareous shales and argillaceous calci- siltites. Section 7 - units 004 to 008 (arrows) appoxi- mately 20 metres thick.

Figure 5. Shale ramp biostromal units overlying shale ramp-clino- them transitional sequence (mainly nodular calisilites). Figure is two metres. Section 8, Bell Creek.

Figure 6. Deep open-marine bituminous shales and limestones of the Carcajou "Marker" deposited during incipient drowning event and marking initiation of the carbonate platform. Scale is two metres. Section 3.

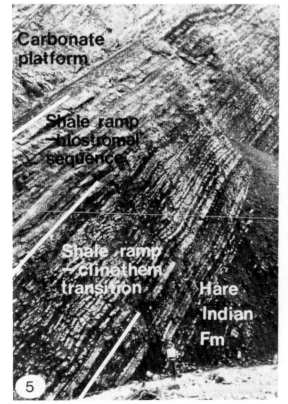

Formation along distal portions of the shale bank clinoforms. These black shales thin basinward and may even pinch out. This would account for the missing stratigraphic section attributed to pre-Canol erosion by other workers (Warren and Stelck, 1962; Bassett and Stout, 1968; Gilbert, 1973; Braun, 1966, 1977 and 1978).

Upper First-Order Cycle

The upper first-order cycle includes the remaining portion of the Ramparts (Kee Scarp) Formation, the Canol shales and the Imperial shales and sandstones.

Ramparts (Kee Scarp) Formation - The base of the upper first-order cycles is marked by the occurrence of a widespread, dark coloured, very shaley and carbonaceous limestone unit known as the "Carcajou Marker" (Figs. 6 and 7). This marker is normally 1-3 metres thick and contains some corals and crinoids. The occurrence of this marker in both the Mackenzie Mountains and in the Norman Wells area is very important because it forms the basis for overlying platform and reefal correlations between the Mackenzie Mountains and Norman Wells described later in this paper.

The "Carcajou" beds grade up into cleaner limestone beds of the Ramparts (Kee Scarp) Formation. The Ramparts (Kee Scarp) consists of a lower widespread platform unit, attaining a thickness of 57 metres in the Mackenzie Mountains and 63 meters in the Norman Wells subsurface, and an upper, more areally restricted reefal unit. The upper reefal unit attains a maximum thickness of 160 meters in the Mackenzie Mountain outcrops and only 90 meters in the Norman Wells subsurface. Facies of the platform and reefal limestones are described later in the paper.

Canol Formation – The Canol Formation consists of black, laminated, partly siliceous, organic-rich shales with thin limestone-turbidite deposits adjacent to the reefal complexes (Fig. 8). These shales have a rare planktonic-nektonic fauna of stylolinids and fish remains. The lower part of these shales intertongue with Ramparts (Kee Scarp) platform and reef limestones; the upper part onlaps the reef in the Mackenzie Mountains outcrop and onlaps onto and drapes over the reef in the Norman Wells subsurface. A deep, euxinic depositional environment, with water depths in excess of one hundred meters, is interpreted.

Imperial Formation – Canol shales grade up into less carbonaceous, greenish coloured shales and sandstones of the Imperial Formation (Fig. 9). These deposits are interpreted as the clinoform-facies equivalent of the basin floor (fondoform) Canol shale. Shales of the Imperial Formation cap the Ramparts reef complex in the Mackenzie Mountains.

FACIES OF THE RAMPARTS (KEE SCARP) CARBONATE PLATFORM AND REEF

Six environmental facies displaying distinctive textures, sedimentary structures, fossils and other constituents are recognized. Some of these facies are divided into subfacies. Depositional interpretation of these facies and subfacies is based on both their lateral and vertical extent and on comparison with recent analogies.

Figure 7. Allochthonous foreslope equivalents (mainly fine-grained turbidites) of the shale ramp and carbonate platform facies. Carcajou "Marker" lithofacies approximately four metres thick. Section 25, Mountain River distributary.

Figure 8. Basinal laminites and limestone concretions in the Canol Formation, Section 1, Powell Creek. The person is two metres high.

Figure 9. Gradational contact between Canol Formation and overlying coarsening- and thickening-upward basin-fill sequences of the Imperial Formation. Section 1, Powell Creek.

Figure 10. Middle-foreslope wafer stromatoporoid rudstones and bindstones. Section 4. Ramparts Formation. Scale in centimetres.

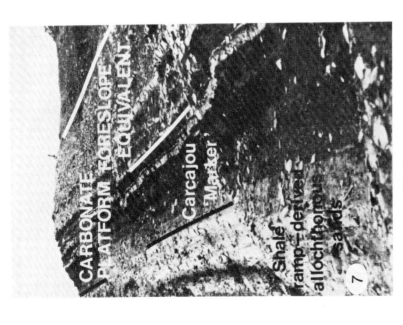
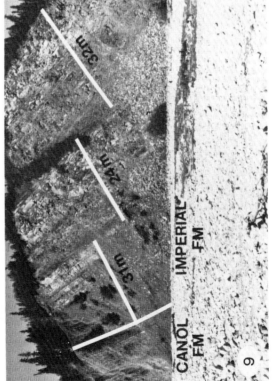

Many more subfacies are recognized in the Norman Wells subsurface than in the Mackenzie Mountain outcrops. This difference is attributed more to the greater ease of discerning and delineating subfacies in cores than any real lithologic differences between the outcrop and subsurface.

The lithofacies and their interpreted depositional environment are described below:

(1) SHOAL LIMESTONES were deposited on well circulated, open-marine areas lacking well-established reef margins. These shoal limestones characterize the majority of carbonate platform deposits and certain reefal intervals. These limestones consists of rudstones and floatstones containing cylindrical stromatoporoids and branching corals.

Three major subfacies are recognized. One is dominated by cylindrical stromatoporoids with subordinate tabular stromatoporoids, a second is dominated by branching corals and the third consists of a mixture of cylindrical stromatoporoids and branching corals. A fourth subfacies, containing distinctively small cylindrical stromatoporoids, is recognized in the Norman Wells subsurface. This subfacies characterizes a more sheltered shoal environment.

(2) BASIN AND FORESLOPE-TOE LIMESTONES AND SHALES were deposited in low-energy, deep-water settings along lower parts of

the foreslope and the adjacent basin floor. Two subfacies are recognized:

(a) Nodular Lime Mudstones consist of bioturbated micritic limestones with a characteristic nodular texture con- taining a sparse to abundant fauna of crinoids and brachiopods. The nodular texture appears to have resulted from a combination of burrowing and selective early cementation accentuated by later compaction. Deposition of these sediments is inferred to have taken place in deeper, low-energy dysaerobic to aerobic environments.

(b) Basinal Laminites are dark unfossiliferous lime mudstones or shales with essentially undisturbed planar millimetre laminations of carbonate or clay and carbonaceous matter. These sediments may contain a sparse pelagic fauna of conodonts, stylolinids and tentaculitids. The sediments represent accumulation in deep-water, oxygen-starved euxinic conditions of the surrounding basin.

(3) FORESLOPE LIMESTONE include both in situ stromatoporoid and coral bearing limestones and allochthonous rubble and sand-size detritus.

(a) In Situ Foreslope Limestones include two major subfacies. Cylindrical Stromatoporiod Foreslope Lime- stones consists of rudstones containing cylindrical stromatoporoids and lesser amounts of Amphipora. Branching corals occur abundantly in places. Desposition occurred in well-circulated, intermediate-energy environments similar to that for the staghorn coral Acropora cervicornis in recent reefs. These limestones occur either downslope of the reef margin or along more sheltered reef margins.

Middle Foreslope Wafer-Cylindrical-Stromatoporoid Limestones characterize deeper water foreslope settings. These limestones consist of micritic rudstones and boundstones with abundant thin-wafer and cylindrical stromatoporoids along with common crinoids and brachiopods (Fig. 10). The wafer stromatoporoids have a growth form similar to some platy corals such as in recent deep forereefs and, like these corals, may have been adapted to reduced light conditions.

(b) Allochthonous Foreslope Limestones include both coarse stromatoporoid rubble and fine lime sands. Stromatoporoid Rubble consists of rudstones containing abraded thick, tabular, encrusting, and cylindrical stromatoporoid fragments and a carbonate sand matrix (Fig. 11). These limestones characterize upper foreslope settings. Foreslope Carbonate Sands are made up of micritized skeletal grainstones and packstones interbedded with micritic basinal sediments (Fig. 12). These lime sands were shed from upslope shallow reef environments where they were formed by the physical and biological breakdown of larger reef building organisms.

(4) REEF MARGIN LIMESTONES formed in high-energy, wave swept, turbulent conditions and consist of alternating in situ thick-tabular stromatoporoids (boundstones) up to a few tens of centimetres thick and broken-up stromatoporoid rubble (Fig. 13). The thick-tabular stromatoporoids were adapted to an encrusting habitat, generally comparable to that of the coralline red algae Lithothamnion in recent reefs.

(5) REEF FLAT LIMESTONES consist of an assemblage of brecciated and/or well-washed Amphipora, cylindrical and more massive stromatoporoid rudstones (Fig. 14). More irregular stromatoporoids commonly encrust abraded stromatoporoid clasts. Deposition occurred on shallow flats, immediately behind reef margins, subject to intensive reworking by major storms.

Two subfacies are recognized in the Norman Wells subsurface. Stromatoporoid Rubble occurs immediately behind reef margins. Bulbous Encrusting, Cylindrical Stromatoporoid Reef Flat Limestones represent more in situ growth and accumulation under less intense wave energy, commonly behind the more rubble reef flat.

(6) REEF INTERIOR LIMESTONES consist of cyclic subtidal lagoonal and tidal-flat limestone deposited in relatively sheltered

Figure 11. Allochthonous foreslope stromatoporoid rubble, Ramparts Formation. Sequence composed mainly of broken cylindrical stromatoporoids and thamnoporids with a carbonate sand matrix. Section 18, unit 038. Scale in decimetres.

Figure 12. Allochthonous foreslope carbonate sand facies, Ramparts Formation. Possible dewatering structure which triggered small-scale synsedimentary faulting (arrows). Section 1, unit 031, Powell Creek. Scale is 15 centimetres.

Figure 13. Typical stromatoporoid morphologies in Ramparts reef-margin boundstones. Robust cylindrical stromatoporoid rudstone matrix. Hammer is 30 centimetres long. Section 4.

Figure 14. Spar-reduced interparticle porosity in reef-flat facies. Note overturned abraded bulbous stromatoporoids and robust cylindrical stromatoporoids. Scale is 15 centimetres. Section 4.

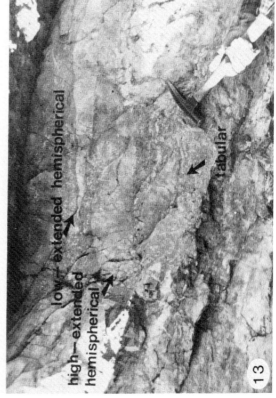

environments behind the reef margin. Subtidal Limestones include either stomatoporoid rudstones and floatstones with a generally micritic matrix or more micrite-lean peloidal packstones. Bulbous and cylindrical stromatoporoids occur in the more open-lagoonal settings; Amphipora predominates in the more restricted lagoons. Tidal-Flat Limestones represent deposition along the shorelines of islands in the interior reef lagoon. These limestones formed along low— energy shorelines, above mean low tide, and generally consist of pelletal mudstones to packstones with crinkly cryptalgal laminations.

In outcrop, subtidal and tidal-flat limestones are mapped together and are divided into two subfacies with differing cycle characteristics. "Shallow" Reef Interior Limestones are characterized by relatively thin cycles, restricted Amphipora subtidal facies and common tidal-flat deposits. "Deep" Reef Interior Limestones are characterized by thicker cycles, more open-marine bulbous and cylindrical stromatoporoids as well as Amphipora, and by uncommon tidal-flat deposits.

CYCLICAL CARBONATE PLATFORM-REEF GROWTH

Major changes in the distribution and type of facies occur during the evolution of the Ramparts (Kee Scarp) platform-reef complex in both the Mackenzie Mountains and in the subsurface at Norman Wells. Major shifts in facies in these complexes occur at distinct levels shown on the facies cross-sections in Figures 15 and 16 (in pocket at end of book). As illustrated on the cross-sections, these levels can be traced from one foreslope to the other within each complex. Stratigraphic units between each level consist of an overall upward-shoaling sequence of facies and represent a single megacycle or second-order cycle of sedimentation. The change from one megacycle to a succeeding megacycle is marked by an abrupt shift to deeper water facies.

Eleven megacycles are recognized within the Kee Scarp Formation in the subsurface at Norman Wells. There are labelled K1A through K4 and form the basis for reservoir zonation in the complex (Fig. 15). Many of the tops of these megacycles have been correlated, on the basis of their relative position to the "Carcajou Marker", to cycle boundaries in the Mackenzie Mountain Ramparts complex (Fig. 16). Cycle boundaries in the Mackenzie Mountain complex are designated by their equivalent subsurface zone.

Comparison of the cross-sections through both complexes show that only the subsurface megacycles with the greatest shift in facies can be recognized in the Mackenzie Mountain complex. In addition, the cross section in Figure 16 also shows that the Mackenzie Mountain complex contains two additional megacycles of reef growth (termed K5 and K6) not present in the subsurface Norman Wells reef complex. These additional two cycles account for the appreciably thicker reef section in the Mackenzie Mountains than in the Norman Wells subsurface.

The recognition of these second-order megacycles is particularily important to understanding the evolution of the Ramparts (Kee Scarp) reef complexes. We believe that the cyclic nature reflects an episodically rising sea level. The initiation of each megacycle corresponds to a marked increase in the rate of sea-level rise. During these times reef growth was generally outpaced by rising sea level. At times of lower sea-level rise (often termed stillstand), reef growth kept pace or exceeded the rate of sea-level rise, producing the characteristic upward-shoaling sequences.

Major changes in sea level for any given megacycle are made up of minor increments most accurately recorded by the response to the interior lagoonal sediments. Sediments in the lagoon form at much lower rates than those in the more actively growing reef margin. Consequently, these sediments may be outpaced by a sea-level rise that has little or no effect on the more competent margin. These minor rises of sea level produced the cyclic pattern of lagoonal sedimentation in the Norman Wells reef complex shown on the cross section in Figure 15. Where completely developed each lagoonal or third-order cycle grades from subtidal limestones of varying nature up to tidal flat cryptalgal laminities. Individual cycles (in many cases) can be traced across the reef interior, but only at megacycle (second-order cycle) boundaries do they extend to the reef and foreslope sediments.

The use of cycles defined in both the Mackenzie Mountain and the subsurface Norman Wells reef complexes enables true growth stages to be defined. By employing this method, distinct phases of platform and reef growth could be discerned. These phases are discussed in the following section.

EVOLUTION OF HARE INDIAN THROUGH RAMPARTS (KEE SCARP)

Figure 17 (in pocket) depicts the evolution of the Mackenzie Mountain and Norman Wells complexes through successive phase from the end of Hume sedimentation through Ramparts and Kee Scarp reef development. Platform and reef phases are labelled according to their equivalent subsurface reservoir zone. Eight phases are recognized and are described below:

(1) DROWNING OF THE HUME LIMESTONE

A pronounced and rapid sea-level rise at the beginning of Givetian time caused drowning of the Hume Formation. Sea level continued to rise, leading to water depths in excess of 150 metres in both the Mackenzie Mountain and Norman Wells areas.

(2) PROGRADATION OF THE HARE INDIAN SHALE BANKS

As the rate of sea-level rise declined, Hare Indian shale banks began to build out into the previously created deep-water basin. Sedimentation occurred on shale clinoforms, both from clay suspension and carbonate turbidite deposition, and on the basin floor from very slow shale suspension, resulting in a thin accumulation of black, organic-rich shales of the Bluefish Member. As shale bank progradation continued, the areas in the present day locations of the Mackenzie Mountains and the Norman Wells field built closer to sea level and became the sites of _in situ_, relatively shallow-water, carbonate "shale ramp" deposition.

(3) DROWNING OF THE HARE INDIAN SHALE BANKS AND DEPOSITION OF THE CARCAJOU MARKER

Another pronounced and rapid sea-level rise led to cessation of Hare Indian shale bank progradation in both the Mackenzie Mountain and Norman Wells areas. This rapid deepening led to the widespread deposition of the very shaly and carbonaceous "Carcajou Marker".

(4) CARBONATE PLATFROM INCEPTION AND INITIAL GROWTH (K1A) AND LOCALIZED UPBUILDING (K1B-K2A)

Following the preceeding sea-level rise, branching corals quickly became established on topographic highs on the underlying "shale ramp". These corals built toward sea level and gradually became replaced by shallow water cylindrical stromatoporoids. The resulting stromatoporoid and coral-bearing limestones also built basinward, filling in lows and creating broad but still submerged platforms by the end of K1A time.

A subsequent rise in sea level led to incursion of relatively deep water and deposition of foreslope limestones on slightly lower areas of the carbonate platforms. Along more positive areas of the carbonate platforms, stomatoporoids and corals continued to grow and keep pace with the sea-level rise. These conditions led to the creation of more areally restricted highs on the carbonate platforms, with surrounding deeper water areas, during K1B and K2A time.

(5) REEF INCEPTION AND PROGRADATION (K2B) AND SUBSEQUENT AGGRADATION (K2C THROUGH K3B)

Another sea-level rise led to temporary cessation of carbonate growth. Reef-forming stromatoporids soon colonized the margins of the previous platform highs and thus initiated reef growth. During K2B time the reefs built very slowly upwards but quickly prograded out and filled adjacent lows, thereby forming more areally extensive reefal deposits. The stromatoporoid reef margins sheltered an interior lagoon, which became the site of deposition of <u>Amphipora</u> limestones and tidal-flat sediments.

Successive sea-level rises following K2B time produced only small increments of backstepping of the reef margins, so that the reef margins essentially grew upwards in the same general positions. Not all the sea-level rises had the identical effect in both buildups, however. One notable difference is the creation of a widespread K3A shoal deposit in the subsurface Norman Wells reef complex and its absence in the Mackenzie Mountain complex. With the exception of the K3A zone at Norman Wells, shallow-water lagoonal and tidal-flat deposits continued to form in the sheltered interiors of both complexes.

(6) REEF BACKSTEPPING AND SUBSEQUENT AGGRADATION (K3C THROUGH K4)

A relatively pronounced rise in sea level led to appreciable backstepping of the reef margins, especially in the subsurface Norman Wells reef complex. Following this rise, the reef margins accreted upwards, with only small increments of backstepping, producing an aggrational style of reef growth again. As a result of the rapid sea-level rise at the onset of this phase, a more open-marine and slightly deeper water style of lagoonal sedimentation became established in the Mackenzie Mountain complex. Shallower, but higher energy peloidal sand deposition occurred in the Norman Wells reef interior.

(7) DROWNING OF THE NORMAN WELLS REEF COMPLEX AND BACKSTEPPING OF THE MACKENZIE MOUNTAIN REEF COMPLEX (K5-K6)

Another relatively rapid sea-level rise at the beginning of K5 time led to drowning of the Norman Wells reef complex and backstepping of the reef in the Mackenzie Mountains. During the first megacycle (K5) of growth in the Mackenzie Mountain complex, open-marine, slightly deeper water lagoonal sedimentation continued in the reef interior. Another sea-level rise at the beginning of K6 time led to the formation of an areally restricted cylindrical stromatoporoid-coral shoal.

(8) DROWNING OF THE MACKENZIE MOUNTAIN REEF COMPLEX (POST K6)

Another rapid rise in sea level caused the ultimate drowning of the Mackenzie Mountain reef complex. Following cessation of reef growth, both complexes remained uncovered by sediments for a considerable period of time. Eventually prograding Imperial clinothem and Canol fondothem sediments onlapped or downlapped onto the reef complexes and finally covered them.

SUMMARY

Study of the Hare Indian through the Ramparts (Kee Scarp) Formations indicate that distinct cycles of sedimentation can be correlated between the Mackenzie Mountain and Norman Wells complexes. Three general levels of cycles at present. Each cycle consists of an upward-shoaling sequence of facies corresponding to a distinct phase of sedimentation initiated by accelerated rates of relative sea-level rise. The correlation and the small-scale periodicity of the second and third-order cycles suggest that eustatic sea-level rises produced the accelerated rates of sea-level rise that initiated the cycles of sedimentation. Alternate mechanisms, such as changes in the rate of either tectonic subsidence or subsidence due to underlying shale compaction, operate on much longer term periods and consequently would not initiate the observed cycles of sedimentation.

First-order cycles consist of thick sequences, at least 200 metres thick, that are on the same scale as "depositional sequence" defined from seismic stratigraphic studies (Vail and others, 1977). These cycles can be correlated throughout the area, from shallow platform to deep basinal settings. In the Mackenzie Mountain and Norman Wells areas, initial parts of these cycles are characterized by relative rapid sea-level rises and by the backstepping and aggradation of carbonate platforms and reefs in shallow-water areas and by condensed basinal sedimentation in offshore or deeper water areas. Later parts of these cycles are characterized by much slower rates of relative sea-level rise and by the progradation of large siliciclastic banks out into the basin.

First-order cycles contain a number of smaller second-order cycles initiated by accelerated rates of relative sea-level rise. These cycles are best defined in shallow-water platforms and reef complexes, where they can be traced across the entire complex, but also can be observed in the Hare Indian shale clinothems. In shallow-water areas, these cycles are commonly characterized by the backstepping of platform and reef margins. Recognition of these second-order cycles is particularly important because it allows growth stages to be defined and delineated within reef complexes and to be correlated between complexes. One important ramification of correlating these cycles between complexes is that the response to a particular sea-level rise can be seen to vary according to setting.

Even smaller third-order cycles are present but are confined to platform and reef lagoons. The reason for this more restricted occurrence is that small increments of sea-level rise may have no appreciable effect on the rapidly accreting platform of reef margin, but may be sufficiently rapid to outpace sedimentation in the more sheltered lagoon, and thus initiate new cycles of sedimentation.

The successful correlation of cycles of varying magnitudes between two widely separated complexes suggests that the same procedure applied elsewhere should aid in unravelling the evolution of other carbonate provinces.

REFERENCES

Bassett, H.G. and Stout, J.G., 1967. Devonian of Western Canada: in D.H. Oswale (ed.), International Symposium on the Devonian system, Calgary, 1967. v. 1, p. 717-752.

Braun, W.K.., 1966, Stratigraphy and microfauna of Middle and Upper Devonian Formations, Norman Wells area, Northwest Territories, Canada: Neues Jahrbuch fur Geologie and Palaontologie Abhandlungen. v. 125, p. 247-264.

Braun, W.K., 1977, Usefulness of ostracods in correlating Middle and Upper Devonian sequences in Western Canada: in M.A. Murphy, W.B.N. Berry, and C.A. Sandberg (eds.), Western North America: Devonian. Univ. of California (Riverside) Campus Museum Contribution No. 4, p. 65-79.

Braun, W.K., 1978, Devonian ostracodes and biostratigraphy of Western Canada: in C.R. Stelck and B.D.E. Chatterton (eds.), Western and Arctic Canadian Biostratigraphy. Geol. Assoc. Can. Spec. Paper 18, p. 259-288.

Gilbert, D.L.F., 1973, Anderson, Horton, Northern Great Bear and Mackenzie Plains, Northwest Territories: in R.G. McCrossan (ed.), The future petroleum provinces of Canada. Can. Soc. Pet. Geol., Mem. 1, p. 213-244.

Vail, P.R., Mitchum, R.M. Jr., Todd, R.G., Widmier, J.M., Thompson, S., III, Sangree, J.B., Bubb, J.N., and Hatelid, W.G., 1977, Seismic stratigraphy and global changes of sea-level: in C.E. Payton (ed.), Seismic stratigraphy - applications to hydrocarbon exploration. Am. Assoc. Petrol. Geol. Mem. 26, p. 49-212.

Warren, P.S. and Stelck, C.R., 1962, Western Canadian Givetian: J. Alberta Soc. Pet. Geol., v. 10, p. 273-291.

Wendte, J.C. and Stoakes, F.A, 1982, Evolution and corresponding porosity distribution of the Judy Creek reef complex, Central Alberta: in W.G. Cutler (ed.), Canada's Giant Hydrocarbon Reservoirs: Am. Assoc. Petrol. Geol.-Can. Soc. Petrol. Geol. Core Conference, Calgary, 1982, p. 63-81.

SEDIMENTOLOGY OF A CARBONATE SOURCE ROCK:

THE DUVERNAY FORMATION OF ALBERTA CANADA

F. A. STOAKES
Exploration Department
Esso Resources Canada Limited
237 - 4th Avenue S.W.
Calgary, Alberta T2P OH6
Canada

S. CREANEY[1]
Exploration Department
Esso Resources Canada Limited
237 - 4th Avenue S.W.
Calgary, Alberta T2P OH6
Canada

ABSTRACT

The Duvernay Formation is an organic-rich basinal carbonate succession that has provided most of the petroleum presently found in Leduc-age reefs of east central Alberta. It accumulated under deep-water anoxic bottom conditions that favoured the preferential preservation of organic material. This, together with slow sedimentation rates resulted in a rich (up to 17 wt. percent TOC) source rock. Rich source intervals occur as dark black laminites interbedded on a fine scale with leaner bioturbated lime mudstones. Intervals sampled had very low amounts of insolubles, rendering them true carbonate source rocks.

Comparison of reservoir oil Level of Organic Metamorphism (LOM) and source rock LOM suggests that long distance secondary migration has taken place within the basin with distances on the order of 100 kms (60 miles) being indicated. Migration took place through a dolomitized aquifer underlying the Leduc reefs, with the amount of petroleum entering each reef being dependant on the nature of vertical permeability barriers separating the two. A portion of the reef-platform system conforms to a classical example of spill-point updip displacement of petroleum, but other portions do not, and an understanding of the mode of migration into these buildups is important. Furthermore, it seems conceivable that the Duvernay Formation could have contributed significantly to hydrocarbons presently found in the giant Athabasca and associated tar sands. The multidisciplinary approach embodied in this study has resulted in a better understanding of carbonate source rocks in general, as well as the apprection of a more fully integrated exploration model.

---

[1] Present Address: Exxon Production Research Company, P.O. Box 2189, Houston, TX 77001

INTRODUCTION

Carbonate rocks are generally considered to provide excellent reservoirs for petroleum but to exhibit very low source potential. This is based, in large part, on the high percentage of the world's oil currently in carbonate reservoirs and their low average analyzed total organic carbon (TOC) content. This has led many authors to seriously question whether carbonate rocks can provide adequate sources for petroleum, and in many cases to seek out alternative more distant or deeply buried "shale" sources.

In this paper we will attempt to show that the majority of the petroleum found in the Upper Devonian of east-central Alberta was derived from the maturation of, and subsequent migration from the Duvernay Formation, an organic-rich basinal carbonate unit.

The core exhibit will illustrate sedimentologic and depositional controls on the preferential accumulation, quality and preservation of organic matter in a carbonate source rock. Evidence for subsequent maturation and migration relies heavily on an understanding of burial history and measurement of a number of geochemical parameters in both the source and reservoir intervals. The availability of cored material from the Duvernay Formation and samples of oils found in adjacent Leduc reefs provides an unparalleled opportunity to study systematic maturation patterns within the source rock and the subsequent migration and entrapment of petroleum.

STRATIGRAPHY AND SEDIMENTOLOGY

Geological Setting Of The Duvernay Formation

The Duvernay Formation (Upper Devonian) of east central Alberta is a typical marine, oil-prone source rock (Stoakes and Creaney, 1984) (Fig. 1). In the East Shale Basin (Fig. 2) the unit overlies platform carbonates of the Cooking Lake Formation with only minor discontinuity, and is the "starved" basin equivalent of surrounding Leduc reefs (Stoakes, 1980). Where the Cooking Lake Formation is absent, as in the West Shale Basin (Fig. 1), the Duvernay conformably overlies rocks of similar basinal aspect referred to as the Majeau Lake Member. The Majeau Lake Member is the basinal equivalent of the Cooking Lake platform and, as such, pre-dates reef growth in this part of the basin and consequently the majority of Duvernay deposition. Lithologically the Majeau Lake Member is identical to the Duvernay Formation and undoubtedly contributed hydrocarbons to the Leduc reservoirs. For convenience it has been grouped with the Duvernay west of the Rimbey-Morinville Leduc reef trend.

Lithofacies Of The Duvernay Formation

Carbonate rocks of the Duvernay Formation represent accumulation in marine, deep water, low-energy, basinal conditions that existed between Leduc reef complexes. Sediments reflect accumulation in water depths of greater than 100 meters (300 feet) based on detailed paleobathymetric reconstructions (Fig. 3 and

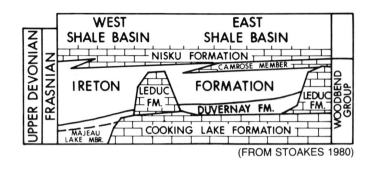

Figure 1. Stratigraphy of east central Alberta.

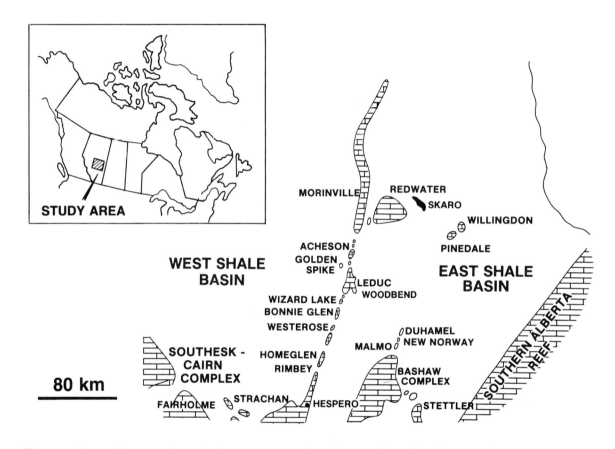

Figure 2. Map of study area showing distribution of Leduc reef complexes and intervening basinal areas.

Figure 3. Paleobathymetric profile of Duvernay and Ireton formations.

Stoakes, 1980, Fig. 42), but may ocur as shallow as 65 meters (200 feet) in areas of poor circulation.

The Duvernay Formation comprises a number of interbedded lithofacies:

Nodular to Nodular-Banded Lime Mudstones

These rocks exhibit an irregular fabric of dense micritic carbonate and more banded argillaceous lime mudstone (Figs. 4a, 4b). This results from bioturbation and subsequent differential cementation of the sediment (Stoakes, 1980). Individual nodules show a variation in morphology from highly irregular to more-or-less banded (Figs. 4a, 4b). Nodules exhibit a gradational contact with the enclosing matrix and often show evidence of bioturbation. More banded lime mudstones (Figs. 4c, 4d) show evidence of horizontal grazing traces.

The presence of only a sparse epifauna and evidence of varying degrees and types of bioturbation suggests oxygen-reduced conditions of the water column (and within the sediments) for many of these nodular mudstones. The predominance of horizontal grazing trails in the more banded sediments suggests that euxinic conditions may have existed below the sediment-water interface.

Analyses of nodular bioturbated sediments generally give total organic carbon (TOC) values lower than 0.5 percent by weight and high isolubles (Fig. 4a). More banded to laminated carbonates give higher TOC's, up to five percent, and much lower values of insolubles (Figure 4d; see also Fig. 6).

FIGURE 4

NODULAR LIME MUDSTONES

		DESCRIPTION	ENVIRONMENTAL INTERPRETATION	T.O.C. (WT. %)	INSOLUBLES (WT. %)
a)		BIOTURBATED IRREGULAR LIGHT GREY CARBONATE-RICH NODULES IN A DARK GREY/GREEN ARGILLACEOUS WACKESTONE MATRIX. SPAR-FILLED FRACTURES CUT ACROSS INDIVIDUAL NODULES. DUVERNAY FORMATION IMP. CDN. SUP. BUCK CREEK 14-29-48-6W5 2640.8 m, 8664 ft	TOE OF SLOPE DYSAEROBIC	0.25	75
b)		BIOTURBATED LIGHT GREY CARBONATE-RICH NODULES IN A DARKER GREY/GREEN ARGILLACEOUS WACKESTONE MATRIX. MARGINS OF NODULES ARE DIFFUSE AND APPEAR TO BE CONTINUOUS LATERALLY INTO THE MORE COMPACTED MATRIX. DUVERNAY FORMATION BANFF IMP. ET. AL. LEDUC 10-34-49-25W4 1763 m, 5784.5 ft	TOE OF SLOPE DYSAEROBIC		

BIOTURBATION

		DESCRIPTION	ENVIRONMENTAL INTERPRETATION	T.O.C. (WT. %)	INSOLUBLES (WT. %)
c)		DARK PLANAR MILLIMETRE LAMINATED CARBONATE MUDSTONE PASSING UP INTO MORE BURROWED CARBONATE MUDSTONE. MOST OF THE BURROWS ARE HORIZONTAL TO SUBHORIZONTAL GRAZING TRAILS PROBABLY REFLECTING THE EUXINIC NATURE OF THE SEDIMENT. DUVERNAY FORMATION IMP. CDN. SUP. BUCK CREEK 14-29-48-6W5 2656 m, 8715 ft	BASINAL DEEP RESTRICTED MARGINALLY EUXINIC		
d)		DARK PLANAR MILLIMETRE LAMINATED CARBONATE MUDSTONE WITH ABUNDANT LIGHT GREY CARBONATE-RICH MASHED HORIZONTAL GRAZING TRAILS. DUVERNAY FORMATION IMP. TOMAHAWK 16-18-52-5W5 2340 m, 7677 ft	BASINAL LAMINITE DEEP RESTRICTED EUXINIC	4.55	9.02

Figure 4.

Laminated Lime Mudstone

This lithofacies comprises dark bituminous lime mudstones with planar millimeter laminations (Fig. 5). Laminations appear to consist of very fine carbonate material and organic-rich layers, essentially undisturbed (Fig. 5). They may contain a sparse pelagic fauna of conodonts, stylolinids, tentaculitids and ostracods, and bear no evidence of bioturbation.

These sediments represent accumulation in deep water oxygen-starved euxinic conditions. Analyses of these sediments provides some of the highest values of TOC in the Duvernay section. They generally average between five and ten percent TOC with low to negligible insolubles (Fig. 5). Analyses adjacent to the Redwater Leduc reef (Fig. 2) have produced individual values of around 17 weight percent. percent TOC, values that would compare favourably with an oil shale!

The low values of insolubles, in compositional analyses, reveal these are indeed, true carbonate source rocks.

Lateral and Vertical Distribution of Rich Source Intervals

Rich organic source intervals appear confined to the dark laminated mudstones described above. In vertical section these source intervals are present as discrete interbeds along with more bioturbated intervals. However, systematic larger-scale changes are apparent (Fig. 6). Dark laminated sediments mark cycle bases of pronounced deepenings resulting from an increased rate of sea level rise (e.g., 8712 ft. on core log, Fig. 6). As the depositional surface shallowed

FIGURE 5

CARBONATE-RICH LAYERS

		DESCRIPTION	ENVIRONMENTAL INTERPRETATION	T.O.C. (WT. %)	INSOLUBLES (WT. %)
a)		PLANAR MILLIMETRE LAMINATED CARBONATE MUDSTONE. FINE CARBONATE-RICH LAYERS SHOW EARLY NODULAR CEMENTATION AND SUBSEQUENT DIFFERENTIAL COMPACTION. IN UPPER PART OF PHOTO CARBONATE LAYERS LOOK LIKE MASHED HORIZONTAL GRAZING TRAILS. DUVERNAY FORMATION IMP. TOMAHAWK 16-18-52-5W5 2334.5 m, 7658.5 ft	BASINAL LAMINITE DEEP RESTRICTED EUXINIC	4.19	10.6
b)		DARK BLACK PLANAR MILLIMETRE LAMINATED CARBONATE MUDSTONE. FINE LIGHT GREY CARBONATE DEFINES MANY LAMINATIONS AND A CARBONATE-ENRICHED BAND IS PRESENT IN THE UPPER CENTRE OF THE PHOTOGRAPH. DUVERNAY FORMATION IMP. CDN. SUP. BUCK CREEK 14-29-48-6W5 2654.3 m, 8708 ft	BASINAL LAMINITE DEEP RESTRICTED EUXINIC	1.6	3.3

LAMINITES

		DESCRIPTION	ENVIRONMENTAL INTERPRETATION	T.O.C. (WT. %)	INSOLUBLES (WT. %)
c)		DARK BROWN/BLACK FINE MILLIMETRE PARALLEL LAMINATED CARBONATE MUDSTONE. SCATTERED LIGHT COLOURED CARBONATE SHELL FRAGMENTS DEFINE LAMINATIONS. MAJEAU LAKE MEMBER IMP. CDN. SUP. BUCK CREEK 14-29-48-6W5 2722.2 m, 8931 ft	BASINAL LAMINITE DEEP RESTRICTED EUXINIC	5.22	
d)		DARK BROWN/BLACK FINE MILLIMETRE PARALLEL LAMINATED CARBONATE MUDSTONE. FINE LIGHT GREY CARBONATE DEFINES MANY OF THE LAMINATIONS. DUVERNAY FORMATION IMP. TOMAHAWK 16-18-52-5W5 2339 m, 7677 ft	BASINAL LAMINITE DEEP RESTRICTED EUXINIC	9.88	3.22

Figure 5.

marginally, more oxygenated conditions allowed the establishment of a burrowing infauna that bioturated and disturbed the laminated fabric. These organisms probably consumed the organic matter as well as introduced oxygenated sea water into the sediment, which resulted in its further destruction through oxidation. Minor interbedding of laminated source and bioturbated non-source intervals may reflect more rapid turnover of the water column, resulting from major storm events. This fine interbedding of source and non-source intervals could lead to dilution of TOC values when dealing with analyses of drill cuttings averaged over ten-foot or three-metre intervals.

MODEL OF SOURCE ROCK ACCUMULATION

Accumulation of organic-rich source intervals, then, appears to be governed by the presence of euxinic conditions at or below the sediment—water interface, a process well documented by Demaison and Moore (1980). This has the effect of inhibiting the burrowing infauna and allowing the accumulation and preferential preservation of organic matter. Stoakes (1980) has proposed a tripartite layering of the water column based on degree of oxygenation within the Duvernay-Ireton basin (Figure 3). This is based primarily on observation of the distribution of shelly fauna and bioturation and follows the general model proposed by Byers (1977). In shallow portions of the basin aerobic well-circulated conditions predominate as evidenced by an abundant shelly fauna and ubiquitous infaunal bioturbation. At intermediate depths dissolved oxygen content decreases and a major change in faunal diversity takes place. These marginally oxygenated waters (dysaerobic, less than 2 ml/l dissolved oxygen) contain communities with lower diversity, fewer calcified species and are dominated by burrowing infauna. The lower stagnant zone is characterized by a complete lack of shelly fauna or burrowing infauna resulting in conspicuously laminated sediments. This results from a complete absence of oxygen. Similar faunal changes may result from salinity rather than oxygen fluctuations but this alternative may be somewhat ruled out by the complete absence of evaporite deposits in this part of the section.

The values of paleo-water depth given in Figure 3 serve as a general guide and may be expected to vary with the stage of basin evolution and circulation patterns. Many of the smaller scale variations in source quality and amount reflect short-term local perturbations of the system.

GEOCHEMISTRY

Abundant available core control within the Duvernay Formation and samples of Leduc Formation oils provide an unparalled opportunity to study systematic maturation changes in the source rock and the progressive secondary migration of oils to adjacent and distant reservoirs.

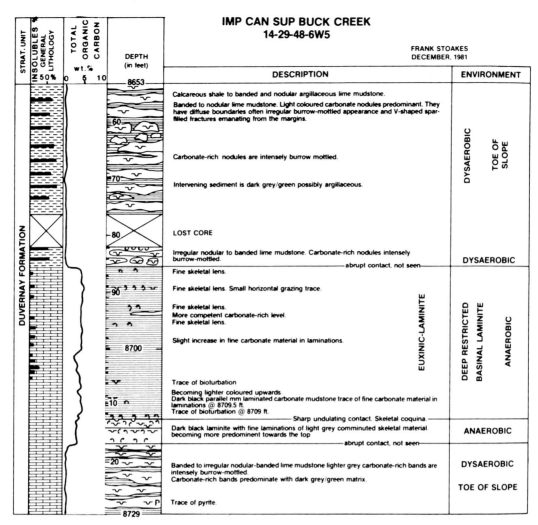

Figure 6. Graphic core log and analyses of Duvernay section.

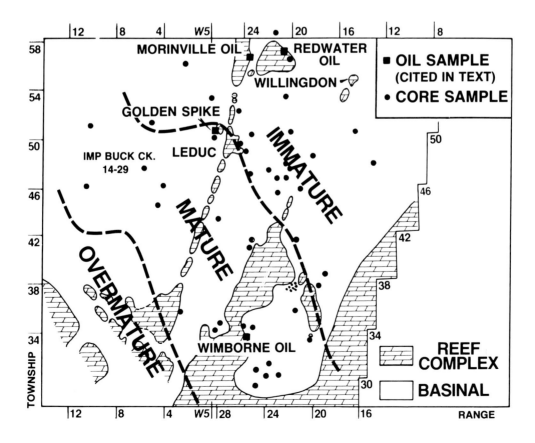

Figure 7. Map of Duvernay source rock maturity showing core control points and selected oil samples.

Duvernay Laminites

This study utilized samples obtained from over 50 cored intervals within the Duvernay Formation, and to a lesser extent the Majeau Lake Member west of the Rimbey-Morinville chain (Fig. 7). A number of analyses were performed to determine (1) Organic matter quality (OMT - organic matter type), (2) Organic matter quantity (TOC) and (3) Source rock maturity (LOM - Level of organic metamorphism).

Before a detailed discussion of Duvernay Formation source rock potential can be understood, a few generalizations must be understood.

(1) Organic matter within the Duvernay Formation comprises almost entirely oil-prone, unstructured, Type II kerogen (classification of Tissot and others, 1974). In the case of the Duvernay this reflects material accumulated in a marine setting probably from a planktonic origin. The Devonian age of these sediments assists this distinction by largely precluding strong influences of terrestrial organic matter.

(2) TOC analyses were performed on washed, dried and ground core following the removal of the carbonate matrix with a solution of hydrochloric acid. Duvernay laminites generally average 5 to 10 percent TOC by weight but individual samples have been measured up to 17 weight percent TOC.

(3) Methods of assessing source rock maturity are many and varied. In this paper an attempt is made to introduce the reader to the

process of hydrocarbon generation via a discussion of the interplay of source rock composition and thermal maturity. Throughout this discussion maturity will be referred to in terms of the LOM (Level of Organic Metamorphism) scale from Hood and others (1975). This scale provides a convenient means of discussing maturity and can be used to correlate maturity data derived from a variety of different analyses. Figure 8 illustrates the correlation of LOM to vitrinite reflectance.

Source Rock Maturation

The effect of increasing maturity on a source rock can be observed as progressive changes in the solid organic residue or as changes in the hydrocarbons being released into source rock porosity from the solid organic matter.

The most convenient way to express the various stages of source rock maturity is via an analysis of the contents of source rock porosity. The onset of maturity can be defined as the first point at which thermogenic hydrocarbons appear in source rock porosity. Analyses which sample pore contents include Total Solvent Extraction, Cuttings Gas analysis, and Rock-Eval pyrolysis (S_1). These analyses also interact with the solid organic matter, to some extent, but in oil-prone source rocks this contribution is overwhelmed by liquid hydrocarbons appearing in source rock porosity. Figure 9 is a composite plot of total extract yield (normalized to total organic carbon) and S_1 (in mg of hydrocarbons per gram of rock) against LOM for Duvernay core samples. It is apparent that hydrocarbons begin to

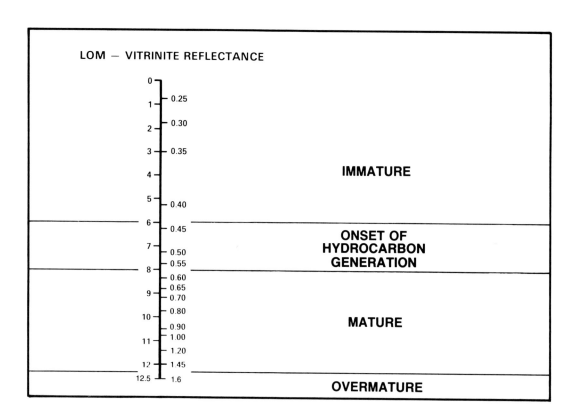

Figure 8. Correlation of LOM (level of organic metamorphism) and vitrinite reflectance (modified after Hood and others, 1975).

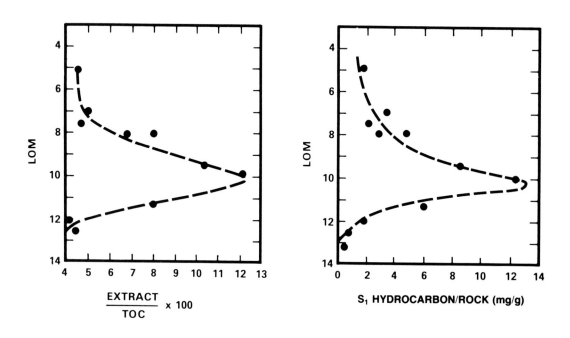

Figure 9. (a) Relationships between total extract yield (expressed as a percent of total organic carbon) and maturity (LOM) for the Duvernay Formation.

(b) Relationship between Rock-Eval pyrolysis S_1 yield (in milligrams of hydrocarbon per gram of rock) and maturity (LOM) for the Duvernay Formation.

appear in source rock porosity at an LOM of 6-8 (vitrinite reflectance = 0.43 - 0.57). A maxima occurs in the curves at an LOM of 10 - 10.5 (vitrinite reflectance = 0.82 - 0.92) indicating maximum oil saturation in the source rock. It is at this maturity that maximum oil expulsion from this type of source rock can be expected. As maturity increases further the amount of hydrocarbon in the porosity decreases, approaching zero at an LOM of 12 - 12.25 (vitrinite reflectance = 1.45 - 1.60). This latter LOM is considered to be the onset of overmaturity and is characterized by the absence of liquid hydrocarbon in porosity and a switch to gaseous hydrocarbon preservation only. This is better observed using cuttings gas analysis, since extract and pyrolysis analyses do not observe hydrocarbon in the $C_1 - C_4$ range. Bailey and others (1974) summarized the effect of overmaturity on marine source rocks. Briefly, the percentage of C_1 + components decreases across the mature/overmature boundary until only C_1 remains. Figure 10 shows the trend in compound depletion with passage into the zone of overmaturity, based on analyses of many wells in the Western Canada Basin. Obviously, the ability to retain liquid hydrocarbons in source rock porosity is gone prior to an LOM of 12.5 (vitrinite reflectance = 1.60) and wet gas (C_2+) is absent by an LOM of 13 (vitrinite reflectance = 1.85).

Figure 7 is a regional maturity map for the Duvernay in east central Alberta. The LOM data are derived from a combination of Rock-Eval pyrolysis (T max) data on the source interval itself plus vitrinite reflectance information projected from overlying Lower Cretaceous coals. The tectonic and thermal history of this basin are relatively simple (Deroo and others, 1977) and hence strong correlations between maturity and present burial depth (which approximates maximum burial depth in this area) are possible (Fig. 11) and allow for contouring of the separate data points.

Oil - Source Correlation

GC-MS representation of steranes and triterpanes from both source and reservoired oils in the basin has allowed correlation of oils from the Leduc Rimbey-Meadowbrook trend reefs to the Duvernay, confirming the geological observation that this was the most likely source of the hydrocarbons.

Reservoir Hydrocarbon Maturity

The data base for this assessment comes from analyses of ninety-three oils from separate accumulations (primarily Leduc-age reefs) whose source rock is known or thought to have been the Duvernay interval.

In recent years a few authors have published work which attempted to assess the maturity (LOM) of reservoir hydrocarbons (Lijmbach and others, 1981; James, 1983). This approach provides the possibility of increasing are insight into the secondary migration pathways taken by oil and gas from source to trap. In the present study the same approach has been used to test the oil generation model outlined above for a marine source rock. If the total extract components removed from a source rock are representative of oil

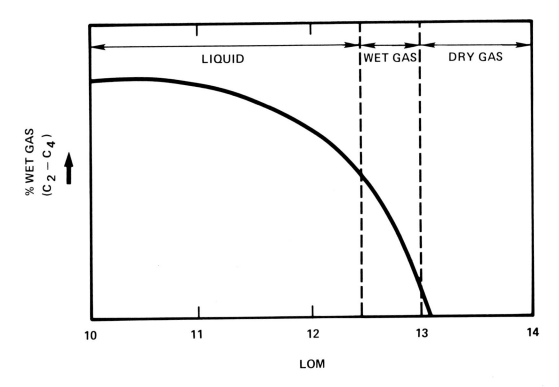

Figure 10. Schematic relationship between cuttings gas composition and advanced maturity (LOM) (based on analyses of many wells in Western Canada - see Bailey and others, 1974 for an example of an individual profile).

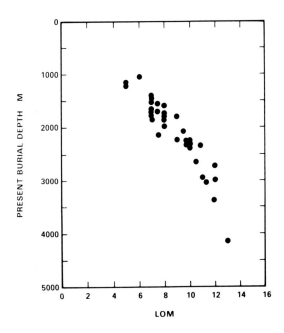

Figure 11. Relationship between maturity (expressed as LOM) and present burial depth for the Duvernay Formation in east-central Alberta.

retained in source rock porosity then they will reflect the maximum maturity attained by that source rock. Figure 12 is a composite of whole extract gas chromatograms from a series of Duvernay cores of increasing maturities. Certain progressive changes can be observed:

(a) Virtually no n-alkanes are present in the immature extract (LOM 5) and the n-alkane spectrum develops with increasing maturity.

(b) The unresolved "hump" (steranes etc.) on the far right of the chromatograph decreases progressively with increased maturity.

(c) At LOM values greater than 12 (vitrinite reflectance = 1.45) the heavy n-alkanes diminish extremely rapidly as thermal degradation "cracks" the long chain hydrocarbons.

Overall comparison of these total extract G.C. profiles with whole oil gas chromatograms of oils from Leduc reef reservoirs (whose source is the Duvernay) allows the assignment of an approximate LOM to the oil. This assignment is made based on n-alkane/isoprenoid ratios, overall n-alkane profile etc. The resulting LOM value is obviously somewhat subjective and, as with all attempts to apply a maturity value to a mobile phase, subject to a number of sources of error. The prime sources of error and an assessment of their applicability would be:

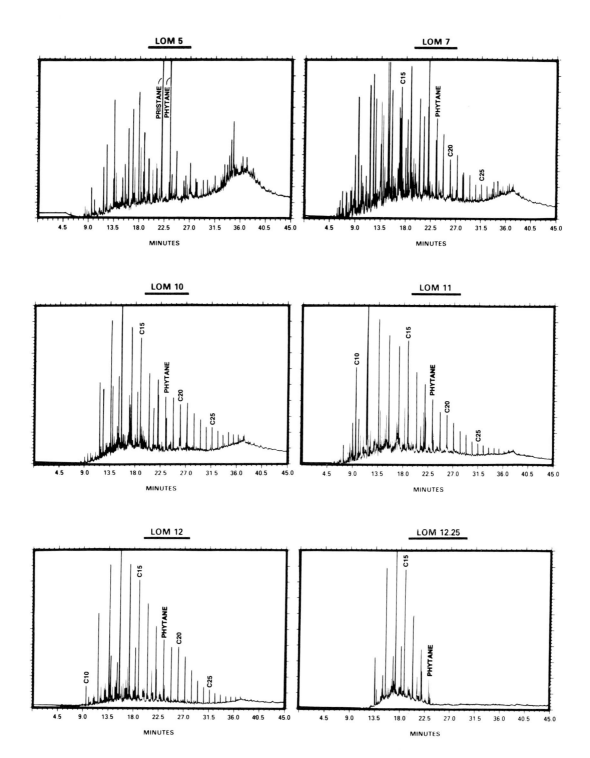

Figure 12. Total extract gas chromatographs from Duvernay core samples at a range of maturity values (expressed as LOM)

GC: Hewlett Packard Model 5710 fitted with an auxillary liquid CO_2 cryogenic attachment

Split: 250 or 300:1

Column: 15 m SE-30 WCOT (0.2mm I.D.) fused silica

Sample: 5L od 5% solution

Program: 0°C-300° at 8°C/min
Hold at 300°C for 8 mins.

(a) Mixing of oils from different sources (no other sources are documented in the area whch could contribute significantly to Leduc reservoirs).

(b) Lateral organic - facies changes within the source (the lack of evidence of siliciclastic or terrigenous input to this carbonate depositional system and the Devonian age of the rocks tend to argue against this).

(c) Mixing of oils from various maturities through time. (Realistically, this must happen in any trap, but the geochemical response from early generated, low maturity oil will subsequently be overwhelmed by that from later, higher maturity oil which has been generated from the source in exponentially increasing volumes (Waples, 1980)).

Figure 13 shows two oils from opposite ends of the maturity spectrum, as examples of the assignment of an oil LOM (see Fig. 7 for pool locations). This process has been repeated on oils from 93 separate pools in the Devonian of east central Alberta. All oils are presumed to have been derived from the Duvernay marine source rock. Figure 14 is a histogram of the total volume of oil in place (samples of which were used in this study) versus their assigned LOM value. The resulting distribution confirms the generation model established for the Duvernay marine source rock (compare to Fig. 9). Small amounts of liquid hydrocarbons occur which have LOM's less than 8 or greater than 12, and the maximum in the distribution occurs at an LOM of 10-11.

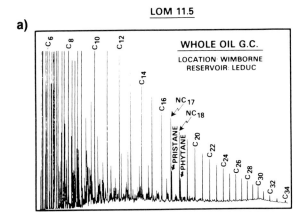
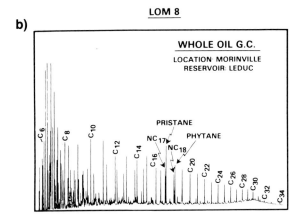

Figure 13. Whole gas chromatograms of two Leduc reservoir oils derived from the Duvernay

    (a)   Wimborne oil - LOM 11.5

    (b)   Morinville oil - LOM 8

    GC conditions as in Figure 12 except sample = 9.5 L

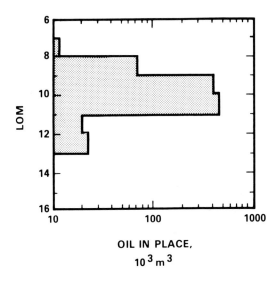

Figure 14. Distribution of Duvernay - derived oil reserves according to the average maturity (LOM) of each pool/field. Ninety-three oil samples were involved in this analysis.

MIGRATION

It is evident from comparison of Figure 7 and Figure 15 that hydrocarbons presently found in the Rimbey-Morinville chain of reefs, and attributed to a Duvernay source, lie beyond the mapped "onset of maturity" contour for the source rock interval. Clearly, conventional wisdom dictates that long distance secondary migration must have taken place. Furthermore, comparison of the LOM of the reservoir and adjacent source rock at the trapping site allows for an estimation of the distance of this secondary migration.

Theory Of Differential Entrapment

In a landmark paper Gussow (1954) first noted the distribution of liquid and gaseous hydrocarbons in these Leduc reefs and proposed his theory of differential entrapment or, as it is more commonly known, the spill-point theory (Fig. 15). This theory involves long distance migration of oil and gas through buoyant forces.

As the adjacent Duvernay source rock matures, oil enters the system at its lower end (Fig. 15). As the trap fills it approaches its spill-point and "spills" updip to the next trap. With further burial the source rock becomes overmature and generates gas, which in turn enters the system and displaces oil to spill-point eventually filling the traps with gas.

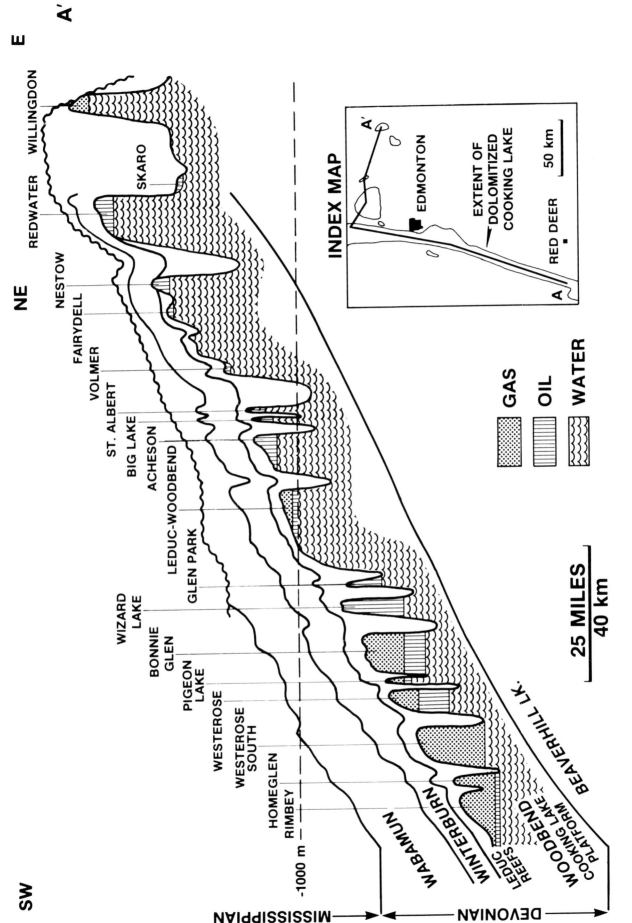

Figure 15. Structural cross-section showing the distribution of oil, gas and water in the Leduc reefs of east-central Alberta.

In the idealized case of this migration pattern we would expect to find the most deeply buried traps full of gas, then gas over oil, oil over water, and then finally all water. As can be seen in Figure 15, this pattern generally holds true for the lower part of the chain betwen the Rimbey and Wizard Lake reefs. However, a gas cap is present at Leduc—Woodbend updip from an oil-filled reef (Wizard Lake). Furthermore, oil occurs at Acheson despite the fact that the Leduc-Woodbend oil column is hundreds of feet from spill-point. The same is also true for the reefs updip from Acheson.

Clearly, then, the mechanism of simple updip displacement does not provide a complete explanation for the distribution of oil and gas in this reef system. In the present study we have identified a different mode of secondary migration via the dolomitized Cooking Lake margin underlying these Leduc buildups (Fig. 16). It is known that reefs on the chain, although separated, share common aquifer systems through the underlying Cooking Lake (Fig. 15). The margin of this platform underwent late, coarse dolomitization as it initially provided an updip conduit for fluids dewatering from the adjacent basinal section. Subsequently, this same conduit was used by similarly adjacent Duvernay source rocks which, as they matured, yielded hydrocarbons which migrated updip through the Cooking Lake. The observed pattern of reservoir fluids suggests that the Cooking Lake aquifer acted as an underlying "leaky pipeline", with the amount of oil each reef received being largely dependent on the presence of permeability barriers between the reef and underlying platform. That is not to say that spill-point displacement was not a factor, as it can be used to elegantly explain gas-oil distribution at the lower end of

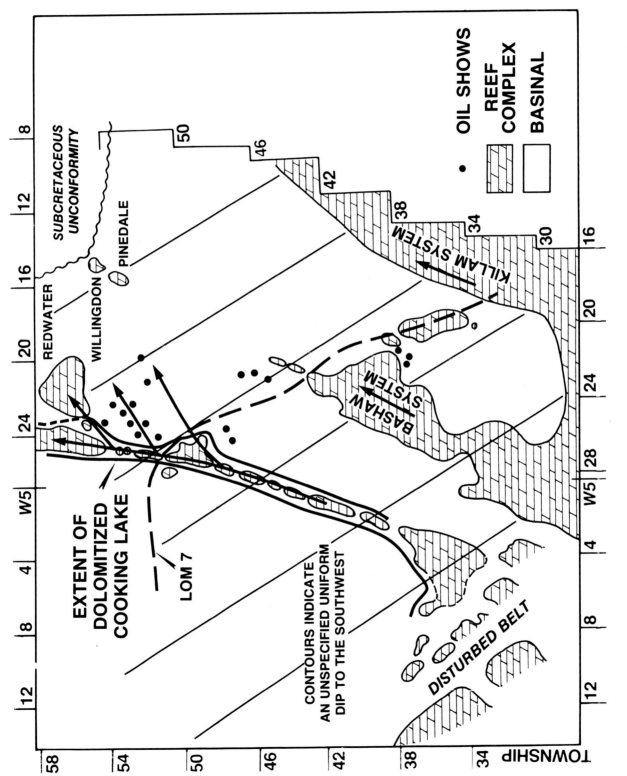

Figure 16. Secondary migration systems. Diagram showing extent of the dolomitized Cooking Lake, structural attitude and reported Cooking Lake oil shows.

the chain. But this was probably a later phenomenon resulting from a sequential emplacement of hydrocarbons into the system, over a long period of time, initially as an oil phase, then followed by entry of overmature gas which was able to penetrate many of the vertical permeability barriers that inhibited oil migration.

Two illustrative examples of variations to this general migration pattern occur at the Golden Spike and Redwater reefs (Fig. 7).

Golden Spike Migration

The Golden Spike reef contains 300 million bbls of oil and provides a good example of short distance (local) migration. Both reservoir oil and adjacent source rock plot at LOM's of 9 (Fig. 17). This buildup lies to the west of the main Rimbey-Morinville reef chain and does not appear to share the same Cooking Lake aquifer. Clearly this explains why this buildup contains oil that is at the same LOM as the adjacent source rock, a similar buildup surrounded by immature source rock would not be considered to be prospective.

Redwater Reef - Long Distance Migration

The Redwater reef, on the other hand, is an example of long distance migration since the reservoir oil and adjacent source rock extract are so different (Fig. 17). The oil appears to have been exposed to an LOM of 9.5-10 whereas the adjacent Duvernay is currently at an LOM of 5 (immature) (Fig. 7). Comparison of source

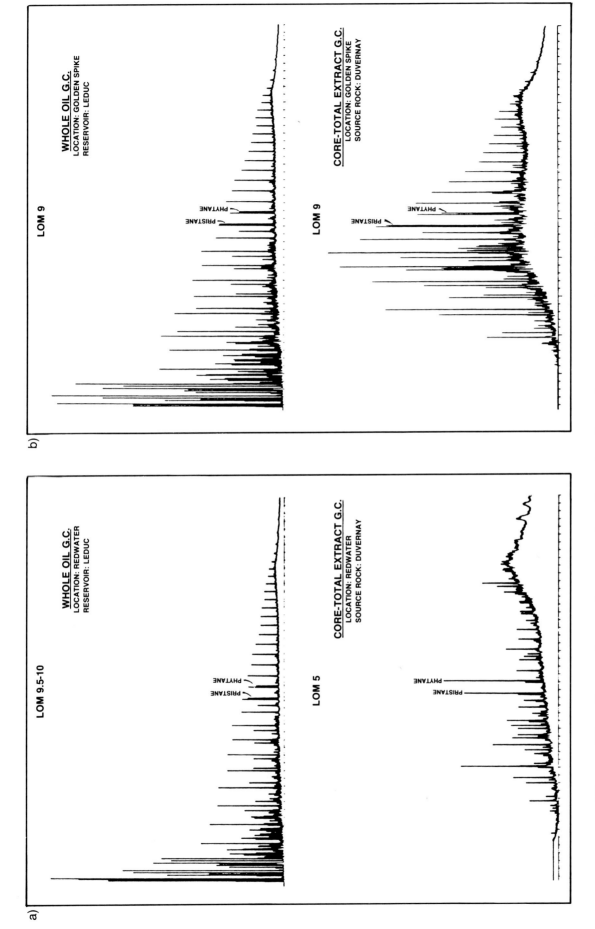

Figure 17. Comparison of reservoir oil and adjacent source rock maturities.

(a) long distance migrated oil - Redwater

(b) locally derived oil - Golden Spike

(Note: light ends in total extract G.C.'s lost during analytical procedure).

rock and oil maturities in the basin suggests that the oil at Redwater (1.2 billion bbls) must have migrated a minimum of 100 kms (60 miles). This value is a minimum because all Duvernay rocks further to the southwest have also passed through this maturation level and could have generated this oil. Like Golden Spike, the Redwater reef is also offset from the continuous Cooking Lake fluid system, but as shown earlier bears evidence of the long distance migration of oil.

Our knowledge of the structural orientation of these sediments during oil migration and emplacement is incomplete; it would appear, however, to be close to its present attitude (dip to the south-west, Fig. 16). Consequently, fluids moving up the reef trend would be proceeding at an angle somewhat oblique to true dip.

Within the Cooking Lake Formation a number of oil shows north and east of the Ledue Woodbend field suggest that oil bled from the dolomitized Cooking Lake system more directly updip through platform limestones and in this way passed beneath and migrated vertically into the Redwater reef (Fig. 16).

Furthermore, oil staining in the Willingdon Leduc reef further east is seen to be heavily biodegraded and Devonian-sourced. It is likely that this reef never trapped Devonian oil but rather acted as a conduit into the overlying Cretaceous and thence to outcrop (Fig. 16). Oil continuing up the main reef chain eventually accessed the Grosmont Formation, an overlying widespread platform carbonate, and in this way migrated updip to subcrop.

The steps illustrated in these two examples can be repeated for all oils in this portion of the basin, and provide a basis for consideration of distance of source and pathways of secondary migration. As such they add to our appreciation of an overall exploration model for this portion of the basin. As the preceding two examples show, however, the explorationist, must exercise extreme caution. Both the Golden Spike and Redwater reefs are removed from the continuous fluid system, the dolomitized Cooking Lake margin. At Golden Spike the surrounding source rock is mature and the reef therefore prospective, but at Redwater knowledge of the regional source rock maturity patterns (Fig. 7) should logically have led to the conclusion that this reef was unprospective! It is only through more detailed investigation of timing of migration and structural attitude of the strata that it would appear prospective.

Implications for the Tar Sands Deposits of Alberta

To the north and east of the study area lie some of the giant oil sands deposits of Alberta. Numerous hypotheses about the origin of these hydrocarbons have been proposed, and recent geochemical and sedimentological studies indicate that they represent the biodegraded remnants of a supergiant conventional oil deposit.

Published work suggests that this oil was derived from the Lower Cretaceous section (see Jones, 1981) with the Mannville Group shales as the principal source rocks (Deroo and others, 1977). However, in a recent study Moshier and Waples (1985) showed that the generative capacity and yield of hydrocarbons in the Lower Cretaceous Mannville Group of central Alberta falls far short of the hydrocarbon volumes found in these deposits. It seems conceivable, from the evidence of the present study, that the Duvernay Formation could have contributed significantly to these deposits. Indeed, Hitchon and Filby (1984) demonstrated that, for trace element characteristics, a sample of Athabasca crude correlates closely with a group of oils associated with Upper Devonian sources.

CONCLUSIONS

The Duvernay Formation is an organic-rich basinal carbonate succession that has provided most of the oil presently produced from Leduc reefs of east-central Alberta. It accumulated in deep water anoxic bottom conditions that allowed preferential preservation of organic material; slow sedimentation rates resulted in a rich (up to 17 weight percent TOC) source rock.

Rich source intervals occur as dark black laminites interbedded on a fine scale with leaner bioturbated lime mudstones. This fine interbedding of source and non-source intervals could lead to considerable inaccuracy when dealing with analyses of drill cuttings averaged over ten-foot or three-meter intervals. Analysis of these rocks reveals a wide range in the amount of insolubles. Generally TOC appears to be independent of clay content, and many of the rich source intervals sampled in this study have very low amounts of insolubles, rendering them true carbonate source rocks.

Comparison of reservoir oil LOM and source rock LOM substantiates that long distance secondary migration has taken place within the basin. Migration distances on the order of 100 kms (60 miles) are indicated. It is possible in many instances to determine migration pathways, and this has significant exploration implications.

Oil generated from mature Duvernay source rocks subsequent to dolomitization of the Cooking Lake platform margin reached this permeable conduit and proceeded updip. This dolomitized margin formed in effect a "leaky pipeline", with the amount of oil accumulating in the overlying Leduc buildups being dependent on the presence or absence of vertical permeability barriers. As source rock burial proceeded into the overmature stage, gas was generated and filled the reefs at the lower end of the chain resulting in a classical example of spill-point updip displacement. North of the Leduc-Woodbend field oil bled laterally in the true updip direction through platform limestones, finally reaching the Redwater reef, with some oil continuing on to subcrop. Evidence presented here suggests that at least a portion of the giant Athabasca and associated tar sands deposits were derived from the rich Duvernay source rock deposits of central Alberta. The Devonian Duvernay source rock and Leduc reservoir oils of central Alberta provides an unparalled opportunity to document maturation, migration and entrapment of hydrocarbons. This can serve as a powerful predictive model for other marine carbonate systems.

ACKNOWLEDGEMENTS

This study was undertaken as a joint geochemistry-sedimentology project in the Exploration Research and Technology Division of Esso Resources Canada Ltd. All analyses were performed by the staff of Esso's Production Research Lab and the authors would also like to thank Esso for permission to publish this work. Dave Milner and Jack

Wendte provided constructive cristicism and assistance at various stages during this study.

REFERENCES

Bailey., N.J.L., Evans, C.R., and Milner, C.W.D., 1974, Applying petroleum geochemistry to search for oil: Examples from Western Canada Basin: Am. Assoc. of Petrol. Geol. Bull,, v. 58, p. 2284-2294.

Byers, C.W., 1977, Biofacies patterns in euxinic basins: a general model: in H.E. Cook and P. Enos (eds.), Deep Water Carbonate Environments, Soc. Econ. Paleont. Mineral. Spec. Pub. 25, p. 5-17.

Demaison, G.J. and Moore, G.T., 1980, Anoxic environments and oil source bed genesis: Am. Assoc. Petrol. Geol. Bull., v. 64, p. 1179-1209.

Deroo, G., Powell, T.G., Tissot, B., and McCrossan, R.G., 1977, The origin and migration of petroleum in the Western Canadian Sedimentary Basin, Alberta: A geochemical and thermal maturation study: Geol. Surv. Canada Bull., 262, 136 p.

Gussow, W.C., 1954; Differential entrapment of oil and gas: A fundamental principle: Am. Assoc. Petrol. Geol. Bull., 138, p. 816-853.

Hitchon, B., and Filby, R.H., 1984, Use of trace elements for classification of curde oils into families - example from Alberta, Canada: Am. Assoc. Petrol. Geol. Bull., v. 68, p. 838-849.

Hood, A., Gutjahr, C.C.M., and Heacock, R.L., 1975, Organic metamorphism and the generation of petroleum: Am. Assoc. Petrol. Geol. Bull., v. 59, p. 986-996.

James, A.T., 1983, Correlation of natural gas by use of carbon isotopic distribution between hydrocarbon components: Am. Assoc. Petrol. Geol. Bull., v. 67, p. 1176-1191.

Jones, R.W., 1981, Some mass balance and geological constraints on migration mechanisms: Am. Assoc. Petrol. Geol. Bull., v. 65, p. 103-122.

Lijmbach, G.W.M., Veen, F.M. Van Der, and Engelhardt, E.D., 1981, Characterization of crude oils and source rocks using field ionisation mass spectrometry: in Bjoroy, M. and others, (eds.), Advances in Organic Geochemistry 1981: New York, Wiley, p. 788-798.

Moshier, S.O. and Waples, D.W., 1985, Quantitative evaluation of Lower Cretaceous Mannville Group as source rock for Alberta's oil sands: Am. Assoc. Petrol. Geol. Bull., v. 69, p. 161-172.

Stoakes, F.A., 1980, Nature and control of shale basin fill and its effect on reef growth and termination: Upper Devonian Duvernay and Ireton Formatios of Alberta, Canada: Bull. Canadian Petrol. Geol., v. 28, p. 345-410.

Stoakes, F.A. and Creaney, S., 1984, Sedimentology of a carbonate source rock: The Duvernay Formation of central Alberta: in Eliuk, L.(ed.), Carbonates in Subsurface and Outcrop, 1984, Canadian Soc. Petrol. Geol. Core Conference, Calgary, Alberta, p. 132-147.

Tissot, B., Durand, B., Espitalie, J., and Combaz, A., 1974, Influence of nature and diagenesis of organic matter in formation of petroleum: Am. Assoc. Petrol. Geol. Bull., v. 58, p. 499-506.

Waples, D.W., 1980, Time and temperature in petroleum formation: Application of Lopatin's method to petroleum exploration: Am. Assoc. Petrol. Geol. Bull., v. 64, p. 916-926.

EVIDENCE FOR RAPID FLUID MIGRATION DURING DEFORMATION, MADISON GROUP, WYOMING AND UTAH OVERTHRUST BELT

JOYCE M. BUDAI
Department of Geological Sciences
The University of Michigan
Ann Arbor, MI 48109-1063

ABSTRACT

The relationship between deformation-induced pressure solution and fracturing in carbonate rocks can be complex and may be interactive. Multiple episodes of pressure solution and fracturing are common in strongly deformed carbonate units such as the Mississippian Madison Group of the western Overthrust Belt. Paragenetic relationships in these rocks suggest that fractures both modify and enhance continued pressure solution by opening the host rock to fluid migration.

The composition of vein and stylolite mineralization may be used to evaluate the history of fluid migration during deformation. In the Madison carbonates, the earliest veins were filled by dolomite or calcite, while all subsequent veins were filled with calcite. Host limestone and dolomite are non-luminescent while filled veins are variably luminescent. The isotopic compositions of vein-filling calcite and dolomite are distinct from host rock compositions and document changes in fluid chemistry during burial and deformation. Taken together, the temporal change in mineralogy, luminescence and isotopic compositions of various vein-filling carbonate cements vs. host rock carbonates are strongly suggestive of rapid allochthonous fluid migration during deformation of the Madison Group in the western Overthrust Belt.

INTRODUCTION

The Overthrust Belt of the western U.S. is a north-south trending structural province which consists of a series of imbricated, west-dipping thrust sheets affecting Precambrian through Tertiary sedimentary rocks (Fig. 1). First movement on the Paris thrust is dated as latest Jurassic and the final stages of movement on the Prospect Thrust occurred near the end of the Paleocene (Wiltschko and Dorr, 1983). In general, deformation began in the west and moved east with time. However, intermittent movement on the older faults may have been synchronous with major movement on the Absaroka thrust (Armstrong and Oriel, 1965; Jordan, 1981; Wiltschko and Dorr, 1983). In the Overthrust Belt of Wyoming and Utah the Madison Group forms resistant cliffs within eroded thrust belt mountains. East of exposed thrust sheets, the upper Madison Group yields hydrocarbons in commercial quantities.

This study focuses on three outcrop sections of the Madison Group in Wyoming and Utah and core material from the Carter Creek-Whitney Canyon gas field in southwest Wyoming. The Haystack Peak section (location 4, Fig. 1) is within the central part of the Absaroka plate. The Carter Creek 1-5 well core (location 3, Fig. 1) is an 80 meter (240 feet) core taken from the upper half of the Madison Limestone. The section is in the southern part of the Absaroka plate. The Laketown Canyon and Crawford Mountains sections (locations 1 and 2, Fig. 1) are within the Crawford thrust sheet, bounded on the west by the Paris thrust fault and separated from the Absaroka plate to the east by the Crawford thrust fault.

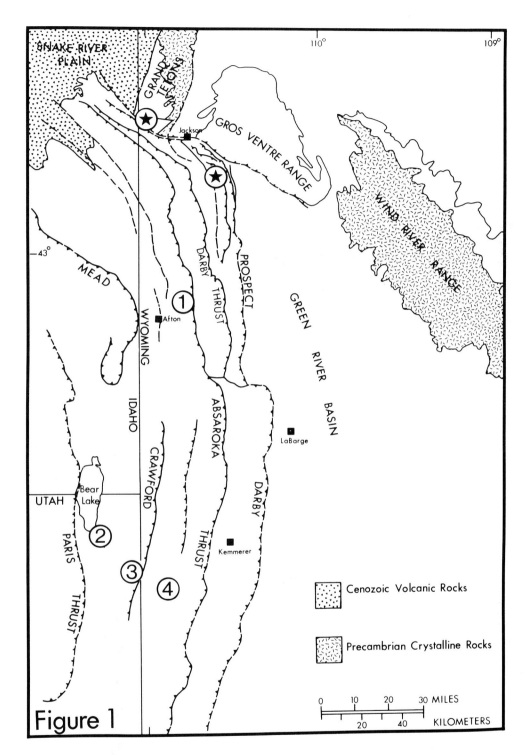

Figure 1. Location map showing generalized tectonic features of study area. Modified from unpublished map by J.A. Dorr, 1983. Locations described in this paper are as follows: (1) Haystack Peak, (2) Laketown Canyon, Utah, (3) Brazer Canyon in the Crawford Mountains, and (4) Carter Creek 1-5 well core. The two starred locations were included in another study.

STRATIGRAPHY AND EARLY DIAGENESIS

The upper half of the Mississippian Madison Group, the Mission Canyon Formation, is a massive, dolomitic and in places evaporitic shallow marine carbonate sequence (Fig. 2). In western Wyoming the Mission Canyon ranges in thickness from 260 to 330 meters (800 to 1000 feet). A regional unconformity following Mission Canyon deposition resulted in some erosion and meteoric alteration of Madison Group sediments. The upper half of the Madison Group contains evaporitic intervals which have been dissolved giving rise to intrastratal collapse breccia beds. Evaporite solution may have occurred during the upper Mississippian unconformity, but the presence of Amsden sandstone clasts within some breccia beds suggests that some amount of collapse occurred following deposition of the overlying Amsden Formation.

The upper part of the Laketown Canyon section is younger than the Mission Canyon Formation at the other locations and is referred to as the Monroe Canyon Limestone (Fig. 2) (Sando and Sandberg, 1977). Because this study is concerned with the burial and deformation history of the Madison group, the difference in age between these locations has no bearing on the results and interpretations. However, there are lithologic differences between Laketown Canyon and the other locations which may have exerted some control on the style of deformation. The upper part of the Monroe Canyon Limestone includes quartz sandstone and sandy dolomite units as well as limestone and dolomite (Fig. 2). There is less massive dolomite at Laketown Canyon than in the other two outcrop locations.

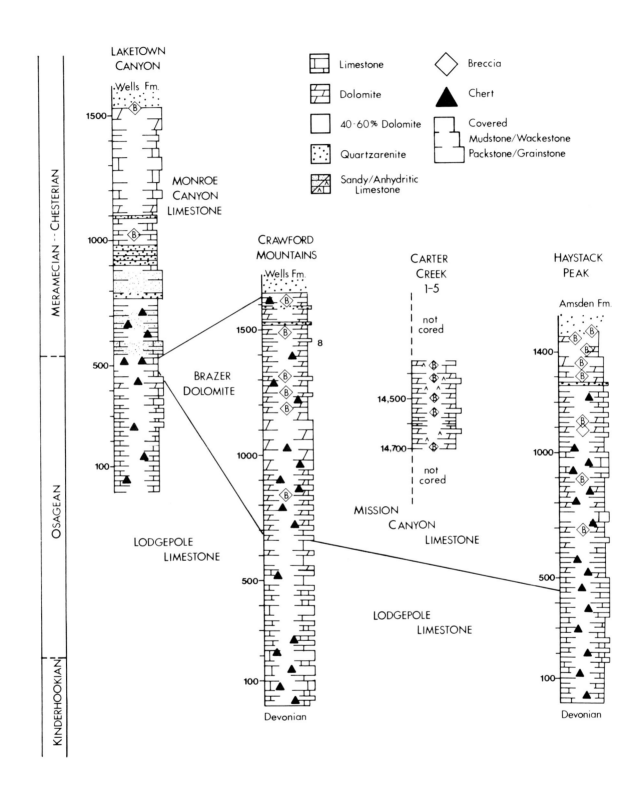

Figure 2. Stratigraphy and lithology of the Madison Group in the sections studied. Stratigraphic correlations from Sando and Sandberg (1977) and Gutschick and others (1980).

as well as limestone and dolomite (Fig. 2). There is less massive dolomite at Laketown Canyon than in the other two outcrop locations.

The lower half of the Madison Group, the Lodgepole Limestone, is thin- to thick-bedded, slightly dolomitic open marine limestone. The Lodgepole Limestone is approximately 200 meters (600 feet) thick at the Crawford Mountains and Haystack Peak sections. The upper part of the Lodgepole Limestone was collected at the Laketown Canyon section. Because the Carter Creek core ends in Mission Canyon dolomite, there is no Lodgepole Limestone from that location.

There is considerable petrographic and isotopic evidence that the Madison Group was dolomitized before deep burial (Budai and others, 1984; 1985). Replacive dolomite is non-ferroan, non-luminescent and predates all veining and stylolitization. Dolomitization was most extensive in upper Madison Group shallow marine to peritidal facies and is rare in the lower Lodgepole Limestone (Fig. 2).

The stable isotopic compositions of Madison Group replacive dolomite, crinoidal calcite and early carbonate cements are representative of host rock composition before burial and thrust belt deformation (Fig. 3). These compositions are compatible with predicted marine to meteoric carbonate values in the Mississippian (Brand and Veizer, 1981; Meyers and Lohmann, 1985; Budai and others, 1985) and are very similar to isotopic compositions reported for early diagenetic components in other Mississippian carbonate formations (Choquette and Steinen, 1980; Meyers and Lohmann, 1985; Banner and others, 1985). Methods used in microsampling, preparation, and sample measurement for carbon and oxygen isotopes were as described in Budai and others (1985).

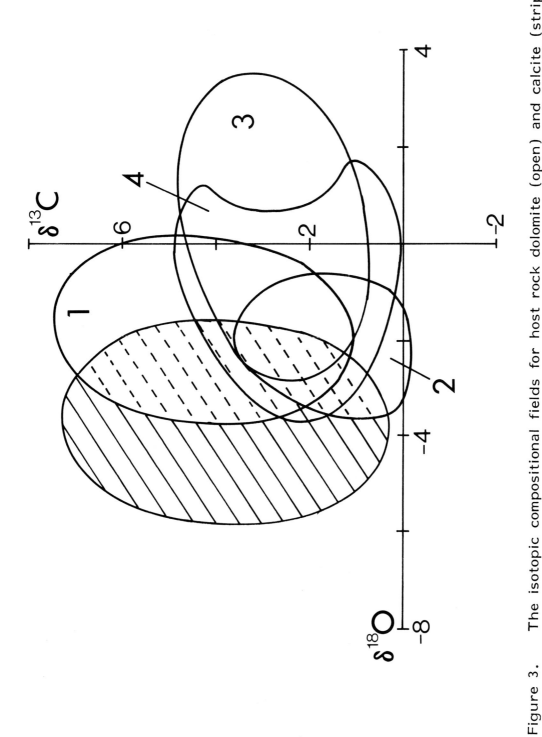

Figure 3. The isotopic compositional fields for host rock dolomite (open) and calcite (striped) in the four study locations. Numbers on dolomite fields refer to locations listed in Fig. 1. Stripped field includes early diagenetic calcite from all sections. Both carbon and oxygen isotopes are relative to the PDB standard.

EVIDENCE FOR DEFORMATION IN THE MADISON GROUP

Superimposed on the petrographic record of early diagenesis is a complicated sequence of fracturing and stylolitization. The severity and complexity of fracturing varies between locations, but in all cases there is clear evidence for multiple episodes of stylolitization, fracturing and fracture-filling, or veining, by dolomite and calcite cement.

A generalized paragenetic sequence can be applied to all sections, though the details vary with location (Fig. 4). The earliest filled fractures, or veins, are roughly normal to bedding and are truncated by bedding parallel stylolites or a second set of veins. These may record deep burial compaction or the onset of thrust belt deformation. Fractures, veins and stylolites which are oriented at variable angles to bedding transect the earlier fracture sets and stylolites. These veins and stylolites are referred to as tectonic because they appear to record thrust belt deformation of the Madison Group. An additional feature shared by all sections is the common reactivation of first and second veins by subsequent fracturing and filling. This is indicated by multiple vein fillings that differ in mineralogy, petrography or cathodoluminescence. Multiple fracturing along the sample plane is also apparent because later veins, which can be sinuous, curve into and bisect pre-existing veins and stylolites.

Several important characteristics of veining and stylolitization change depending on geographic location of the section, stratigraphic position of the sample and lithologic composition of the surrounding host rock. These characteristics include: (1) mineralogy of vein

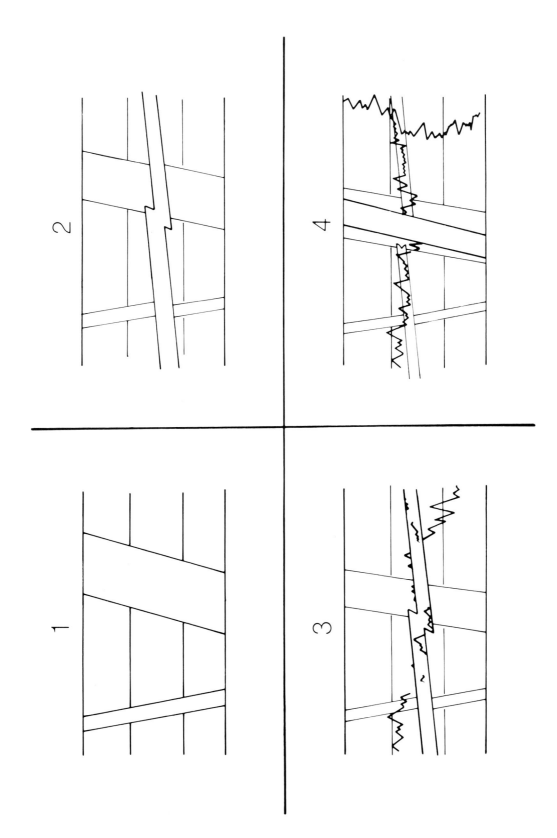

Figure 4. Idealized sequence of deformation events observed in the Madison Group. Cartoons 2 and 3 are interchangeable because in some samples bedding parallel stylolites follow horizontal veins, and in other cases horizontal stylolites are ruptured by bedding parallel veins.

material, (2) cathodoluminescence of veins, (3) isotopic composition of vein-filling cement and (4) the complexity of fracturing, vein-filling and stylolitization.

The importance of section location relative to major thrust faults is difficult to assess because the detailed structural history of the Overthrust Belt is poorly understood. However, the large scale deformation events have been well-documented. First movement and much of the displacement on the Paris thrust occurred in Early Cretaceous, well before first movement on the Crawford thrust (Royse and others, 1975; Wiltschko and Dorr, 1983). First movement on the Absaroka thrust is thought to have been in the Late Cretaceous, following the last major movement on the Crawford thrust (Wiltschko and Dorr, 1983). In a qualitative way, comparison of the Madison Group from one location to another suggests that there are mechanical as well as compositional differences between the deformation history of the Crawford and Absaroka plates. Veining and stylolitization are generally more complex in Crawford plate sections suggesting a longer and perhaps more intense deformation history. Veins exhibit greater mineralogic and cathodoluminescent variability in the Crawford plate than in the Absaroka plate. The stable isotopic differences between locations are discussed below, but preliminary data suggest there are systematic changes in composition of vien-filling cement based on geographic location.

Stratigraphic position controls physical parameters such as bedding thickness and lithologic composition. Proximity to changes in either of these may be equally important in affecting the style of deformation recorded in the Madison Group. For example, the lowest

Lodgepole Limestone at Haystack Peak is composed of thin-bedded lime mudstones and wackestones. In this interval slip planes with fibrous calcite ornamentation have developed parallel and normal to bedding. These appear as simple filled veins unless the slip surface is examined. Calcite fibers have grown in the direction of layer movement which is approximately parallel to the regional transport direction of the Absaroka thrust plate. These fibrous cements formed during layer movement and are considered to be syndeformational. In the Madison Group such slip surfaces are best-developed in thin-bedded limestones where anisotropy was high. In the massive dolomite beds characteristic of the upper Lodgepole Limestone and Mission Canyon Formation, brittle fractures and well-developed stylolite seams are more common.

Contrasts in bedding thickness and lithology appear to enhance the development of solution cleavage, slip surfaces, stylolitization and veining in adjacent strata. This was observed at the transition between the Mission Canyon Formation, which is dolomitic and massively bedded at its base, and the upper Lodgepole Limestone which has variable bed thickness and is slightly dolomitic. Because the two formations responded somewhat differently to the same stress, deformation styles change vertically, and lithologic transition zones have the most complicated structural history. This, in turn, affects the degree of host rock interaction with vein-filling fluids and the final composition of cements within slip surfaces and viens.

In all sections there is an apparent host rock control on the mineralogic composition of early tectonic veins. Dolomite veins are more likely to occur in dolomitic samples. Pure limestone intervals have some dolomite-filled veins, but, with minor exceptions, the veins are small and often fragmented by subsequent calcite veins and stylolitization. Most limestone samples contain only calcite-filled veins whereas dolomite samples with dolomite veins also contain calcite-filled veins. In addition, in all samples the early dolomite veins are volumetrically reduced through corrosion and replacement by calcite that fills later veins. In some cases only partial rhombs lining a vein wall remain as evidence of a former dolomite vein. The cause(s) for this pattern may include: (1) crystallographic substrate control on vein mineralization or (2) cannibalization of adjacent wall rock during vein-filling. There is clearly no host rock control on late veining because all are filled with calcite, in both dolomite and limestone samples.

CRAWFORD PLATE

Laketown Canyon Section

The Laketown Canyon section within the western part of the Crawford plate has been complexly faulted and folded (Fig. 1). Beds are now oriented vertically and the Madison section is incomplete due to faulting and loss of section. This section differs from the others in that the upper unit is lithologically heterogeneous. Quartz sandstones are interbedded with dolomitic limestone and sandy dolomite. Two perpendicular orientations of well-developed stylolite seams are visible on the outcrop. Veins are large, often split and filled by a second or third carbonate phase. Veining and stylolitization are more complicated in limestone and dolomitic units than in sandstone intervals.

In thin section the cross-cutting relationships are extremely complex. Generally the earliest tectonic veins were filled with dolomite. Subsequent veins were filled with calcite (Fig. 5). Stylolitization occurred intermittently during or between veining events. Where dolomite veins were split and filled with calcite, dolomite margins adjacent to the younger vein are corroded and replaced by the later calcite. Under fluorescent light vein dolomite is blue and in some cases displays zoned intensity. Host rock dolomite is yellow to tan and in some cases has fluorescent yellow cores and blue rims. Calcite is uniformly blue, probably reflecting the fluorescent color of the underlying epoxy. Under cathodoluminescence each vein has two to four different generations of filling cement. Vein dolomite is generally zoned from dull to bright red, while vein calcite ranges from dead to bright yellow or orange. With few exceptions, host limestone and dolomite are non-luminescent. In several extensively veined samples the fluid that filled veins with luminescing cement clearly interacted with adjacent host dolomite, corroding crystal margins and imparting later generations of luminescent dolomite cement. Fluorescent zones in host dolomite are consistent with this observation.

The first and second vein filling cements in Laketown Canyon samples have an oxygen composition depleted by 5.0 to 8.0 per mil relative to the host rock dolomite (Fig. 6). Dolomite veins have the same range in composition as early calcite veins. Calcite veins have widely variable oxygen and carbon compositions. All later veins are depleted by at least 10 per mil in oxygen and 3 per mil in carbon relative to the host rock composition. The fine-scale compositional differences suggested by luminescent variations within each vein may

Figure 5. Multiple veining episodes in a dolomitic sample from the upper part of the Laketown Canyon section. Slab has been stained for calcite with Alizarin Red. S. White veins are dolomite (d), dark veins are calcite (c). Black bar scale = 1 cm.

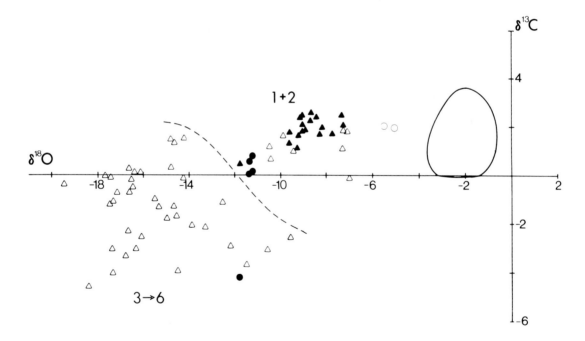

Figure 6. Stable isotopic composition of host dolomite and limestone, and vein-filling calcite and dolomite in the Laketown Canyon section. All values are relative to the PDB standard in per mil notation. Host dolomite = open field, host limestone = open dots, host dolomite in extensively veined samples = solid dots, vein dolomite = solid trianagles, vein calcite = open triangles. Numbers by fields indicate vein generations with 1 being the oldest vein.

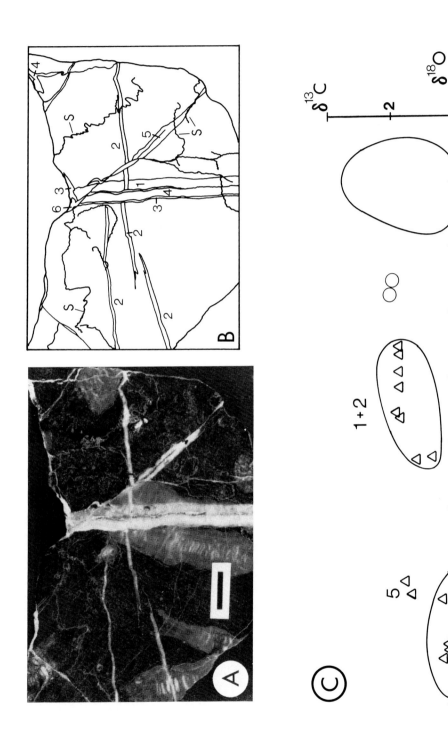

Figure 7. (a) A complexly veined limestone sample from the upper part of Laketown Canyon section.
(b) Line drawing indicates relative vein sequence with 1 being oldest.
(c) The isotopic composition of successive veins in this sample. Numbers correspond to the line drawing. Isotopic symbols and notation the same as in Figure 6.

be contributing to the apparent scatter in later vein analyses. Vein compositions for a single sample exhibit and isotopic range approaching that of all samples analyzed (Fig. 7). This suggests that fluids associated with vein mineralization migrated through the entire section, independent of lithology or stratigraphic position.

Based on the multiple fracturing and vein-filling events preserved in this section, a comparatively long deformation history has been recorded. Changes in mineralogy, luminescence and isotopic composition of successive vein-fills indicate the composition of fluids associated with deformation changed dramatically through time.

Crawford Mountains Sections

The Crawford Mountains section is located on the eastern edge of the Crawford plate, 25 kilometers (15 miles) southeast of the Laketown Canyon section (Fig. 1). The upper half of this section is more extensively dolomitized than the other outcrop locations. Fracturing and stylolitization are common throughout the section but are generally less complicated than in the Laketown Canyon section. Dolomite-filled veins are less abundant in this location, but always predate calcite veins. There are no distinctive fluorescent carbonate cements at this location. Luminescence of vein-filling cements is variable and considerably less common than in Laketown Canyon samples. Vein dolomites range from dead to brightly luminescent. Luminescent calcite veins are younger than veins filled with non-luminescent calcite based on consistent cross-cutting textures where both occur in the same sample. All host limestone and dolomite are non-luminescent. In dolomite samples the wall rock adjacent to calcite-filled veins is commonly corroded and partially replaced by calcite (Budai and others, 1984). Dedolomitization is also observed within dolomite veins cut by younger fractures filled with calcite.

The isotopic composition of vein dolomite is enriched by 3.0 per mil in oxygen compared to vein dolomite in the Laketown Canyon section (Fig. 8). The host rock dolomite in the Crawford Mountains has a mean oxygen composition approximately 3.0 per mil enriched relative to the other sections (Fig. 3). Vein-filling dolomite has the same carbon composition as that observed in Laketown Canyon dolomite veins. In both sections the carbon and oxygen isotopic compositions of dolomite veins are considerably less variable than those in calcite veins.

Calcites exhibit a range of oxygen and carbon compositions, all depleted relative to the mean host rock composition. The veins with the lightest oxygen ($-14.5\ \delta O^{18}$) are significantly less depleted than the lightest veins at Laketown Canyon ($-19.5\ \delta O^{18}$). Carbon compositions in calcite veins are slightly more depleted, but comparable to those in the Laketown Canyon section. The two sections exhibit overlapping vein compositions, but are distinguished by differences in the depleted range of their respective oxygen and carbon trends.

ABSAROKA PLATE

Carter Creek 1-5 Well Core

The Carter Creek well core penetrates the southern part of the Absaroka plate and is separated from the first two locations by the Crawford thrust (Fig. 1). The core consists of anhydritic dolomitic

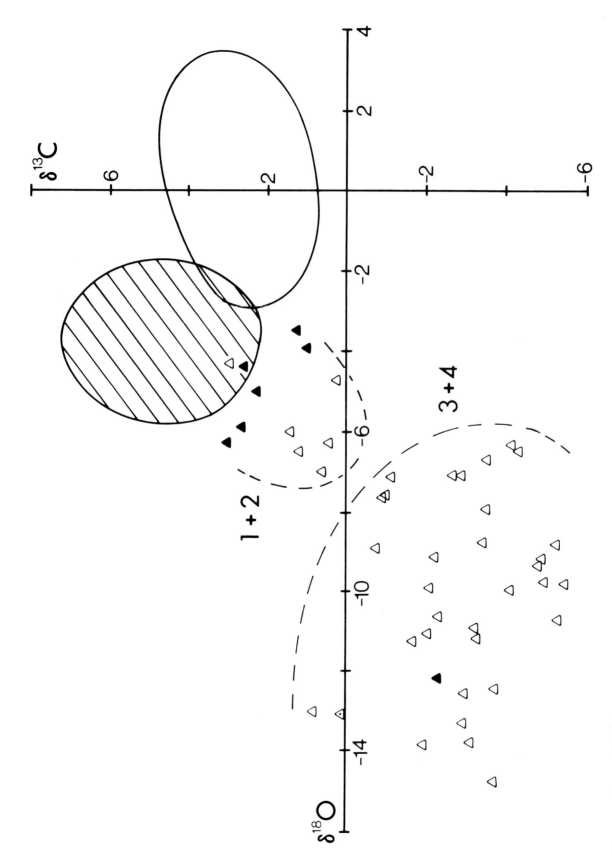

Figure 8. Stable isotopic composition of host rock and veins from the Crawford Mountains section. Striped field = crinoidal calcite and early calcite cements. All other symbols the same as in Figure 6.

and minor limestone from the upper half of the Madison Group. The core begins approximately 80 meters (250 feet) below the top of the Madison Group. Fractures, veining and stylolitization are common through the cored interval. Dolomite-filled veins are rare. Most veins are lined with calcite and/or anhydrite. Less commonly bitumen and calcite fill veins. The Carter Creek core has no luminescent cement in veins and the host rock is non-luminescent. Preliminary examination indicates there are no dolomite or calcite phases with distinctive fluorescent zoning. Therefore it is petrographically impossible to separate vein-filling generations except where cross-cutting fabrics exist.

There are two stages of stylolitization in the Carter Creek core. Bedding-parallel, low-amplitude stylolites are truncated by fractures, veins and highly irregular tectonic stylolites (Fig. 9). Many calcite veins appear to pre-date tectonic stylolites. The morphology and filling material in the tectonic stylolites suggests a fairly complex history. Most are lined with bitumen and filled with calcite or, less commonly, calcite plus anhydrite. Small fragments of adjacent host dolomite are incorporated into enlarged seams. Fine blebs of bitumen also occur surrounded by stylolite calcite. Veins or tension gashes originate at tectonic stylolites seams and are filled with calcite and bitumen. Calcite-filled veins truncated by tectonic stylolites do not contain bitumen, but calcite and bitumen do occur within fractures and pores of permeable, brecciated intervals. Textural evidence for calcitization of host rock dolomite and anyhdrite is common adjacent to calcite veins and within tectonic stylolite seams (Budai and others, 1984). This suggests that fluids associated with deformation and

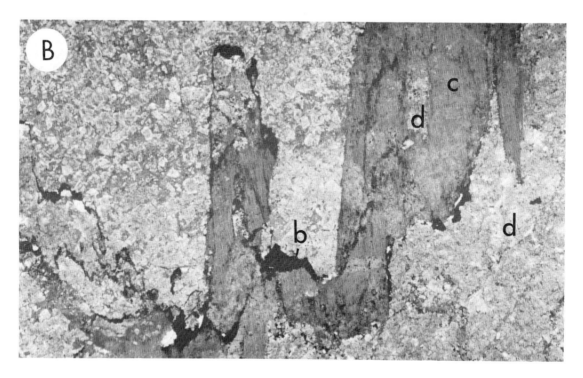

Figure 9. (a) Photomicrograph of tectonic stylolite seam in the Carter Creek 1-5 well core. Bedding parallel sytlolites (1) are cut by an irregular tectonic stylolite (2). B = bitumen, c = calcite, d = dolomite. Field of view = 10 mm.

(B) Calcite and bitumen within a tectonic stylolite seam. Letters the same as in Figure 9a. Field of view = 3.5 mm.

hydrocarbon migration were brines high in calcium, but low enough in magnesium and sulfate to dissolve dolomite and anhydrite.

The isotopic compositions of Carter Creek veins and stylolites are helpful in confirming successive filling cement generations that are petrographically indistinguishable (Fig. 10). Those calcite-filled veins that appeared to pre-date tectonic stylolites exhibit a carbon composition similar to host dolomite and have an oxygen composition depleted by 2.0 to 8.0 per mil relative to the mean host rock oxygen. Dolomite-filled veins are isotopically the same as early calcite veins. In contrast to vein-filling cements at other locations, calcite associated with bitumen in stylolites and veins has a small range of oxygen compositions (-7.0 to -10.0 δO^{18}) and a broad range of depleted carbon compositions (-4.0 to -12.0 δO^{18}). There are no veins with intermediate isotopic signatures. This may suggest a significant break in time between the two stages of veining and certainly a change in the composition of fluids associated with deformation. The depleted carbon in the stylolitic calcite may record the effects of hydrocarbon degradation and release of light carbon dioxide or bicarbonate during calcite filling of veins. The relatively invariant oxygen suggests a single fluid source and rapid precipitation during this stage of deformation.

Haystack Peak Section

This is the only location that has been studied from a rigorous structural and well as diagenetic point of view. The following

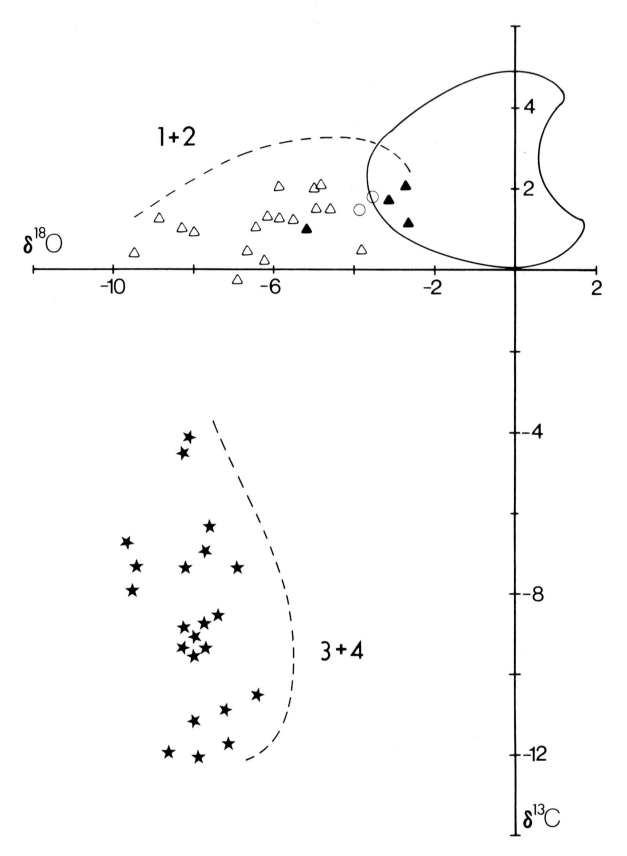

Figure 10. Isotopic composition of host dolomite, vein dolomite and calcite, and stylolite calcite in the Carter Creek 1-5 well core. Stylolite calcite = stars, all other symbols and notation the same as in Figure 6.

observations are based on preliminary data gathered during a reconnaissance field trip and will be refined in future work.

The Haystack Peak section is in the middle of the Absaroka plate, approximately 140 kilometers (85 miles) north of the Carter Creek-Whitney Canyon gas field (Fig. 1). As noted earlier, the detailed burial and structural history for his region has not been established so there may be no reason to anticipate diagenetic or structural similarities between this and other locations. However, the depositional and early diagenetic history is identical for the Haystack Peak, Carter Creek and Crawford Mountains sections. Thus, the pre-deformation lithologies and bulk chemical compositions are roughly the same (Budai and others, 1985). Contrasts in deformation style or composition of cements in veins and stylolites should reflect differences in the chemical and mechanical conditions during development of the Overthrust Belt.

The complete Mississippian section is exposed on the eastern flank of Haystack Peak. Beds are dipping approximately 25° to the northwest and the entire outcrop is part of a larger fold in the Absaroka thrust plate formed over a lateral ramp in the underlying Absaroka thrust fault (D. V. Wiltschko, pers. comm., 1985). The lowest Lodgepole Limestone is thin-bedded, fine-grained limestone. Veins are both normal and parallel to bedding and generally filled with calcite. Bedding and fault plane slip surfaces are common. The cross section of these surfaces is parallel to bedding. In some cases slip surfaces are ornamented by multiple layers of syntaxial calcite fibers which formed during movement of adjoining beds. Fibers are oriented parallel to the direction of layer movement. Higher in the section bedding thickness increases and the Lodgepole Limestone is more dolomitic. Both dolomite and quartz-filled veins occur, but calcite-filled veins are more common. In some cases quartz veins have been split and filled by later calcite veins. Unlike other sections, there is a rare, later generation of dolomite that fills the centers of some calcite veins.

Reconnaissance isotope analyses of veins from both the Lodgepole Limestone and Mission Canyon Formation display a somewhat different pattern than those of other sections (Fig. 11). Veins and slip surface calcite fibers in the Lodgepole have highly variable oxygen compositions ranging from host rock oxygen to strongly depleted values (-2.0 to -20.0 δO^{18}). In contrast to other locations, the carbon composition of Lodgepole Limestone veins is only slightly depleted and in a few samples is enriched relative to mean host rock carbon. In such veins the high carbon is identical to that in early calcite and dolomite cements, and host limestone or dolomite in the same samples. These enriched carbon intervals are restricted stratigraphically to the middle Lodgepole Limestone at both Haystack Peak and the Crawford Mountains section and are interpreted to record early diagenesis during shallow burial anaerobic fermentation (Irwin and others, 1977; Budai and others, 1985). The occurrence of the same isotopoc composition in veins that clearly postdate early diagenesis strongly suggests that during at least one stage of deformation host rock carbonate has been incorporated into veins and slip surface. It is evident that there are multiple sources of vein-filling carbonate because individual samples contain veins with isotopic compositions that range from that of the surrounding host rock

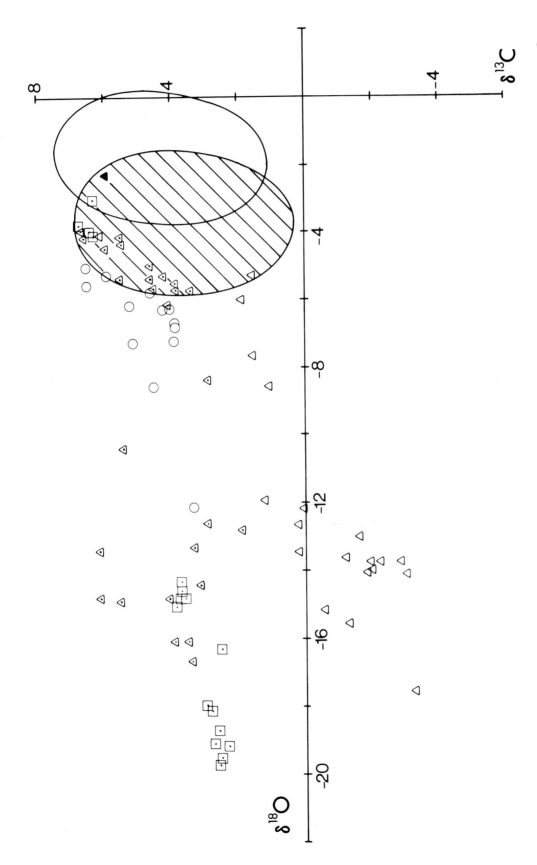

Figure 11. Isotopic compositions of host rock and veins in the Haystack Peak section. Open squares = calcite fibers on slip surfaces, open dots = host rock limestone adjacent to veins. Open triangles = calcite veins from the Mission Canyon Limestone, dotted triangles = calcite veins in the Lodgepole Limestone. All other symbols and notation the same as in Figure 8.

to the most depleted compositions characteristic of late veins in other locations (Fig. 11). It is likely that there are multiple veining events in these Lodgepole samples as has been inferred at other locations.

Vein-filling calcite in the Mission Canyon Formation exhibits a carbon-oxygen compositional range similar to that observed at Laketown Canyon. These veins are strongly depleted relative to mean host rock compositions. There are no light carbon veins like the stylolite-filling calcite in the Carter Creek core. The vertical isotopic differences in vein-filling carbonate observed at Haystack Peak may point to a strong lithologic influence on the mechanism of veining in carbonate rocks. In the thin-bedded limestones of the lower formation some veins are apparently filled with remobilized host rock carbonate. The identical isotopic composition suggests extremely high rock to water ratios during this veining episode. Vein generations with highly depleted compositions in both the Lodgepole Limestone and Mission Canyon Formation may record greater fluid contribution to vein filling. Haystack Peak is the only location wher veins have been systematically analyzed from the bottom to top of the section. In the other sections most vein analyses are from the upper half of the Group. Therefore, based on a preliminary data set, it is impossible to confirm whether the Lodgepole Limestone veins at Haystack Peak are typical or unusual.

DISCUSSION

There are some general observations that can be made at this stage in the study. Although it is not possible ot place each vein and stylolite generation in a precise temporal sequence for each location, there are similar textural patterns and compositional trends that occur on a regional scale. Earliest veins were filled with either dolimite or calcite, but where dolomite veins occur they always predate calcite veins. In all locations dolomite veins are less abundant than calicte veins, but are most common in the Laketown Canyon section and decrease sharply to the east. The isotopic compositions of earliest dolomite and calcite veins in each section appear to be influenced by the mean isotopic composition of the Madison host rock. Dolomite veins from Laketown Canyon have a more depleted oxygen composition than dolomite veins from the Crawford Mountains, which corresponds to a more enriched host rock oxygen composition in the Crawford Mountains section.

There are marked changes in luminescence when locations in the Crawford plate are compared to those in the Absaroka plate. Dolomite and calcite veins are commonly luminescent and complexly in Laketown Canyon samples, less commonly so in the Crawford Mountains section, and non-luminescent in the Carter Creek core. Based on preliminary observations, luminescent veins are rare in the Haystack Peak section. Because host limestone and dolomite in each section are non-luminescent, the fluid source of luminescing vein cements may be allochthonous. In a general way, the luminescent veins become less abundant from west to east, perhaps reflecting a fluid source on the west and gradual depletion of trace metals with fluid migration eastward during deformation.

In all locations the calcite which fills later veins also corrodes and replaces host rock dolomite and earlier dolomite veins. Later veins are grouped together because petrographically they are difficult to separate, but these clearly record multiple stages of fracturing and filling. Based on the range of isotopic compositions in younger veins, fluids associated with the later stages of deformation were highly variable and chemically distinct from surrounding host rock and older veins. The earliest dolomite and calcite veins have isotopic compositions that are depleted by a small amount relative to host rock carbonate. In contrast, the later veins are strongly depleted relative to host rock carbon or oxygen. This change in composition with time of veining may indicate a fundamental change in fluid sources, degree of host rock interaction, or rates of deformation and fluid flow.

The timing of stylolitization relative to veining is often ambiguous suggesting that both processes occur on an intermittent basis. Veins and stylolites are commonly reactivated by later deformation. This has been observed in all locations and indicates that once movement has occurred along planes of weakness, forming veins or stylolites, subsequent fluid migration or renewed pressure solution is channeled through these pre-existing conduits. Thus, a vein filled by several cement generations preserves a picture of episodic fluid compositional changes through time.

The petrographic and compositional differences observed between the two locations in the Crawford plate suggest that there was a gradual evolution in composition as burial fluids migrated through the Madison Group during thrusting. The similarity of sequential changes in mineralogy, luminescence and isotopic composition between early and later veins in different thrust plates indicates that there was some fluid migration between plates during various stages of the Overthrust Belt deformation.

ACKNOWLEDGEMENTS

Permission to use core samples from the Carter Creek-Whitney Canyon gas field was kindly granted by Amoco Oil Company and Chevron U.S.A, both in Denver. Financial support for the study was provided by research grants from Gulf Oil Company, Casper, Wyoming; Marathon Oil Company, Denver Research Center and Champlin Petroleum Company, Denver. The author would like to recognize David V. Wiltschko and John P. Craddock, Center for Tectonophysics, Texas A&M University, for their considerable assistance collecting samples at the Haystack Peak section and teaching me about the importance of structural analysis in the Madison Group. This study is only a preliminary report on an integrated structural/diagenetic project with them. Finally, Karen Rose Cercone and William J. Meyers at SUNY, Stony Brook are thanked for providing the author with office space and access to microscopes, photographic equipment and a word processor. The manuscript would not have been completed without their support.

REFERENCES

Armstrong, R.L., and Oriel, S.S., 1965, Tectonic development of Idaho-Wyoming thrust belt: Am. Assoc. Petrol. Geol. Bull., v. 49, p. 1847-1866.

Banner, J., Hanson, G., Meyers, W., and Prosky, J., 1985, Geochemistry of regionally extensive dolomites, Burlington and

Keokuk Formations (Mississippian) Iowa and Illinois (Abs): Am Assoc. Petrol. Geol. Bull., v. 69, p. 236.

Brand, U., 1982, The oxygen and carbon isotope composition of carboniferous fossil component: Sea-water effects: Sedimentology, v. 29, p. 139-147.

Brand, U., and Veizer, J., 1981, Chemical diagenesis of a multicomponent carbonate system--2: Stable isotopes: Jour. Sed. Petrol., v. 51, p. 987-997.

Budai, J.M., 1984, Evidence for rapid fluid migration during thrust belt deformation (Abs): Am. Assoc. Petrol. Geol. Bull., v. 68, p. 458.

Budai, J.M., Lohmann, K.C., and Owen, R.M., 1984, Burial dedolomitization in the Mississippian Madison Group, Wyoming and Utah Thrust Belt: Jour. Sed. Petrol., v. 54, p. 276-288.

Budai, J.M., Lohmann, K.C., and Wilson, J.L., 1985, Synsedimentary and early mixed water dolomitization of the Madison Group, Wyoming and Utah Overthrust Belt: submitted to Am. Assoc. Petrol. Geol. Bull.

Choquette, P.W., and Steinen, R.P., 1980, Mississippian non-supratidal dolomite, St. Genevieve Limestone, Illinois Basin: Evidence for mixed water dolomitization: in Zenger, D.J., Dunham, J.B., and Ethington, R.L. (eds.), Concepts and Models of Dolomitization: Soc. Econ. Paleont. Mineral. Spec. Pub. 28, p. 163-196.

Gutschick, R.C., Sandberg, C.A., and Sando, W.J., 1980, Mississippian shelf margin and carbonate platform from Montana to Nevada: in Fouch, T.D., and Magathan, E.R. (eds.), Rocky Mtn. Sect., Soc. Econ. Paleont. Mineral., West-Central U.S. Paleog. Symp. 1, p. 111-128.

Irwin, H., Curtis, C., and Coleman, M.L., 1977, Isotopic evidence for source of diagenetic carbonates formed during burial of organic-rich sediments: Nature, v. 269, p. 209-213.

Jordan, T.E., 1981, Thrust loads and foreland basin evolution, Cretaceous, western United States: Am. Assoc. Petrol. Geol. Bull., v. 65, p. 2506-2520.

Meyers, W.J., and Lohmann, K.C., 1985, Isotope geochemistry of regionally extensive calcite cement zones and marine components in Mississippian limestones, New Mexico: in Schneidermann, N., and Harris, P. M. (eds.), Carbonate Cements Revisited: Soc. Econ. Paleont. Mineral. Spec. Pub. 36, p. 233-239.

Royse, F., Jr., Warner, M.A., and Reese. D.L., 1975, Thrust belt structural geometry and related stratigraphic problems, Wyoming-

Idaho-Northern Utah: Deep Drilling Frontiers of the Central Rocky Muntains, Rocky Assoc. Geol., Symp. Vol., p. 41-54.

Sando, W.J., and Sandberg, C.A., 1977, Old Laketown Canyon Section, Utah: in Dutro, J.T., Jr., Gilmour, E.H., Gutschick, R.C., Sandberg, C.A., and Sando, W.J. (eds.), Carboniferous of the Northern Rocky Mountains: Am. Geol. Inst. Selected Guidebook Series No. 3, p. 46-48.

Wiltschko, D.V., and Dorr, J.A., Jr., 1983, Timing of deformation in Overthrust Belt and Foreland of Idaho, Wyoming, and Utah: Am. Assoc. Petrol. Geol. Bull., v. 67, p. 1304-1322.

PENNSYLVANIAN PHYLLOID-ALGAL MOUND PRODUCTION AT TIN CUP MESA FIELD PARADOX BASIN, UTAH

Wilson H. Herrod
Marathon Oil Company
Casper, Wyoming 82601

Mike H. Roylance
Marathon Oil Company
Casper, Wyoming 82601

Elizabeth C. Strathouse
Marathon Oil Company
Casper, Wyoming 82601

ABSTRACT

Tin Cup Mesa field in the Utah portion of the Paradox Basin produces from the Upper Ismay Zone of the Pennsylvanian Paradox Formation. The principal depositional and diagenetic facies within the reservoir are visible in cores from two wells. The Tin Cup Mesa #3-26 (NW NE Sec. 26, T28S-R25E) was drilled approximately 1160 feet northeast of, and updip from, the #3-26 and encountered only 30 feet of carbonate with a commensurate increase in anhydrite to a thickness of 90 feet.

The five major lithofacies within the field area include: (1) the phylloid-algal mound; (2) mound cap; (3) mound flank; (4) intermound; and (5) an evaporitic facies. All of these are present in the Marathon #3-26 producer, while in the Marathon #1-23 dry hole, facies numbers 4 and 5 above predominate. All lithofacies are potentially productive with the exception of the intermound and evaporitic facies found in the #1-23.

Diagenetic processes which have enhanced the reservoir include dolomitization and carbonate dissolution. These two processes, along with primary shelter porosity, are responsible for all of the reservoir present in the field. Cementation was the primary process by which porosity was destroyed, and at least six different types of cement are present. Compaction, bioturbation, and internal sedimentation were additional agents of reservoir destruction. It should be emphasized that repeated early exposure of the reservoir at Tin Cup Mesa field was of paramount importance in the creation of secondary porosity and in the enhancement of primary porosity.

INTRODUCTION

Tin Cup Mesa field is a northwest-southeast trending carbonate stratigraphic trap located in the southeastern Utah portion of the Paradox Basin (Fig. 1). The field produces from phylloid-algal limestones and dolomites of the Pennsylvanian (Desmoinesian) Paradox Formation. The purpose of this paper is to illustrate and discuss the principal depositional and diagenetic facies present in the field, with emphasis on their relationship to the distribution of porosity and permeability, and their control on petroleum entrapment. Two cores will be described in detail to illustrate these relationships.

Regional Setting and Stratigraphy

The Paradox Basin was a silled Pennsylvanian depocenter bordered by the Uncompahgre Uplift to the north and east, the Defiance-Zuni Uplifts to the south, and the Monument-Kaibab, Circle Cliffs-Emery Uplifts to the west (Fig. 2). Several passages through the surrounding uplifts have been postulated to have provided access to the basin, of which the Four-Corners Platform is the best defined. Normal marine waters entered the restricted basin over this platform, and via several possible channels to the northwest termed the Fremont and Oquirrh accessways (Elias, 1963). Evaporite deposition prevailed in the more distal northern parts of the basin, while carbonates accumulated contemporaneously on a shallow shelf to the south, east, and possibly to the west. Hite (1970) and Roylance (1984) have summarized the hydrographic setting which encouraged the

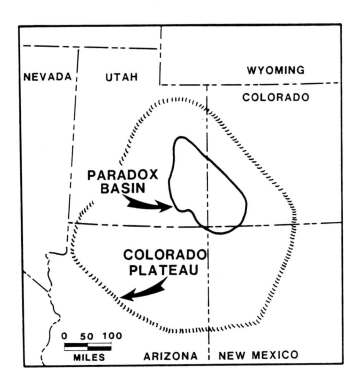

Figure 1. Location of the Pennsylvanian Paradox Basin in the Four Corners area of the Colorado Plateau.

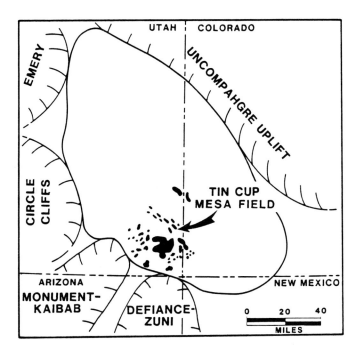

Figure 2. The paleostructural setting of the Paradox Basin restricted the access of normal marine waters resulting in the cyclic deposition of carbonates, sapropelic dolomites, and evaporites (reservoirs, source rocks, and seals).

synchronous deposition of such diverse lithologies in the Paradox Basin.

As illustrated in Figure 3, the Paradox Formation is the middle member of the Hermosa Group, and has been subdivided into four informal zones of which the Ismay is the youngest. Each zone represents at least one transgressive-regressive pulse which generally emanated from the south over the Four-Corners Platform. Each successive transgressive-regressive episode extended the carbonate to evaporite facies transition progressively northward until, by Upper Ismay time, most of the basin was covered by carbonates with thick evaporites relegated to the extreme northern part of the basin (Reid and Berghorn, 1981).

Phylloid-algal mounds and meadows proliferated during the regressive phase of these transgressive-regressive episodes, particularly during the Ismay and Desert Creek cycles. The relatively thin sapropelic dolomites of the Hovenweep, Gothic, and Chimney Rock Zones, as well as lesser thin "shales", are probably in part transgressive deposits which may be utilized as approximate time-lines.

The bulk of Paradox Basin production comes from Ismay and Desert Creek phylloid-algal carbonate stratigraphic traps in the southeastern Utah portion of the basin. Approximately 50 fields have produced from such traps, the most notable being Aneth field (estimated ultimate recovery of 350 million barrels largely from the Desert Creek). A recent resurgence of exploration activity has resulted in eight Ismay/Desert Creek discoveries in the past five years.

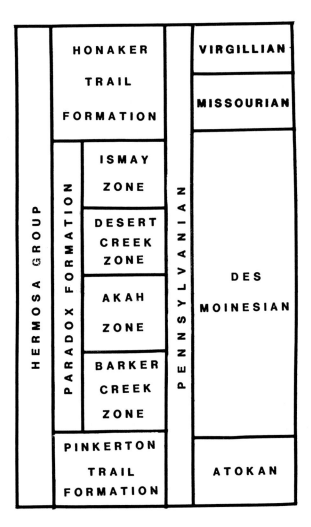

Figure 3. Pennsylvanian stratigraphic column for the Paradox Basin.

TIN CUP MESA FIELD

Tin Cup Mesa field was discovered in April, 1982 on a prospect which was defined by traditional geologic structural and isopach mapping, and refined utilizing high-resolution seismic data. The discovery well, Marathon's Tin Cup Mesa #3-26 (NW NE Sec. 26, T38S-R25E) was completed in the Upper Ismay for 528 BOPD, 401 MCFGD, and no water. This well encountered 120 feet (36 m) of Upper Ismay carbonate capped by 10 feet (3 m) of anhydrite. A subsequent offset, the Marathon Tin Cup Mesa #1-23, (SE SE Sec. 23, T38S-R25E) was drilled approximately 1160 feet (353 m) northeast of and updip from the discovery well largely for acreage reasons. The resulting dry hole encountered only 30 feet (9 m) of carbonate with a commensurate increase in anhydrite to a thickness of 90 feet (27 m). A total of eight wells have been drilled to date in the field, seven of which produce oil. Cumulative gross production as of January 31, 1985 totaled 337,687 BO, 390.5 MMCFG, and 49,960 BW.

Reservoir Distribution and Trap

The reservoir at Tin Cup consists of a heterogeneous series of stacked phylloid-algal mounds separated by rocks representing intermound and evaporitic facies. Local source rocks such as the sapropelic dolomites of the underlying Hovenweep, Gothic and Chimney Rock "shales" charged the reservoir with approximately 4.5 million barrels of oil (stock-tank barrels in place). The vertical seal is provided by the anhydrite cap which thins over the field. The porous buildup grades laterally into impermeable carbonate mudstones and shales, which together with the anhydrite may serve as a lateral barrier.

Although the hydrocarbons at Tin Cup are trapped stratigraphically, the field reflects a southwestern regional dip and the distribution of gas, oil, and water in the reservoir is structurally controlled (Fig. 4). As shown in Figures 5 and 6, the gas-oil contact is at -362 feet (-110 m) subsea and the top of the oil-water transition zone is at -442 feet (-135 m). Reservoir energy is predominately solution-gas drive.

The field's best porosity and permeability development occurs in the vicinity of the Tin Cup Mesa #4-26 where the buildup is the thickest and structurally highest. Figure 7 illustrates the distribution of net hydrocarbon-bearing reservoir (i.e., net pay above the oil-water contact) as determined by production data. Table I provides specific information on producing wells.

Of the five major lithofacies to be discussed in the following section, the phylloid-algal facies exhibits the best reservoir based upon porosity (which ranges from 2% to 20%, averaging 8%) and permeability (which ranges from 0.01 md to 486 md, averaging 10 md). However, this algal facies is not continuous across the field. Several relatively thin dolomites do correlate throughout the field, but are less productive than the algal carbonates. Porosities range from 2% to 30% and average 12%, while permeabilities range from 0.01 md to 60 md and average 5 md. In general, these dolomites dominate the reservoir in the northwestern end of the field (wells #2-23 and #3-23), while algal-carbonates provide the bulk of production from the #3-36 and

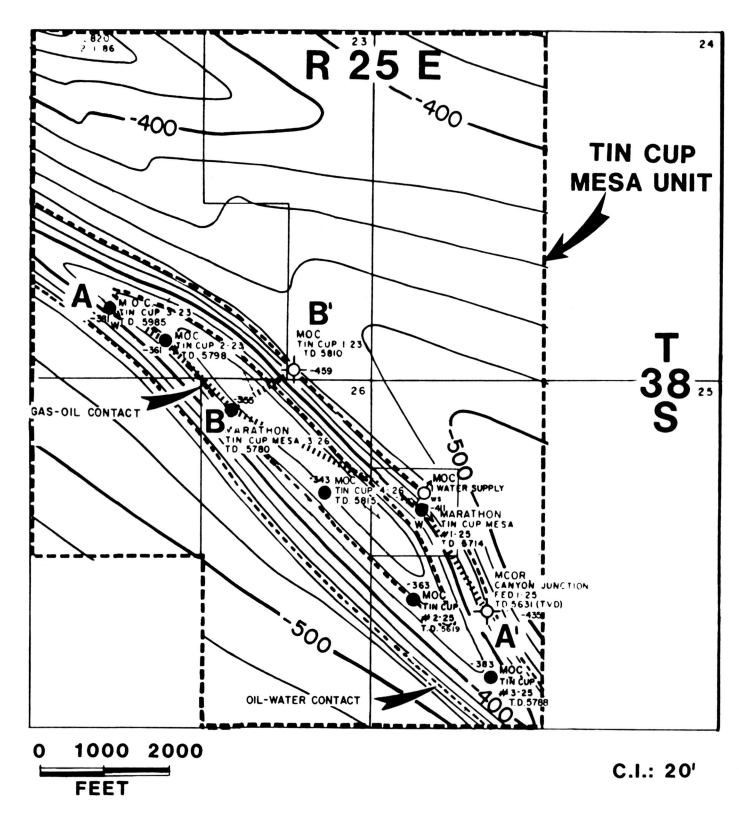

Figure 4. Upper Ismay structure map (at top of carbonate) in the Tin Cup Mesa Field area illustrating southwestern regional dip with stratigraphically induced closure. Note the location of the #3-26 and #1-23 wells as well as the A-A' and B-B' cross sections.

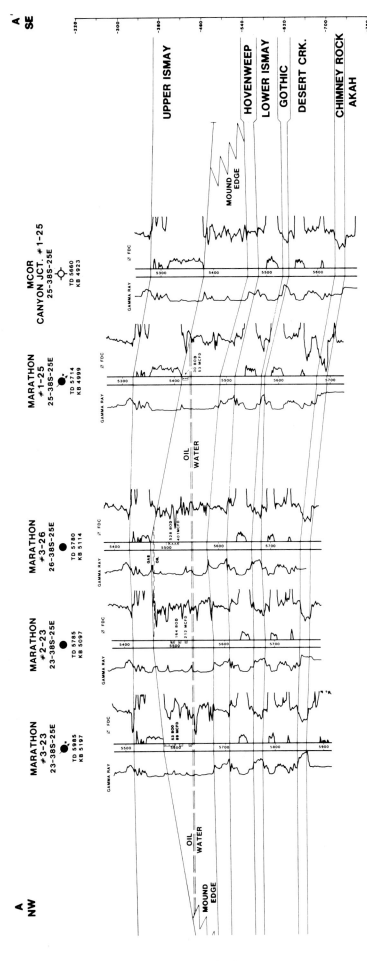

Figure 5. Sturcture cross section A-A' along strike at Tin Cup Mesa Field. Note oil-water and gas oil contacts.

417

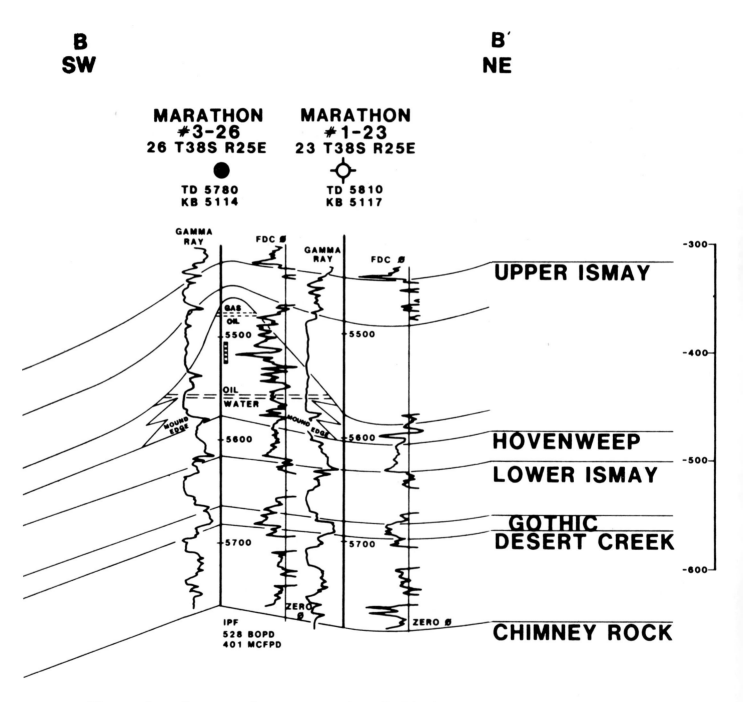

Figure 6. Structural cross section B-B' (dip section) at Tin Cup Mesa Field. Note fluid levels and the dramatic facies transition between the cored wells on display here.

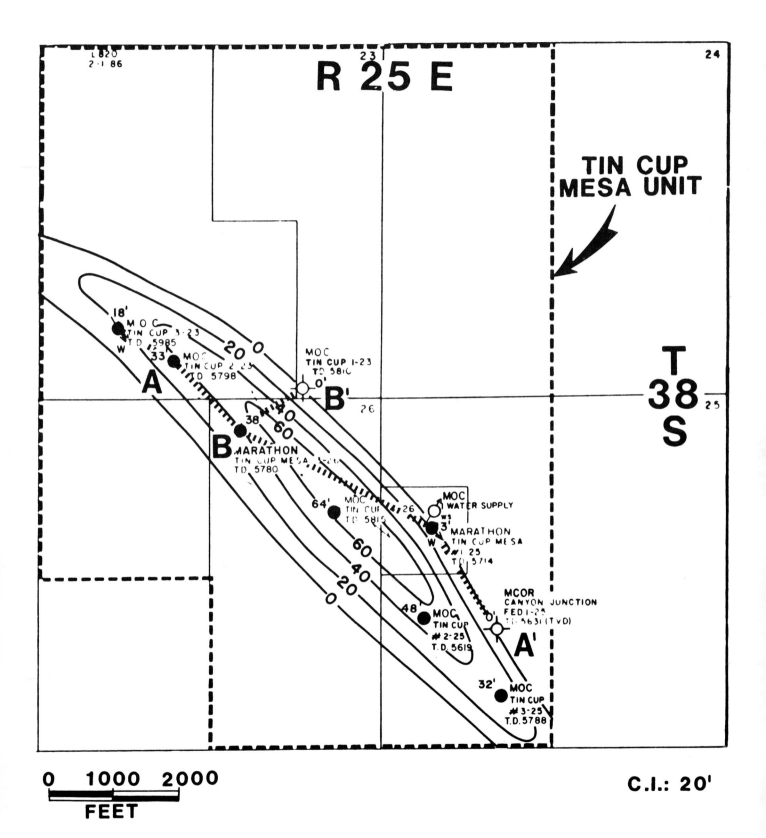

Figure 7. Isopach map of Upper Ismay net pay above the oil-water contact at Tin Cup Mesa Field.

TABLE I

TIN CUP MESA FIELD

T38S, R25E

San Juan County, Utah

OPERATOR	WELL NAME	LOCATION	TOTAL DEPTH	DATE COMPLETED	STATUS	PRODUCTIVE INTERVAL (ISMAY)	INITIAL PRODUCTION BOPD
MCOR	#1-25 Canyon Junction Federal	NE SW Sec. 25	5,660'	2-1981	D&A	0	0
MARATHON	#1-25 Tin Cup Mesa	SW NW Sec. 25	5,714'	1-1983	Oil	5,416' - 26'	30
MARATHON	#3-26 Tin Cup Mesa	NW NE Sec. 26	5,780'	4-1982	Oil	5,504' - 24'	528
MARATHON	#1-23 Tin Cup Mesa	SE SE Sec. 23	5,810'	9-1982	D&A	0	0
MARATHON	#2-23 Tin Cup Mesa	SE SW Sec. 23	5,798'	1-1983	Oil	5,492' - 5525'	164
MARATHON	#3-23 Tin Cup Mesa	SW SW Sec. 23	5,895'	9-1983	Oil	5,578' - 5632'	53
MARATHON	#4-26 Tin Cup Mesa	SE NE Sec. 26	5,815'	11-1983	Oil	5,504' - 62'	310
MARATHON	#2-25 Tin Cup Mesa	NW SW Sec. 25	5,619'	4-1984	Oil	5,300' - 42'	307
MARATHON	#3-25 Tin Cup Mesa	SE SW Sec. 25	5,788'	8-1984	Oil	5,440' - 87'	393

Table I. Current chronology of drilling in the Tin Cup Mesa Field area.

wells to the southeast. Production tests in the #3-23 indicate that rocks with less than 2 md permeability and 5% porosity do not produce.

LITHOFACIES

From cores taken in and around Tin Cup Mesa field, five major lithofacies may be recognized: (1) phylloid-algal mound; (2) mound cap; (3) mound flank; (4) intermound; and (5) evaporitic facies. All of these facies are present in the Marathon #3-26 discovery well, while in the Marathon #1-23 dry hole, facies numbers 1, 2, and possibly 3 are absent (Figs. 8, 9, and 10).

Phylloid-Algal Facies

This facies represents the heart of the algal mound development. It consists largely of gray, phylloid-algal lime wackestones to grainstones (bafflestone). The phylloid algae are presumed to be *Ivanovia* (an extinct codiacean green algae) but are invariably too altered for positive identification. Additional biotic constituents include fenestrate bryozoans, encrusting (tubular or opthalmidid) forams, pelecypods, and occasional crinoids, ostracods, and gastropods. Non-skeletal components include peloids, intraclasts and variety of coated grains, some of which are ooliths. This facies lacks distinct bedding. In general, mud content and the number of crinoids increase downward in a given mound, while phylloid algae commonly increase in abundance upwards, as the rock becomes increasingly

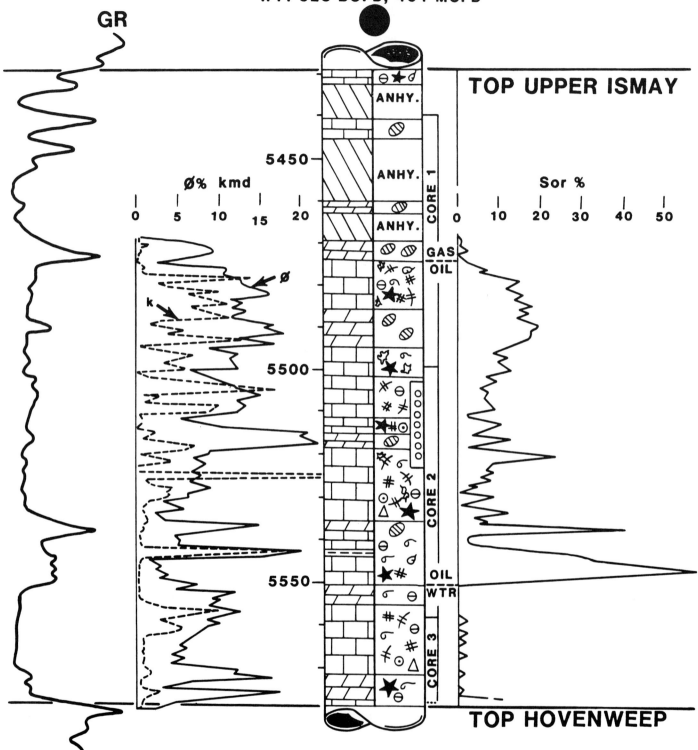

Figure 8. Schematic core description for the Marathon Tin Cup Mesa #3-26 discovery well. Core porosity, permeability, residual oil saturation, and gamma-ray response are illustrated. Note that only about 10 feet of capping anhydrite was encountered. See Fig. 10 for key.

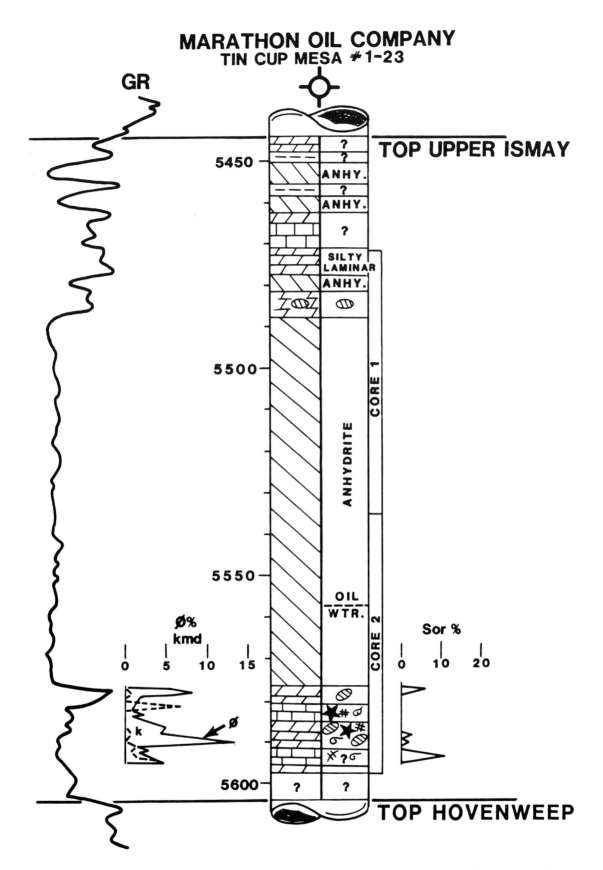

Figure 9. Schematic core description for the Marathon Tin Cup Mesa #1-23. Core porosity, permeability, residual oil saturation, and gamma-ray response are illustrated. Note that about 90 feet of capping anhydrite was encountered. See Fig. 10 for key.

grain-supported (Figs. 11, 12, and 13). Porosity and permeability can be excellent in this facies, with the latter being either primary, solution-enhanced shelter porosity, and/or solution-moldic porosity.

Figure 10. Key for symbols used in Figures 8 and 9.

Figure 11. Phylloid-algal lime wackestone to packstone illustrating solution-modified shelter porosity. Note light-colored calcite cement. Core slab from Tin Cup Mesa #3-26 (5,502 ft.).

Figure 12. Phylloid-algal lime wackestone illustrating relatively tight, compacted algal plates. Lack of early cementation and buttressing by botryoidal aragonite precluded the preservation of primary shelter porosity. Core slab from Tin Cup Mesa #3-26 (5,562 ft.).

Figure 13. Phylloid-algal lime packstone. Note compacted algal plates and occulsion of interparticle porosity by anhydrite. Photomicrograph of this section from Tin Cup Mesa #3-26 (5,561.5 ft.).

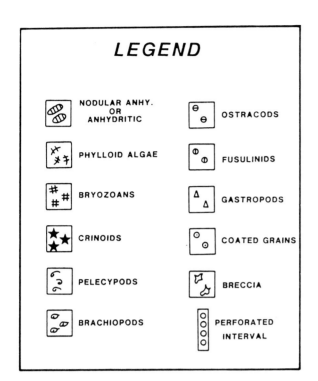

10

11

12

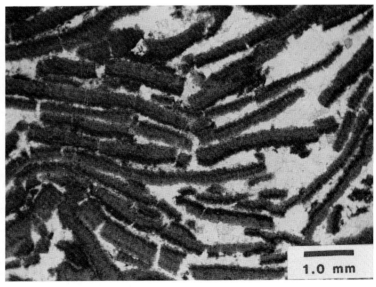

13

The Upper Ismay at Tin Cup Mesa field exhibits three, and perhaps four, individual mound developments (Fig. 8). These stacked mounds vary from about 10 feet to 20 feet (3 to 6 m) in thickness. However, these thicknesses are subjective in that individual mounds often shoal upwards into exposure surfaces, and the resulting brecciation, solution, and dolomitization obscures the transition from a mound to a mound-cap facies. The base of the phylloid-algal facies is gradational with the underlying intermound rocks.

Mound-Cap Facies

Individual mounds generally shoal upwards into capping facies which consist of light-gray to light-brown anhydritic, dolomitic mudstones or dolomitic lime mudstones. Various peritidal features such as nodular anhydrites, mudcracks, intraclastic pavements, fenestral and stromatolitic fabrics, brecciation, and calichification may be present (Figs. 14, 15, and 16). A strong evaporitive, diagenetic overprint obscures much or all of the original texture. Therefore the foraminiferal, peloidal, ooid calcarenites described by Choquette and Traut (1963), Elias (1963), Pray and Wray (1963), and Choquette (1983) as capping phylloid-algal buildups in the Ismay-Flodine field area may be obscured. Due to leaching and dolomitization, porosity in the cap facies can be substantial, whereas permeability is usually reduced relative to the mound proper.

Again, the local effects of brecciation, calichification and porosity-occluding cements are all important. The contact between this capping facies and the overlying intermound, or mound-flank facies, is usually abrupt.

Mound-Flank Facies

This facies is postulated to account for the syndepositional breccias common to many of the cores from Tin Cup Mesa field. These breccia grainstones usually comprise a conglomerate of individual, pebble-sized, angular clasts each consisting of gray mudstone to packstone bound on one side by a phylloid-algal plate. This arrangement simply reflects brecciation along lines of primary shelter porosity. The matrix between breccia fragments often consists of brown to light-brown peloidal wackestone to packstone, and the angular interstices are usually plugged by a combination of calcite spar and anhydrite. The angular nature of these breccias suggests early induration of the mound followed by possible slumping from the crest onto the flanks of the buildup. Certainly, the 90+ feet (27 m) of relief on the Tin Cup mound was sufficient to encourage instability, perhaps triggered by storms. Of course, in situ brecciation due to compaction or exposure is also probable, and is suggested where individual clasts can be refitted, or if vadose (as opposed to marine) cements occupy the angular voids. The reservoir quality of these breccias is highly variable, but can be excellent (Figs. 17 and 18).

Figure 14. Fossiliferous anhydritic, dolomitic wackestone with fossil-moldic and vuggy porosity. Most fossil material is unrecognizable with the exception of crinoidal debris. Some porosity occulsion by anhydrite. Probably represents mound-cap facies. Core slab from Tin Cup Mesa #3-26 (5,514 ft.).

Figure 15. Fossil-moldic porosity formed by leaching crinoid grains (C), foraminifera (FO), and other assorted fossils. Voids have been subsequently filled by anhydrite. Note several large shapeless voids (V) that are probably enlarged fossil molds. Plane-polarized light. Photomicrograph from Tin Cup Mesa #3-26 (5,487 ft.).

Figure 16. Brecciated, fenestral, pelloidal, pseudo-oolitic, slightly fossiliferous, anhydritic dolomitic wackestone to packstone. Most textural elements are best visible in thin-section. Probably represents mound-cap facies. Core slab from Tin Cup Mesa #3-26 (5,516 ft.).

Figure 17. Brecciated, fossiliferous (?), pelloidal, lime wackestone to mudstone. Internal sedimentation and anhydrite occludes most voids. Possible phylloid-algal blades present in this interval. May represent a slump breccia from the mound proper. Core slab from Tin Cup Mesa #1-23 (5,587 ft.).

14

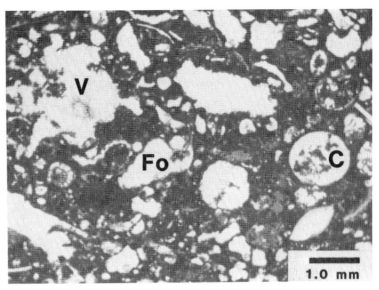

15

16

17

Figure 18. Brecciated, bioturbated (?), fossiliferous, pelloidal, dolomitic wackestone. Probably represents slumping from mound cap, or post-mound (inter-mound) bioturbation. Core slab from Tin Cup Mesa #3-26 (5,473 ft.).

Figure 19. Fossiliferous (crinoidal), lime wackestone from intermound facies. Note wispy compaction features and semi-articulated crinoid column. Core slab from Tin Cup Mesa #3-26 (5,547 ft.).

Figure 20. Fossiliferous (crinoidal), highly bioturbated lime wackestone from intermound facies near contact with Hovenweep interval. Core slab from Tin Cup Mesa #3-26 (5,576 ft.).

Figure 21. Anhydrite, massive to enterolithic with a dolomitic mudstone matrix from evaporitic cap. Core slab from Tin Cup Mesa #1-23 (5,559 ft.).

18

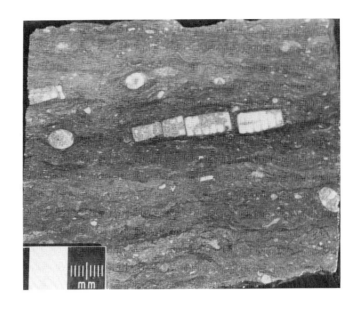

19

20

21

Intermound Facies

The intermound facies is probably the most varied. Depositionally it represents sedimentation between existing mounds, and may transgress those mounds due to minor sea level fluctuations. Generally these are brown-gray to black crinoidal, brachiopodal, spiculitic lime wackestones to sapropelic, dolomitic mudstones (Figs. 9, 19, and 20). These were initially euxinic transgressive deposits which are organic-rich at the base, and shoal upward into lighter-colored fossiliferous (though still restricted), bioturbated limestones. These rocks formed the substrate upon which algal mounds subsequently developed. This facies is generally tight, and unless dolomitized, may serve as a lateral seal.

Evaporitic Facies

This facies caps the Upper Ismay in the field and overprints virtually the entire cored section (Figs. 8 and 9). It is dominated by dark to light-gray anhydrite exhibiting a variety of displacive/replacive, or primary textures including nodular or mosaic (chickenwire) fabrics, a massive texture (with occasional palmate structures), or laminar features (Figs. 21 and 22). These anhydrites are generally interbedded or intermingled with gray to brown-gray dolomitic mudstones to wackestones. Lighter colored dolomites which are interbedded with the anhydrite are generally featureless whereas the darker, organic-rich dolomites with nodular anhydrites may exhibit scattered fossils, burrows, and current laminations. The transition from a primary evaporitic facies into the underlying carbonate buildup is gradual and early dolomitization and anhydritization often mask primary features.

DIAGENESIS

The carbonates at Tin Cup Mesa field exhibit a complex diagenetic overprint. The phylloid-algal lithofacies has undergone the most intense alteration, while the intermound lithofacies has suffered the least. The diagenetic processes that have affected these carbonates can be divided on the basis of whether they enhance or damage the reservoir, or had no effect on reservoir quality.

Reservoir Enhancement

Dolomitization and dissolution of carbonate minerals are the main diagenetic processes responsible for the creation of porosity. These two processes, along with primary porosity, are responsible for all of the reservoir present in the field. The basic porosity types include intercrystalline, moldic (Fig. 15), and solution-enhanced primary porosity (Fig. 23). Conversely, where these processes were inactive, such as in the intermound facies, porosity is negligible.

Dolomitization and dissolution are interrelated processes which are both associated with periods of restricted shoaling conditions and subarial exposure. Several likely mechanisms were involved in the dolomitization/leaching process. One involves the mixing of fresh and marine waters to create a dolomitizing fluid as in the Dorag model

(Badiozamani, 1973). As fresh water entered the system during times of exposure, it would dissolve the less stable carbonate phases. Subsequent mixing with marine fluids within the mound would result in dolomitization. As expected, the most intense leaching is found in the uppermost mound and overlying mound-cap facies, where the likelihood for prolonged exposure was the greatest. The patchy occurrence of dolomite within the buildup probably reflects several periods of exposure and complex pathways for dolomitizing fluids. A second model for dolomitization involves the formation of dense Mg-rich brines, which as they moved through carbonate sediment, could both dolomitize some carbonate phases while dissolving the less stable (aragonitic) grains (Runnels, 1969).

Reservoir Destruction

A large amount of both primary and secondary porosity was occluded as the result of a number of diagenetic processes. These processes included compaction, bioturbation, the formation of botryoidal aragonite cement, plus additional cementation, brecciation, and internal sedimentation.

Compaction and bioturbation were processes by which much of the primary porosity was destroyed. Additional loading throughout the burial history produced stylolites, further reducing porosity. Botryoidal aragonite (now calcite or dolomite) grew syndepositionally in the phylloid-algal mound facies (Fig. 24). This cement is relatively rare in these phylloid-algal mounds, much of it having been leached

(Fig. 25). Botryoidal aragonite grew into existing void space or at the sediment-water interface (Roylance, 1984).

Cementation was the primary process by which porosity was destroyed. There are at least six different types of cement present in the phylloid-algal and associated facies. They are, from oldest to youngest: (1) isopachus fibrous carbonate (Fig. 26); (2) coarse calcite-spar (Fig. 27); (3) coarse crystalline anhydrite (originally gypsum) (Fig. 38); (4) halite (Fig. 23); (5) calcite overgrowths on echinoderm fragments (Fig. 29); (6) baroque dolomite (Fig. 30). Volumetrically, coarse crystalline anhydrite is the most common cement. Coarse calcite-spar is second in abundance. Most of the cementation probably occurred relatively early in the diagenetic history, although with the exception of isopachous-fibrous and botryoidal cements, it generally post-dates the formation of solution porosity. Many large solution pores are completely filled by cement (Fig. 27). Baroque dolomite is the only cement that formed much later (Moore and Druckman, 1981). It always occurs in solution voids and forms large crystals that are sometimes a centimeter in diameter.

Another diagenetic process related to, and probably concurrent with the formation of coarse, crystalline anhydrite cement is the replacement of carbonate by microcrystalline anhydrite (Fig. 31). This process of anhydritization undoubtedly resulted in reduction in porosity, as it occluded any intercrystalline or interparticle porosity as well.

Brecciation is prevalent in the phylloid-algal facies and mound flank facies. Brecciation was probably a consequence of exposure and/or compaction acting on loosely packed fossil grains such as

Figure 22. Nodular to chicken-wire anhydrite with a dolomitic mudstone matrix from evaporitic cap. Core slab from Tin Cup Mesa #3-26 (5,462 ft.).

Figure 23. Solution-enhanced porosity (P). Halite crystal (H) partially fills void. Round "hole", to left and light spot above it, are where colored epoxy did not completely fill void. Plane-polarized light. Photomicrograph from Tin Cup Mesa #3-26 (5,505 ft.).

Figure 24. Botryoidal calcite (originally aragonite) associated with a phylloid-algal blade (P) which is completely leached and altered except for a micrite rind. Note that the botryoid grew into a void that has been completely filled by calcite cement (C) and anhydrite cement (A). Plane-polarized light. Photomicrograph from Tin Cup Mesa #3-26 (5,501.4 ft.).

Figure 25. Original botryoidal aragonite that has been completely leached and the void filled by coarse crystalline anhydrite. Note that the serrated edge of the botryoid is still preserved due to its micritic composition. Plane-polarized light. Photomicrograph from Tin Cup Mesa #3-26 (5,507.6 ft.).

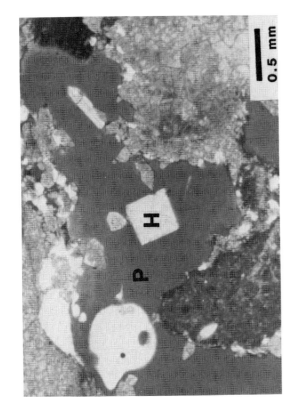

23

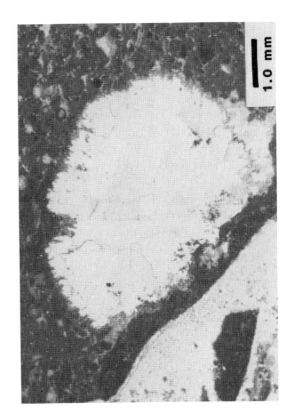

25

22

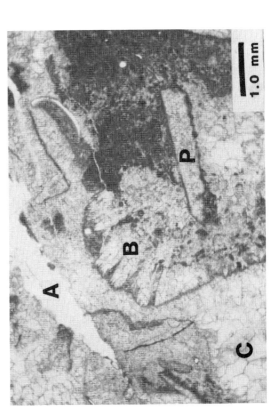

24

Figure 26. A shelter void formed by a phylloid-algal blade (P) that has been lined by isopachous fibrous carbonate cement and subsequently filled by coarse crystalline anhydrite cement. Note the phylloid-algal blade has been leached and replaced by anhydrite. Plane-polarized light. Photomicrograph from Tin Cup Mesa #3-26 (5,500.8 ft.).

Figure 27. Large solution void that has been completely filled by coarse calcite-spar. Note several phylloid-algal blades that are preserved only as micrite rinds. Plane-polarized light. Photomicrograph from Tin Cup Mesa #3-26 (5,501.4 ft.).

Figure 28. A large solution void that has been rimmed by coarse calcite spar and then mostly filled by coarse crystalline anhydrite. Dark spots (P) are residual porosity filled by dyed epoxy. Plane-polarized light. Photomicrograph from Tin Cup Mesa #3-26 (5,525.2 ft.).

Figure 29. Calcite overgrowth (O) on crinoid grain (C). Note angular nature of overgrowth which is in optical continuity with the crinoid grain. Plane-polarized light. Photomicrograph from Tin Cup Mesa #3-26 (5,480.3 ft.).

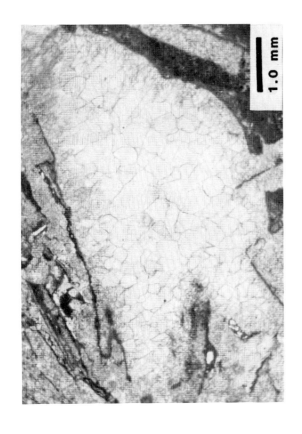

26

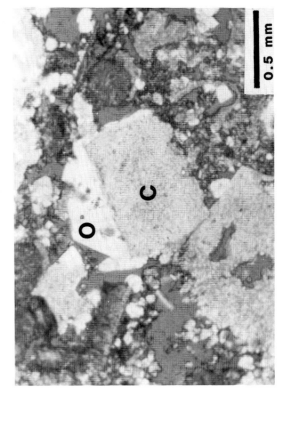

27

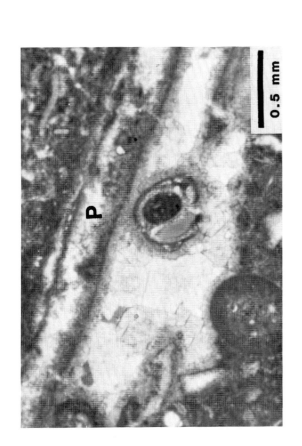

28

29

Figure 30. A single baroque dolomite crystal that almost fills a fossil mold. Note the undulatory extinction characteristic of baroque dolomite. Crossed nichols. Photomicrograph from Tin Cup Mesa #3-26 (5,560.8 ft.).

Figure 31. Replacement of carbonate (originally probably internal sediment) by microcrystalline anhydrite (A). Two completely leached botryoids overlie this microcrystalline anhydrite. Crossed nichols. Photomicrograph from Tin Cup Mesa #3-26 (5,507.6 ft.).

Figure 32. Internal sediment surrounding breccia clasts in mound-cap facies. Plane-polarized light. Photomicrograph from Tin Cup Mesa #3-26 (5,496.5 ft.).

Figure 33. Large solution void with a geopetal structure of internal sediment that has been partially replaced by anhydrite. The rest of the void has been filled by coarse crystalline anhydrite. Plane-polarized light. Photomicrograph from Tin Cup Mesa #3-26 (5,480.3 ft.).

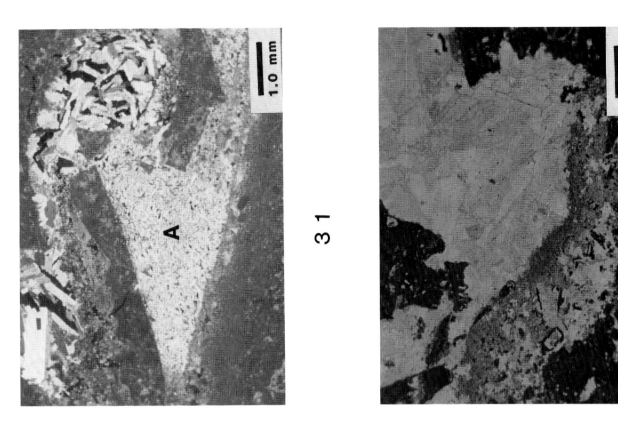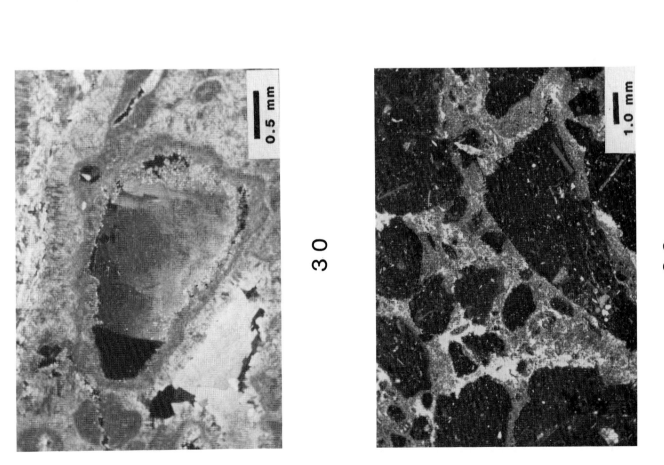

phylloid-algal blades. Brecciation tends to decrease the total amount of porosity (due to the high initial primary porosity) even though it might add some intergranular voids as well as open up channel-ways for dolomitizing and leaching fluids.

Internal sedimentation was another of the major causes of porosity reduction, although in areas of the core with large solution vugs, it is noticeably absent. Internal sediment is most commonly found in the mound-flank or cap facies (Fig. 32). Occasionally it floors some of the large solution voids in the phylloid-algal facies (Fig. 33).

Other Diagenetic Processes

Diagenetic processes that neither enhanced nor reduced porosity include the formation of caliche crusts, silification, recrystallization, and micritization. Caliche crusts (Fig. 16) formed very early in the history of the phylloid-algal mounds due to periodic exposure. Silicification effects only fossil grains, especially crinoid ossicles. Most of the original carbonate sediment has been recrystallized. Phylloid algae were particularly susceptible. Many fossil grains and intraclasts have been partially or completely micritized. In some cases this is fortuitous, since a micrite rind is all that remains of many phylloid-algal grains (Fig. 27).

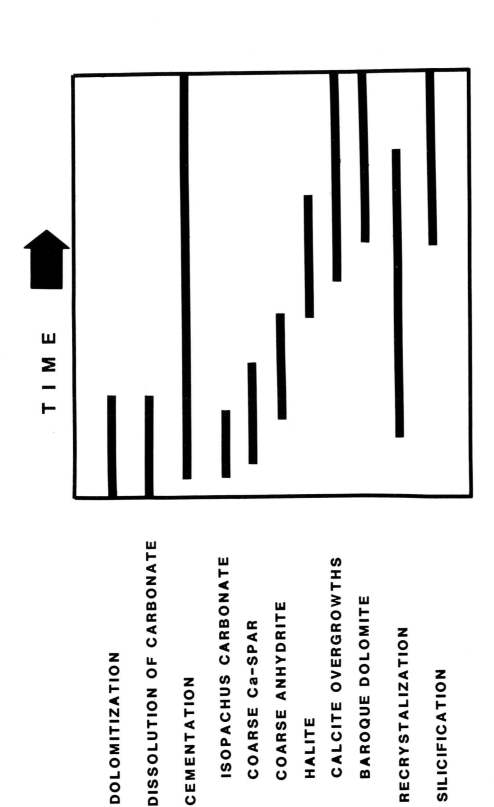

Figure 34. Paragenetic sequence for Tin Cup Mesa field. The diagenetic processed that occurred syndepositionally or nearly so, such as compaction, bioturbation, the formation of botryoidal aragonite, brecciation, internal sedimentation, calichification, and micritization, are not included in the figure.

Diagenetic Sequence

The sequence of diagenetic events is summarized in Figure 34. As illustrated, much of the diagenetic alteration that affected the reservoir occurred concurrently, and very early in the history of the mound. Some events were essentially syndepositional. Overall, it should be emphasized that the repeated early exposure of the reservoir at Tin Cup Mesa field was of paramount importance in the creation of secondary porosity via solution and dolomitization, and the enhancement of primary porosity. In addition, the evaporitic conditions associated with the termination of mound growth, did much to alter and obscure the primary depositional features while obliterating a considerable volume of porosity.

CONCLUSIONS

Phylloid-algal mounds in the Paradox Basin are complex, heterogenous reservoirs characterized by rapid lateral and vertical facies changes. The distribution of algal material is not uniform within a mound and individual buildups may be stacked and/or shingled. These primary facies variations are further complicated by a strong diagenetic overprint resulting from the early and repeated exposure of individual mounds. Due to this complexity, reservoir behavior is very difficult to predict.

REFERENCES

BADIOZAMANI, K., 1973, The Dorag dolomitization model - application to the Middle Ordovician of Wisconsin: Jour. Sed. Petrol., v. 43, p. 965-984.

CHOQUETTE, P. W., 1983, Platy algal reef mounds, Paradox basin: in P.A. Scholle, D. G. Bebout, and C. H. Moore (eds.), Carbonate Depositional Environments: Am. Assoc. Petrol. Geol. Memoir 33, p. 454-462.

CHOQUETTE, P. W., and J. D. TRAUT, 1963, Pennsylvanian carbonate reservoirs, Ismay field, Utah and Colorado: in R. O. Bass and S. L. Sharp (eds.), Shelf Carbonates of the Paradox Basin, A Symposium: 4th Field Conf., Four Corners Geol. Soc., p. 157-184.

ELIAS, G. K., 1963 Habitat of Pennsylvanian algal bioherms, Four Corners area: in R. O. Bass and S. L. Sharp (eds.) Shelf Carbonates of the Paradox Basin, A Symposium: 4th Field Conf., Four corners Geol. Soc., p. 185-203.

HITE, R. J., 1970, Shelf carbonate sedimentation controlled by salinity in the Paradox basin, southeast Utah: in J. L. Ran, and L. F. Dellwig (eds), Third Symposium on Salt: N. Ohio Geol. Soc., v. 1, p. 48-66.

MOORE, C. H., and Y. DRUCKMAN, 1981, Burial diagenesis and porosity evolution, Upper Jurassic Smackover, Arkansas and Louisiana: Am. Assoc. Petrol. Geol. Bull., v. 65, p. 597-628.

PRAY, L. C., and J. L. WRAY, 1963, Porous algal facies (Pennsylvanian), Honaker Trail, San Juan Canyon, Utah: in R. O. Bass and S. L. Sharp (eds.), Shelf Carbonates of the Paradox Basin, A Symposium: 4th Field Conf., Four Corners Geol. Soc., p. 204-234.

REID, F. S., and C. E. BERGHORN, 1981, Facies recognition and hydrocarbon potential of the Pennsylvanian Paradox Formation: in D. L. Wiegand (ed.), Geology of the Paradox Basin: Rocky Mtn. Assoc. Geol. 1981 Field Conf., p. 111-117.

ROYLANCE, M. H., 1984, Depositional and diagenetic control of petroleum entrapment in the Desert Creek interval, Paradox formation, southeastern Utah and southwestern Colorado: unpub. MS. Thesis, Univ. Kansas.

RUNNELS, D. D., 1969, Diagenesis, chemical sediments, and the mixing of natural waters: Jour. Sed. Petro., v. 39, p. 1188-1201.

SEDIMENTOLOGY AND RESERVOIR CHARACTERISTICS OF THE NIOBRARA FORMATION (UPPER CRETACEOUS), KANSAS AND COLORADO

Peter A. Scholle
Gulf Research and Development
P. O. Box 37048
Houston, TX 77236

Richard M. Pollastro
U.S. Geological Survey
Denver Federal Center,
MS 960
P. O. Box 25046
Denver, CO 80225

ABSTRACT

The Niobrara Formation of the Western Interior consists of 100 to 250 m of interbedded chalks and calcareous shales. The formation is divided into two members: (1) a thin, basal, dominantly chalk-bearing sequence, the Fort Hays Limestone Member, and (2) an upper, thick, calcareous shale unit, the Smoky Hill Chalk Member. Both units represent widespread pelagic to hemipelagic sedimentation in an epicontinental seaway during a relative highstand of sea level. During this highstand, the shoreline migrated westward and large areas of the Western Interior were covered by relatively deep water and received little terrigenous influx. Productivity of nanno- and microfossils was sufficiently great to yield moderately thick sequences of regionally homogeneous chalk and marl. These strata are very fine grained, have high primary porosity, and contain few macrofossils other than bivalves.

Sedimentary structures in the Niobrara consist mainly of laminations, fecal pellets, and burrows. Chalk-shale depositional cycles are found at scales ranging from millimeters to tens of meters. Cyclic sedimentation may have been related to variations in climatic patterns which, in turn, influenced terrigenous sediment influx and (or) biological productivity in the region. Climatic fluctuations probably influenced the salinity of the surface waters in the Western Interior seaway by altering circulation patterns and basinal water turnover. This resulted in periods of dysoxic and even anoxic bottom-water conditions within the seaway. Stagnation events are reflected by intervals with virtually no benthic fauna or burrows and with good preservation of millimeter-scale laminations and abundant organic matter.

The primary properties of the chalks and calcareous shales of the Niobrara have been greatly modified by burial diagenesis. Increased burial brought about rapid reduction of porosity by a combination of mechanical and chemical compaction and associated calcite overgrowth cementation. Authigenic pyrite is widespread in association with organic matter. Clay minerals of detrital origin and altered volcanogenic material originally underwent transformation from disordered, predominantly smectitic mixed-layer clays to ordered, predominantly illitic mixed-layer clays. During diagenetic alteration, brittle chalks and calcareous shales were deformed and significant fracturing occurred. Finally, thermal maturation associated with burial led to hydrocarbon generation from the organic-carbon-rich chalks and calcareous shales of the Niobrara. In areas of shallow burial, formation of biogenic methane was widespread and has proven to be economically important.

INTRODUCTION

Geologic studies of the Niobrara have been conducted for nearly a century (Gilbert, 1894; Williston, 1893; Cragin, 1896) and hydrocarbon production from the formation extends back to at least 1919 (Lockridge, 1977). However, a surge in both exploration activity and scientific investigation of this unit in the past two decades was prompted by several external factors. Major advances in paleo-oceanography, spurred largely by the Deep Sea Drilling Project, have led to better understanding of, and heightened interest in, chalk deposition, anoxic events, cyclic sedimentation, paleoclimatology, and related subjects. The Niobrara Formation has been an excellent testing ground for paleo-oceanographic concepts, especially as they apply to sedimentation in an epicontinental seaway. At the same time, major increases in market prices for oil and natural gas coupled with success in finding vast reserves of oil in North Sea chalk reservoirs led to a quantum increase in exploration activity in North American chalks. Although the early stages of the chalk exploration boom focused on the Gulf Coast (Austin Chalk, Selma Group, Annona Chalk, and related units), the play eventually included the Niobrara chalks from western Kansas to the Canadian border (e.g. Lockridge, 1977; Scholle, 1977a; Smagala, 1981; Brown and others, 1982).

Studies of cores and electric logs from chalks and calcareous shales have revealed a radically different kind of carbonate reservoir from those encountered in most exploration situations. Chalk reservoirs have proven to be finer-grained, more porous, less permeable, more uniform and widespread, and considerably more predictable in terms of petrophysical properties than reservoirs in typical shallow-water limestones. Thus, chalks provide an interesting contrast to the other carbonate units discussed at this core workshop.

STRATIGRAPHY AND REGIONAL DISTRIBUTION

The Niobrara Formation is a widespread unit extending over much of the U.S. Western Interior (Reeside, 1944). It was deposited in a broad, shallow, asymmetrical seaway bordered on the west by the rising Cordilleran orogenic belt from which much clastic terrigenous debris was shed. Thus, the unit becomes progressively sandier and shalier to the west and northwest. In the central and eastern parts of the Western Interior seaway (Fig. 1), however, a stable, shallow-water platform existed for long time intervals during which clastic influx was minimal, particularly at times of maximum transgression (Kauffman, 1969; Hancock, 1975; Hancock and Kauffman, 1979). In these areas, chalks and calcareous shales of the Niobrara Formation, Carlile Shale, and Greenhorn Limestone were deposited.

The Niobrara Formation (Fig. 2) includes two members: (1) the basal Fort Hays Limestone Member, a relatively pure chalk ranging in thickness from about 10 to 40 m (Frey, 1972; Hann, 1981), and (2) the overlying Smoky Hill Chalk Member which varies from 100 to 220 m and contains chalks, marls, and shales. Thickness variations in the Niobrara section are partially controlled by synsedimentary movement on broad structural features such as the Transcontinental Arch (Weimer, 1978).

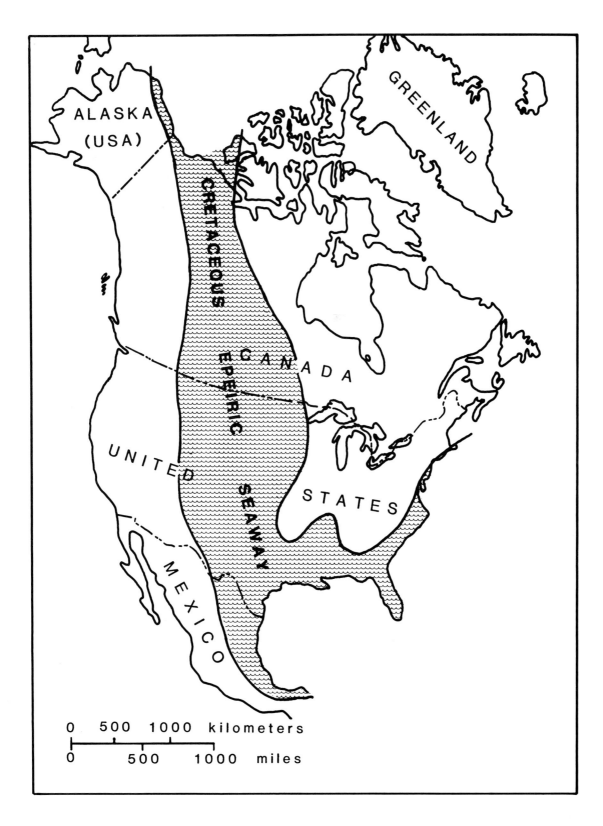

Figure 1. Map showing the location of the Upper Cretaceous seaway in North America (modified from Kaufmann, 1975).

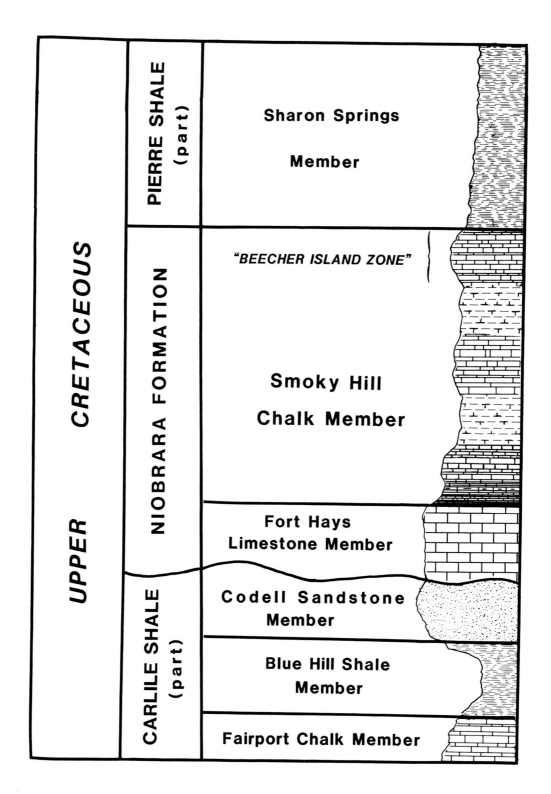

Figure 2. Generalized stratigraphic column for the Denver basin area showing the Niobrara Formation and associated units.

The contact of the Niobrara Formation with the underlying Codell Sandstone (Fig. 2) or Juana Lopez Members of the Carlile Shale is typically unconformable and time-transgressive. Thus, the age of the basal beds of the Fort Hays Limestone Member ranges from late Turonian near Pueblo, Colorado (Scott and Cobban, 1964) to early and middle Turonian to the north and east on the Transcontinental Arch (see review in Hann, 1981). The upper contact of the Niobrara locally is unconformable, but most commonly the Niobrara grades upward into the marine shales of the Sharon Springs Member of the Pierre Shale (Rice and Shurr, 1983). The youngest part of the Niobrara is early Campanian in most of the study area.

The Smoky Hill Chalk Member has been divided into 6 or 7 distinct units consisting of relatively pure, thin chalks and thicker calcareous and noncalcareous shales (Scott and Cobban, 1964; Hann, 1981). Variations in the carbonate content of these units correspond to changes in the relative inputs of terrigenous clastic and volcanogenic detritus versus pelagic carbonate material. Variations in organic carbon content are related to the ratio of clastic to carbonate input as well as to temporal fluctuations in the degree of oxygenation of bottom waters. Lithologic changes, in turn, are linked to both generation and production of hydrocarbons. For example, chalk-rich units generally have higher porosity and permeability; shale-rich units typically have higher organic carbon contents and are excellent potential hydrocarbon source beds.

Widespread drilling activity in the Niobrara during the past decade has led to extensive coring of the unit. Generally, coring has been restricted to the upper part of the formation, particularly the upper chalk unit of the Smoky Hill Chalk Member. In this workshop we present parts of two unusually long Niobrara cores (Fig. 3). The Coquina Oil Berthoud State #3 core (Fig. 4) from Larimer Co., Colorado (Sec. 16, T4N, R69W), includes a virtually complete, although thin, Niobrara section from an area once covered by considerable overburden (probably at least 3 km). Tertiary uplift and erosion have stripped some of this cover and have brought the section to its present depth of about 1 km. The Plains Resources Schock Errington #1 core (Fig. 5) from Sherman Co., Kansas (Sec. 11, T7S, R42W), on the other hand, represents portions of a considerably thicker Niobrara section which has never been buried to depths significantly greater than its current 500 to 600 m.

These are the two longest and most continuous Niobrara cores available and they best illustrate the large-scale sedimentological features of both reservoir and non-reservoir intervals in the formation. Supplemental information from outcrop sections and shorter cores from a variety of localities in and around the Denver basin has also been used in this paper.

SEDIMENTOLOGY

The Niobrara Formation consists of chalks and calcareous shales deposited at low sedimentation rates in areas of the Western Interior seaway far removed from the main sources of terrigenous detritus. Sedimentation of chalks and related units primarily occurred at transgressive maxima or during the gradual regressive phases following such maxima. At such times, sea level highstands led to decreased

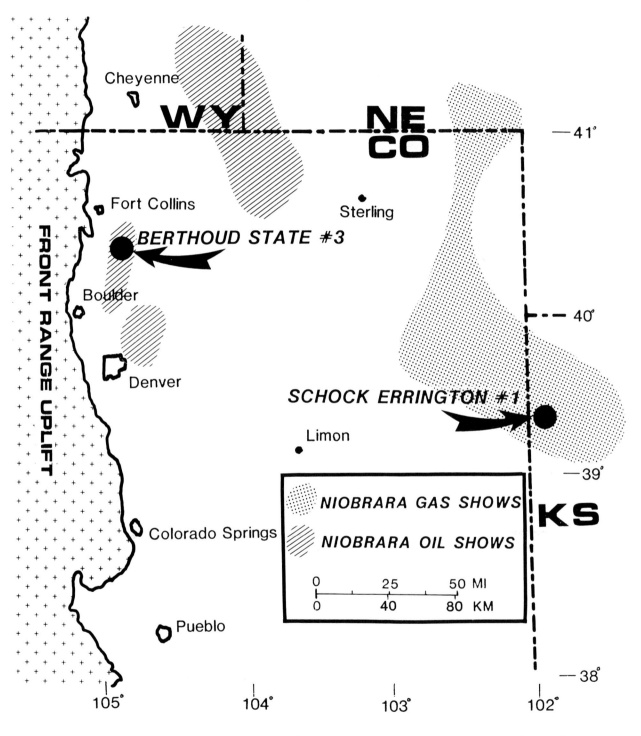

Figure 3. Map of the Denver basin area showing the location of the two cored wells discussed in this paper as well as the general areas of oil and gas exploration.

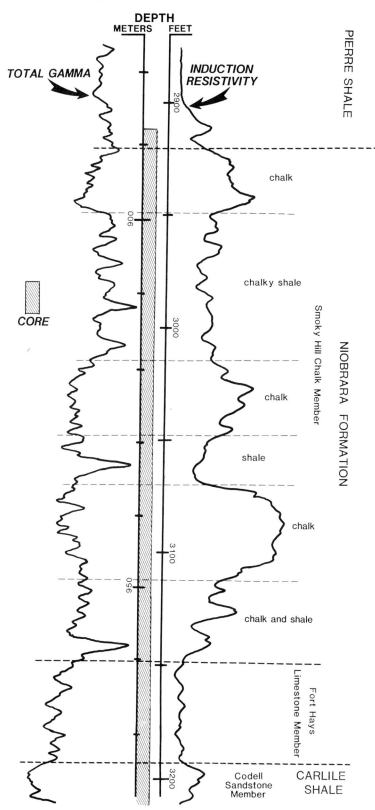

Figure 4. Log section of the Berthoud State #3 well, Larimer Co., Colorado showing total gamma and induction resistivity logs, stratigraphic interpretations, and cored intervals.

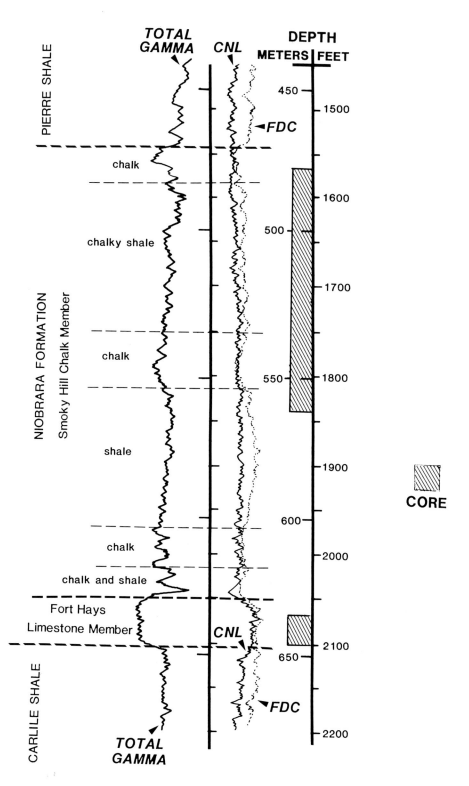

Figure 5. Log section of the Schock Errington #1 well, Sherman Co., Kansas showing Total Gamma, Compensated Neutron (CNL), and Formation Density Compensated Formation Density (FDC) logs, stratigraphic interpretations, and cored intervals.

exposure of source areas, raised erosional base levels, and broadened shelf environments (Kauffman, 1969; Scholle and others, 1983). Pelagic and hemipelagic sediment accumulated in waters between 25 and 200 m deep (Hancock, 1975); that is, below wave base and presumably also below the effective photic zone. Thus, Niobrara strata have remarkably uniform lithology throughout the Kansas, Nebraska, and Colorado area with no biohermal buildups, grainstone bodies, or other evidence of shoal-water sedimentation.

Niobrara strata are composed of three primary end members: calcite of biogenic origin, detrital and volcanogenic silicate minerals, and organic matter with associated pyrite. Variations in the relative proportions of these three constituents account for the wide variability in Niobrara lithologies. Chalk intervals, particularly in the Fort Hays Limestone Member, can have as much as 90% calcium carbonate. This material is now entirely calcitic and was derived primarily from remains of nannoplankton (Fig. 6A). Coccoliths and rhabdoliths are the predominant carbonate components of all Niobrara chalks, but planktic and benthic foraminifers, Inoceramus prisms and shells, and oysters also contribute significant volumes of skeletal calcite (Figs. 7-9). Nannofossils are concentrated in small fecal pellets probably produced by copepods and related plankton-grazing organisms (Hattin, 1975).

The non-carbonate components of the Niobrara Formation consist primarily of clays, detrital quartz and subordinate feldspar, pyrite and organic matter (Pollastro and Martinez, 1985). Clays are mainly illite and mixed-layer illite/smectite with minor amounts of kaolinite and chlorite. Most, but not all, clay minerals in the Niobrara appear to be derived from altered and reworked volcanic ash. Indeed, intact

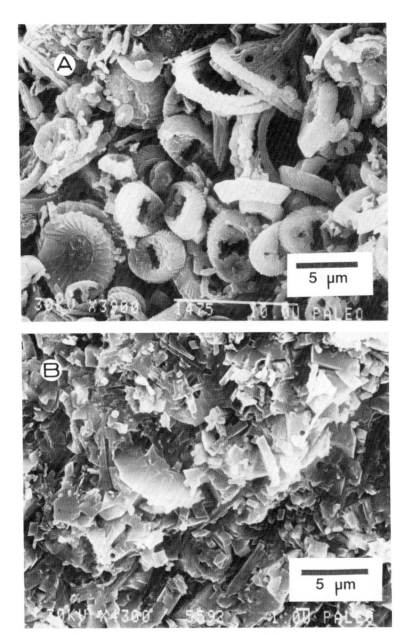

Figure 6. A: Scanning electron micrograph of chalk from the Smoky Hill Chalk Member (Beecher Island zone) showing numerous coccoliths with little diagenetic alteration. Sample from Kansas-Nebraska Natural Gas Whomble #1-32 well in Beecher Island field, Yuma Co., Colorado. Sample is from 1475 ft (450 m) depth and has about 15% insoluble residue and 38% porosity.

B: Scanning electron micrograph of chalk from the Smoky Hill Chalk Member (Beecher Island zone) showing diagenetically altered, overgrown and "welded" coccolith fragments. Sample from Excelsior Oil Co. #1 Alice Nay well, Weld Co., Colorado. Sample is from 5600 ft (1707 m) depth and has about 15% insoluble residue and 9-10% porosity.

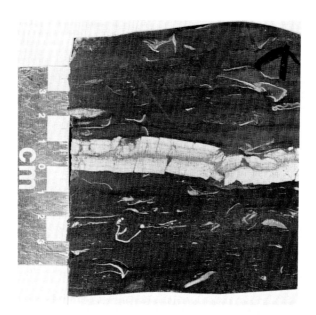

Figure 7. Core photograph from Berthoud State #3 well, 3013 ft (918 m) depth. Shows articulate inoceramid bivalve and scattered oyster shells in calcareous shale matrix.

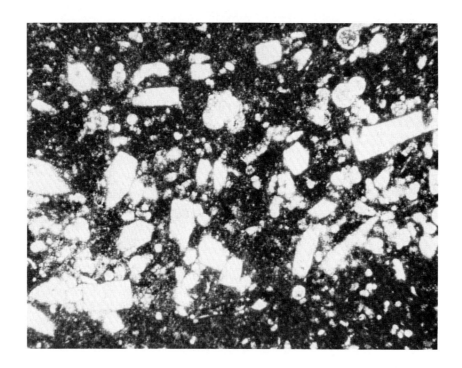

Figure 8. Thin-section photomicrograph showing Inoceramus prisms and foraminiferal tests in Niobrara chalk. Outcrop sample from Boettcher (Ideal Cement) Quarry, north of Fort Collins, Larimer Co., Colorado. Long axis of plots is about 1.2 mm.

bentonite layers are present throughout the Niobrara section (Fig. 10). Quartz, feldspar and some clay minerals were derived from distant terrigenous sources, and carried into the seaway as wind-blown grains or as dilute suspensions of fine-grained sediment in marine currents. Pyrite, which occurs as discrete nodules (Fig. 11) and disseminated framboidal aggregates, formed authigenically in close association with organic matter, reaching levels as high as 16% (by weight) of the insoluble fraction in strata with high total organic carbon (TOC) values (Pollastro and Martinez, 1985).

Organic matter in the Niobrara is fine-grained and therefore difficult to characterize by visual analysis. Geochemical analysis indicates a predominantly hydrogen-rich, sapropelic, Type II kerogen (typical amorphous marine-pelagic organic material) with a lesser component of transported terrigenous organic matter (Rice, 1985). TOC levels generally exceed 1% and in several intervals rise above 5% (by weight). The highest levels of TOC occur in the least calcareous intervals in the Niobrara; thus, TOC is inversely proportional to $CaCO_3$ content (Pollastro and Martinez, 1985).

All three primary constituents of Niobrara strata are very fine grained. Although coccospheres may be several tens of microns across, most disintegrate into individual coccoliths (2 to 10 microns in diameter) or even into single-crystal calcite laths (0.2 to 1 micron), the fundamental architectural elements of coccoliths. Planktic foraminiferal tests and inoceramid prisms, although considerably coarser grained, rarely make up a large (over 10%) proportion of the sediment. Detrital clay and silt as well as disseminated organic matter are also fine-grained. Thus, the Niobrara consists mainly of biomicrite

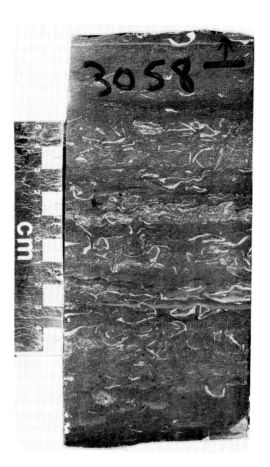

Figure 9. Core photograph from Berthoud State #3 well, 3058 ft (932 m) depth. Shows abundant scattered oyster shells in burrow-homogenized calcareous shale.

Figure 10. Core photograph from Berthoud State #3 well, 3153 ft (961 m) depth. Illustrates one of many bentonites preserved in the unburrowed or only slightly burrowed intervals of the Smoky Hill Chalk Member.

and terrigenous mudstone and, where not greatly altered by diagenesis, retains the high porosity (40-50%) typical for such fine-grained sediment (Scholle, 1977b).

Distribution of sedimentary structures in the predominantly pelagic and hemipelagic sediments of the Niobrara is controlled by two main factors: (1) variations in the input of siliciclastic components and (2) changes in the circulation patterns and oxygenation levels of the water column in the epicontinental seaway. These parameters apparently varied cyclically throughout deposition of the Niobrara Formation. Indeed, some of the earliest studies of cyclic marine sedimentation in the world were conducted in this unit (Gilbert, 1894, 1900).

Cyclic sedimentation is evident on several scales. Mega-cycles consist of packages of predominantly shale or predominantly chalk strata that are tens of meters thick. These cycles are reflected in most informal stratigraphic subdivisions of the Niobrara (Scott and Cobban, 1964; Hann, 1981; Hattin, 1982; Rice and Shurr, 1983). Likewise, oil and gas production from the Niobrara is largely from the chalk megacycles, especially the uppermost chalk bench, informally termed the "Beecher Island zone" (Fig. 2) (Lockridge, 1977). Cyclic alternations of clay-rich and calcite-rich beds also occur on a scale of 20 cm to 1 m (Figs. 12, 13). These cycles have been interpreted to reflect Milankovich-type variations in the earth's orbital parameters with frequencies of 20,000 to 30,000 years (Gilbert, 1894; Fischer, 1980; Arthur and others, 1984). Intermediate-scale rhythms, combining approximately five small-scale units into bundles that presumably represent 100,000- to 120,000-year climatic variations also

Figure 11. Core photograph from Berthoud State #3 well, 3180 ft (969 m) depth. Shows two authigenic pyrite nodules in laminated calcareous shale. Note continuation of laminae through one of the pyrite nodules. Core width is about 10 cm.

Figure 12. Core photograph from Berthoud State #3 well, 3194 ft (974 m) depth. Shows thin chalk-shale cycles and burrows in the Fort Hays Limestone Member of the Niobrara Formation. Mechanical and chemical compaction have reduced the thickness of the shale intervals relative to the carbonate ones (see Scholle and others, 1983).

Figure 13. Core photograph of part of the Fort Hays Limestone Member from the Schock Errington #1 well, 2076-2096 ft (633-639 m) depth. Note repeated chalk-shale cycles and presence of shell fragments and burrows throughout Fort Hays sequence.

are common. Finally, in intervals in which burrows are rare or absent, very small scale cyclic alternations of chalk and calcareous shale occur as millimeter laminations (Fig. 14).

The fundamental causes of these cycles of sedimentation are not completely understood. Changes in temperature and rainfall can affect both the influx of fine-grained terrigenous debris and the productivity of nannoplankton and other calcareous organisms. Changes in the amount of rainfall and fresh-water runoff may also have affected the overall salinity and water-mass stratification of the Western Interior seaway (Scholle, 1977a; Pratt, 1984). Evidence for the possibility of low-salinity intervals comes from oxygen isotopic analysis of whole-rock chalks as well as individual faunal constituents from the Niobrara (Hattin, 1965; Kauffman and Scholle, 1977; Scholle and Kaufman, 1977; Scholle, 1977a). In sections which have not been deeply buried, typical whole-rock δO^{18} values range from -4.5 to -5.5 per mil in the Fort Hays and -5 to -7 per mil in the Smoky Hill (all values relative to the PDB standard). Inoceramids have oxygen isotopic values about 1 to 1.5 per mil heavier than the whole rocks presumably because, as benthic organisms, they lived in cool, deep, and perhaps high-salinity waters. Whole-rock analyses, on the other hand, reflect primarily the composition of coccoliths which lived in the upper part of the water column. The very "light" oxygen isotopic composition of bulk samples of these strata (at least 3 per mil lighter than bulk samples of age-equivalent, fully marine chalks from Europe or the U.S. Gulf Coast), implies either precipitation at remarkably high temperatures (~40°C) from normal marine waters or percipitation at normal continental temperatures from surface water of considerably reduced

Figure 14. Core photograph of part of the Smoky Hill Chalk Member from the Schock Errington #1 well, 1626-1649 ft (496-503 m). Shows typical sequence of millimeter—laminated shaly chalk rich in fecal pellets but devoid (with the exception of two large inoceramid shells) of benthic macrofauna.

salinity (Scholle and Kauffman, 1977). Between one-third and one-half of the North American continent drained into the Western Interior seaway, so it is likely that at times of increased rainfall a low-density, fresh to brackish water layer may have spread over the depositional area of the Niobrara. The relatively low diversity of flora and fauna in Niobrara chalks when compared with their age-equivalent European counterparts (Hancock, 1975) may also be related to abnormal, low salinity conditions in the Western Interior seaway.

Variations in the thickness, salinity, and extent of brackish-water surface layers in the seaway may have led to significant temporal differences in circulation and turnover of the Western Interior water mass (Pratt, 1984). This could have been a major cause of some of the chalk-shale compositional cycles due to coupling of circulation patterns, salinity, biological productivity in surface waters, and rate of sedimentation of terrigenous clastic material. Changes in water-mass stratification may also have significantly influenced oxygenation levels of bottom waters in the seaway. Strata of the Niobrara Formation, as well as those in the underlying Carlile Shale and Greenhorn Limestone, show extreme variability in the numbers of benthic organisms as well as in the degree of bioturbation within medium- and large-scale cycles (Hattin, 1965; Pratt, 1984). Thus, some intervals, especially the chalkier zones, contain common benthic organisms (mainly foraminifers, inoceramids, and oysters) and numerous burrows (Figs. 15, 16), including _Thalassinoides_, _Chondrites_, _Planolites_, and _Teichichnus_ (Frey, 1970). In these intervals, the sediment commonly was completely homogenized and aerated; consequently, primary depositional lamination has been largely

Figure 15. Core photograph from Schock Errignton #1 well, 2096 ft (639 m) depth, showing details of intensely burrowed chalk bed in Fort Hays Limestone Member.

Figure 16. Core photograph from Berthoud State #3 well, 3148 ft (960 m) depth. Note extensive bioturbation and mixing of dark shale into underlying light-colored chalk.

or completely obliterated and TOC levels are generally below 1%. Other intervals, particularly the shalier ones, have little or no trace of burrow activity, retain remarkably well-preserved primary laminae (Figs. 14, 17) and delicate zooplankton fecal pellets, have virtually no benthic macro- or microfossils, and have TOC levels between 3 and 6%.

DIAGENESIS

An understanding of diagenesis is critical to any attempt to predict porosity and reservoir potential of the Niobrara Formation. Studies conducted in the past 15 years have indicated that burial depth is the single most important factor in the diagenesis of chalks and shales (Schlanger and Douglas, 1974; Scholle, 1977a, 1977b; Lockridge and Scholle, 1978; Hower, 1981; Pollastro and Scholle, 1985). This is largely a consequence of the fact that the chalks are composed of calcitic grains which had a high degree of mineralogical stability from the time of sedimentation. Furthermore, deposition took place in relatively deep water. Thus, processes which radically affect the diagenesis of typical shallow-water carbonate strata, such as subaerial exposure and fresh-water alteration, are unlikely to occur in chalks and will have minimal effects when they do occur.

Chalks and shales are deposited as soupy sediments containing up to 80% water-filled porosity (Schlanger and Douglas, 1974). Although in some chalks seafloor cementation can greatly reduce this initial porosity, at least locally, there is no evidence that hardground formation significantly affected Niobrara strata. There is also no evidence that the Niobrara was subaerially exposed during or immediately after sedimentation. Thus, these sediments were probably buried with marine pore fluids and very high initial porosities.

Mechanical compaction during early stages of burial induced such effects as rapid dewatering of the sediment and grain reorientation and breakage. Mechanical compaction effects appear to be most extreme in areas of rapid sediment loading, particularly where abnormally high pore-fluid pressures may have been dissipated rapidly by faulting and fracturing (Fig. 18). Continued burial led to chemical compaction with dissolution of calcite along solution seams and microstylolites (Fig. 19), especially in intervals with high primary clay content. Pressure solution also took place at grain-to-grain contacts. Calcium carbonate dissolved by pressure solution was reprecipitated in nearby areas of lower differential stress and lower clay content. Calcite precipitiation occurred in chalky horizons mainly as small overgrowths on coccolith fragments, eventually yielding a "welded" fabric of interlocking crystals (Fig. 6B).

Shales likewise underwent a series of transformations with progressive burial. In addition to the mechanical effects (mainly grain reorientation) shown by chalk strata, shales also progressed through a series of mineralogical transformations. Although the initial compositions of detrital and volcanogenic clays in the Niobrara were complex, consistent mineralogic changes are observed with depth (Pollastro and Scholle, 1985). Thus, expandibility of mixed-layer illite-smectite (I/S) clay decreases, the proportion of illite-type layers increases, and there is a shift from random to ordered I/S with increasing burial. Regularly interstratified chlorite-smectite was also

Figure 17. Core photograph of laminated chalk from Beecher Island zone chalk in State #1-29 well (Yuma Co., Colorado) at about 1470 ft (448 m) depth. Lamination results from alternations of clay-rich and clay-poor chalk; speckled texture is due to presence of numerous fecal pellets packed with coccoliths. Long axis of photo is approximately 5 cm.

Figure 18. Thin-section photomicrograph of the Fort Hays Limestone Member from the Boetcher (Ideal Cement) Quarry north of Fort Collins, Larimer Co., Colorado. Compaction of fecal pellets and foraminiferal tests, although widespread may be most intense in areas in which sedimentation rates were highest and in which later faulting may have rapidly relieved overpressures. Long axis of photograph is about 1.2 mm.

formed in aluminous bentonites at the elevated temperatures encountered under deep burial (~100°C).

One result of these diagenetic transformations in both chalks and shales is a continuous loss of porosity with depth (Fig. 20). Similar plots have proven useful in predicting porosity loss in chalks throughout the world (Scholle, 1977b). The slight deviations of Western Interior chalks from the global model for porosity reduction may be related to two main factors (Lockridge and Scholle, 1978). First, Western Interior chalks generally have higher insoluble residue and TOC contents than age-equivalent chalks from the Gulf Coast, Europe, the Middle East, and many Deep Sea Drilling Project sites. Because the presence of clays and other minerals (and perhaps also organically-derived compounds) can accelerate the burial-related chemical compaction of chalks, Niobrara chalks may simply lose porosity at a faster rate than their "cleaner" equivalents elsewhere. Alternatively, the present-day burial depths of the Niobrara may not equal their maximum burial depths; that is, overburden may have been removed during Tertiary and Quaternary erosion of the region.

PETROLEUM PRODUCTION AND POTENTIAL

The loss of porosity in chalks and associated calcareous shales at relatively shallow depths limits possible exploration plays in these units. High porosity and moderate to low permeability reservoirs can be anticipated in areas where Niobrara chalk has not been buried to depths of more than about 1 km (Fig. 21). Unfortunately, such areas are generally thermally immature. Thus, hydrocarbons must have

Figure 19. Core photograph from Berthoud State #3 well, 3192 ft (973 m) depth, showing numerous thin solution seams along which carbonate dissolution has taken place during burial. Calcium carbonate dissolved within these clay-rich zones is reprecipitated in adjacent carbonate-rich intervals.

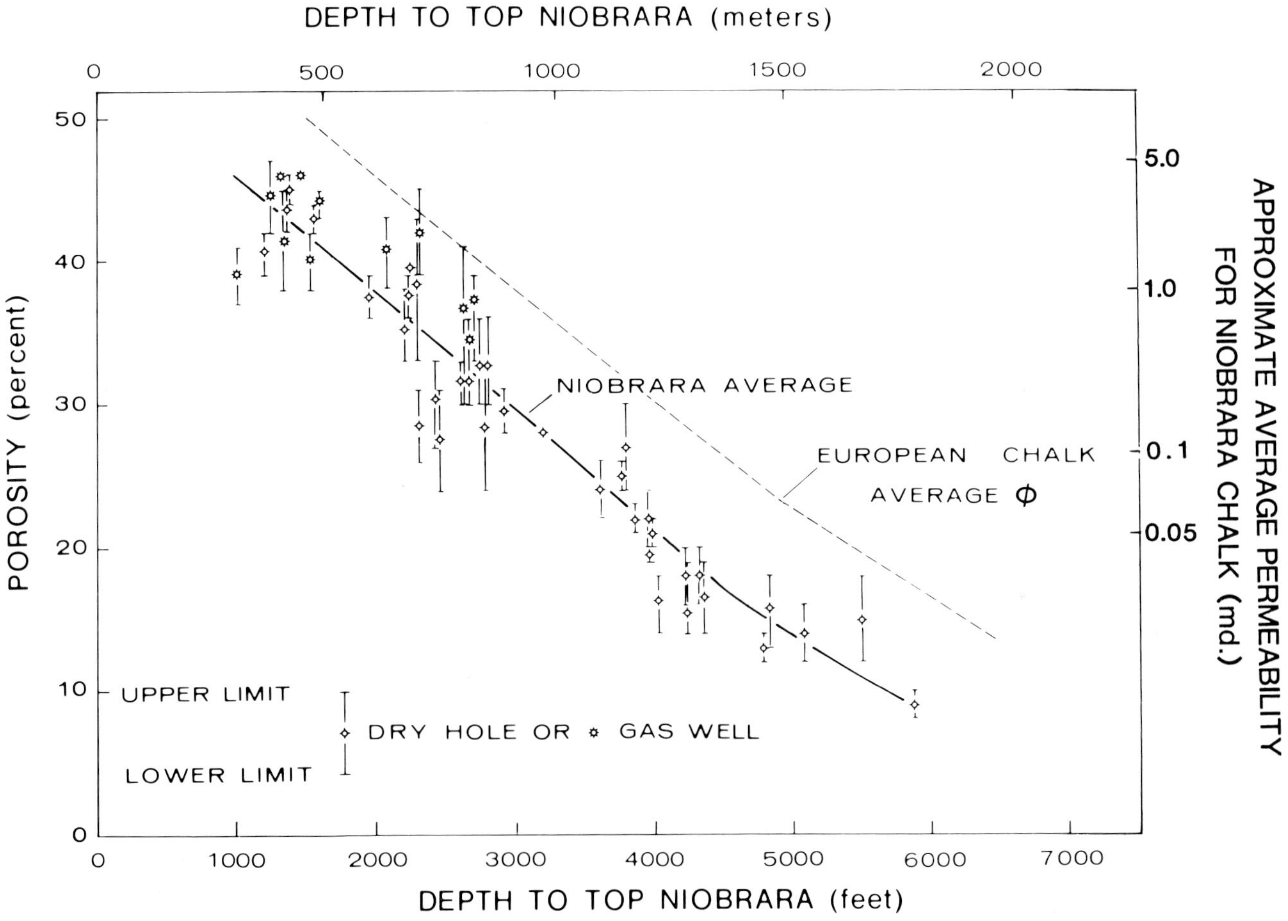

Figure 20. Plot of porosity and permeability versus depth in the Niobrara Formation. European chalk average line is a curve for chalks with less than 1% insoluble residue. Porosities determined from electric logs and checked with direct core plug measurements. Modified from Lockridge and Scholle (1978).

migrated from more deeply buried sections or exploration objectives will be limited to biogenic gas deposits. Drilling during the past decade has proven that exploration for biogenic gas in the Niobrara can be economically viable and that such deposits are widely distributed in the eastern part of the Western Interior region (Lockridge, 1977; Scholle, 1977a, Rice and Claypool, 1981).

Low matrix porosities and permeabilities resulting from burial-related diagenesis have reduced the quality of chalk reservoirs in normally pressured areas which have had more than 1 to 1.5 km of overburden. On the other hand, in areas of significant faulting or flexuring, brittle, chalky limestone can undergo intense fracturing (Figs. 22, 23). Where such fractures have remained open and uncemented, and where the organic-carbon-rich shales of the Niobrara have reached thermal maturity, excellent oil- or gas-production has been obtained. This parallels exploration results in Gulf Coast, especially Austin, chalks (Scholle, 1977a).

The remarkable rates and volumes of oil production which have been found in the North Sea are related to the high pore-fluid pressures found in the Central Graben of the basin. Chalk porosities in the 35 to 45% range are preserved at 3 to 4 km depth where anomalously high pore fluid pressures are present (Scholle, 1977b). Because the post-Niobrara Upper Cretaceous section of the Denver Basin was deposited rapidly and is fine-grained, it is possible, even likely, that abnormally high pore fluid pressures once existed in the basin. However, because the sediment loading in the Denver Basin took place mainly in the Late Cretaceous, later deformation and the passage of 65 million years have allowed the dissipation of these high

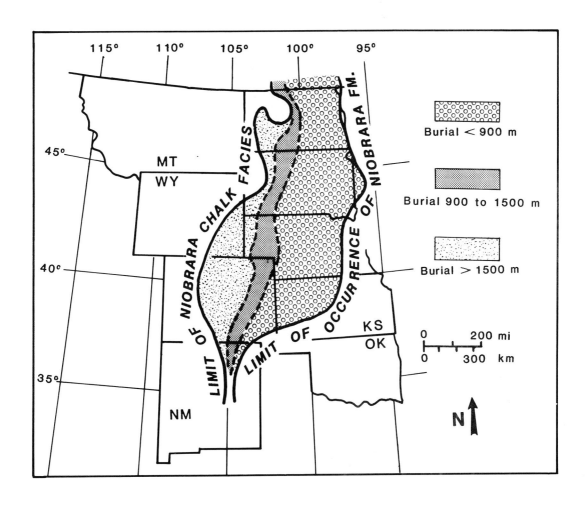

Figure 21. Map showing present-day distribution of Niobrara Formation chalk facies and their approximate maximum burial depth (modified from Scholle, 1977a).

Figure 22. Core photograph from Berthoud State #3 well, 3105 ft (946 m) depth, showing network of small, calcite-filled fractures in calcareous shale from the Smoky Hill Chalk Member. Such fractures, where unfilled, can be very important in enhancing effective permeability of chalk reservoirs.

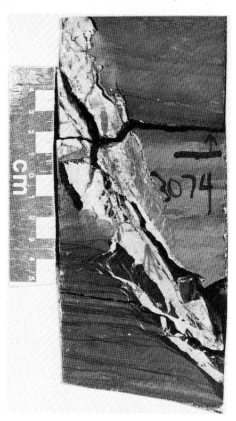

Figure 23. Core photograph from Berthoud State #3 well, 3074 ft (937 m) depth, showing major fracture associated with a high-angle fault.

pore pressures in most areas. In the North Sea, rapid loading continued through Tertiary time, maintaining abnormal pressures to the present. Thus, no chalk oil fields even marginally comparable in production capacity to those of the North Sea have yet been discovered in the Western Interior.

ACKNOWLEDGEMENTS

We thank W. D. Wiggins and R. A. Garber of Gulf Research and Development Company and F. B. Zelt of the U.S. Geological Survey for their careful and constructive reviews of the manuscript. As we followed their suggestions for revision, any remaining errors should be attributed to them.

REFERENCES

ARTHUR, M. A., DEAN, W. E., BOTTJER, D., and SCHOLLE, P. A., 1984, Rhythmic bedding in Mesozoic-Cenozoic pelagic carbonate sequences: the primary and diagenetic origin of Milanovitch-like cycles: in Berger, A. L. and others (eds.), Milankovich and Climate, Part 1: D. Riedel Publ. Co., p. 191-222.

BROWN, C. A., CRAFTON, J. W., and GOLSON, J. G., 1982, The Niobrara gas play: exploration and development of low-pressure, low-permeability gas reservoir: Jour. Petrol. Technology, v. 34, p. 2863-2870.

CRAGIN, F. W., 1896, On the stratigraphy of the Platte series, or Upper Cretaceous of the Plains: Colorado College Studies, v. 6, p. 49-52

FISCHER, A. G., 1980, Gilbert--bedding rhythms and geochronology, in Yochelson, E. L. (ed.), The Scientific Ideas of G. K. Gilbert: Geol. Soc. Am. Spec. Publ. 183, p. 93-104.

FREY R. W., 1970, Trace fossils of Fort Hays Limestone Member of Niobrara Chalk (Upper Cretaceous), west-central Kansas: Univ. Kansas Paleontological Contributions, Art. 53, 41 p.

FREY, R. W., 1972, Depositional environment of Fort Hays Limestone Member of the Niobrara Chalk (Upper Cretaceous) west-central Kansas: Univ. Kansas Paleont. Contrib., Art. 58, 72 p.

GILBERT, G. K., 1894, Sedimentary measurement of Cretaceous time: Jour. Geol., v. 3, p. 121-127.

GILBERT, G. K., 1900, Rhythms and geologic time: Amer. Assoc. Adv. Sci., Proc., v. 49, p. 1-19.

HANCOCK, J. M., 1975, The sequence of facies in the Upper Cretaceous of northern Europe compared with that in the Western Interior: Geol. Assoc. Canada Spec. Publ. 13, p. 83-118.

HANCOCK, J. M., and KAUFFMAN, E. G., 1979, The great transgressions of the Late Cretaceous: Jour. Geol. Soc. London, v. 136, p. 175-186.

HANN, M. L., 1981, Petroleum potential of the Niobrara Formation in the Denver Basin: Colorado and Kansas: Unpublished M.S. Thesis, Colorado State University, Fort Collins, Colorado, 260 p.

HATTIN, D. E., 1965, Upper Cretaceous stratigraphy, paleontology, and paleoecology of Western Kansas: Geol. Soc. Am. Ann. Meeting, (Kansas City), Field Conf. Guidebook, 69 p.

HATTIN, D. E., 1975, Petrology and origin of fecal pellets in Upper Cretaceous strata of Kansas and Saskatchewan: Jour. Sed. Petrol., v. 45, p. 686-696.

HATTIN, D. E., 1982, Stratigraphy and depositional environment of Smoky Hill Chalk member, Niobrara Chalk (Upper Cretaceous) of the type area, western Kansas: Kansas Geol. Surv. Bull. 225, 108 p.

HOWER, J., 1981, Shale diagenesis: in F. J. Longstaffe (ed.), Short Course Handbook No. 7: Mineral. Assoc. Canada, p. 60-80.

KAUFFMAN, E. G., 1969, Cretaceous marine cycles of Western Interior: Mountain Geologist, v. 6, p. 227-245.

KAUFFMAN, E. G., 1975, The value of benthic Bivalvia in Cretaceous biostratigraphy of the Western Interior: in W. G. E. Caldwell (ed.), The Cretaceous System in the Western Interior of North America; Selected Aspects: Geol. Assoc. Canada Spec. Paper 13, p. 163-194.

KAUFFMAN, E.G., and SCHOLLE, P. A., 1977, Abrupt biotic and environmental changes during peak Cretaceous transgressions in

Euramerica: in North American Paleontological Convention II, Abstracts: Jour. Paleont., v. 51, no. 2 suppl., p. 16.

LOCKRIDGE, J. P., 1977, Beecher Island field Yuma County, Colorado: in Exploration Frontiers of the Central and Southern Rockies: Rocky Mtn. Assoc. Geologists, Field Conference Guidebook, p. 271-279.

LOCKRIDGE, J. P., and SCHOLLE, P. A., 1978, Niobrara gas in eastern Colorado and northwestern Kansas: in Pruit, J. D., and Coffin, P. E. (eds.), Energy Resources of the Denver Basin: Rocky Mtn. Assoc. Geol. Symposium, p. 35-49.

POLLASTRO, R. M., and MARTINEZ, C. S., 1985, Mineral, chemical and textural relationships and their implications for hydrocarbons in rhythmic-bedded chalk of the Niobrara Formation, Denver Basin, Colorado: Mountain Geologist, v. 22, in press.

POLLASTRO, R. M. and SCHOLLE, P. A., 1985, Diagenetic relationships in a hydrocarbon-productive chalk: The Cretaceous Niobrara Formation: U.S. Geol. Surv. Bull., in press.

PRATT, L. M., 1984, Influence of paleoenvironmental factors on preservation of organic matter in Middle Cretaceous Greenhorn Formation, Pueblo, Colorado: Am. Assoc. Petrol. Geol. Bull., v. 68, p. 1146-1159.

REESIDE, J. B., Jr., 1944, Map showing thickness and general character of the Cretaceous deposits in the Western Interior of the United States: U.S. Geol. Survey Oil and Gas Investig. Map OM-10.

RICE, D. D., 1985, Occurrence of indigenous biogenic gas in organic-rich immature chalks of Late Cretaceous age, eastern Denver Basin: in Palacas, J. G. (ed.), Geochemistry and Source-Rock Potential of Carbonate Rocks: Am. Assoc. Petrol. Geol. Memoir, in press.

RICE, D. D. and CLAYPOOL, G. E., 1981, Generation, accumulation and resource potential of biogenic gas: Am. Assoc. Petrol. Geol. Bull., v. 65, p. 5-25.

RICE, D. D., and SHURR, G.W., 1983, Patterns of sedimentation and paleogeography across the Western Interior seaway during time of deposition of Upper Cretaceous Eagle Sandstone and equivalent rocks northern Great Plains: in Reynolds, M. W., and Dolly, E. D. (eds.), Mesozoic Paleogeography of the West-Central United States: Rocky Mountain. Sec., Soc. Econ. Paleont. Mineral., Rocky Mountain. Paleogeography, Symposium 2, p. 337-358.

SCHLANGER, S. O., and DOUGLAS, R. G., 1974, The pelagic ooze-chalk-limestone transition and its implications for marine stratigraphy: in Hsu, K. J., and Jenkyns, H. C. (eds.), Pelagic Sediments--On Land and Under the Sea: Internat. Assoc. Sedimentol. Spec. Pub. 1, p. 117-148.

SCHOLLE, P. A., 1977a, Current oil and gas production from North American Upper Cretaceous chalks: U.S. Geol. Surv. Circ. 767, 51 p.

SCHOLLE, P. A., 1977b, Chalk diagenesis and its relation to petroleum exploration: oil from chalks, a modern miracle?: Am. Assoc. Petrol. Geol. Bull., v. 61, p. 982-1009.

SCHOLLE, P. A., ARTHUR, M. A., and EKDALE, A. A., 1983, Pelagic environment: in Scholle, P. A. Bebout, D. G., and Moore, C. H. (eds.), Carbonate Depositional Environments, Am. Assoc. Petrol. Geol. Memoir 33, p. 629-691.

SCHOLLE, P. A., and KAUFFMAN, E.G., 1977, Paleoecological implications of stable isotope data from Upper Cretaceous limestones and fossils from the U.S. Western Interior: in North American Paleontological Convention II, Abstracts: Jour. Paleont., v. 51, no. 2 suppl., p. 24-25.

SCOTT R., and COBBAN, W. A., 1964, Stratigraphy of the Niobrara Formation at Pueblo, Colorado: U.S. Geol. Surv. Prof. Paper 454-L, 30 p.

SHURR, G. W., 1984, Regional setting of Niobrara Formation in northern Great Plains: Am. Assoc. Petrol. Geol. Bull., v. 68, p. 598-609.

SMAGALA, T., 1981, The Cretaceous Niobrara play: Oil and Gas Jour., v. 79, no. 10, p. 204-218.

WEIMER, R. J., 1978, Influence of Transcontinental Arch on Cretaceous marine sedimentation: A preliminary report: in Energy Resources of the Denver Basin: Rocky Mtn. Assoc. Geol. Symposium, p. 211-222.

WILLISTON, S. W., 1893, The Niobrara Cretaceous of western Kansas: Kansas Academy of Science, Trans., v. 13, p. 107-111.